T0309761

**Power System Control
Under Cascading Failures**

Power System Control Under Cascading Failures

Understanding, Mitigation, and System Restoration

Kai Sun
University of Tennessee, USA

Yunhe Hou
University of Hong Kong, Hong Kong

Wei Sun
University of Central Florida, USA

Junjian Qi
University of Central Florida, USA

Registered Offices
John Wiley & Sons, Inc., 111 River Street, Hoboken, NJ 07030, USA
John Wiley & Sons Ltd, The Atrium, Southern Gate, Chichester, West Sussex, PO19 8SQ, UK

Editorial Office
The Atrium, Southern Gate, Chichester, West Sussex, PO19 8SQ, UK

For details of our global editorial offices, customer services, and more information about Wiley products visit us at www.wiley.com.

Wiley also publishes its books in a variety of electronic formats and by print-on-demand. Some content that appears in standard print versions of this book may not be available in other formats.

Library of Congress Cataloging-in-Publication Data

Names: Sun, Kai, 1976– author. | Hou, Yunhe, 1975– author. | Sun, Wei, 1982– author. | Qi, Junjian, 1985– author.
Title: Power system control under cascading failures : understanding, mitigation, and system restoration / Kai Sun, Yunhe Hou, Wei Sun, Junjian Qi.
Description: Hoboken, NJ : John Wiley & Sons, 2019. | Includes index. |
Identifiers: LCCN 2018023661 (print) | LCCN 2018042297 (ebook) | ISBN 9781119282051 (Adobe PDF) | ISBN 9781119282068 (ePub) | ISBN 9781119282020 (hardcover)
Subjects: LCSH: Electric power systems–Control. | Electric power system stability. | Electric power failures.
Classification: LCC TK1007 (ebook) | LCC TK1007 .S86 2018 (print) | DDC 621.31/7–dc23
LC record available at https://lccn.loc.gov/2018023661

Cover Design: Wiley
Cover Image: © Jim Richardson/Getty Images

Set in 10/12pt Warnock by SPi Global, Pondicherry, India

Printed in Singapore by C.O.S. Printers Pte Ltd

10 9 8 7 6 5 4 3 2 1

To:
Fang and Yi
* -Kai Sun*
Lijuan and Jesse
* - Yunhe Hou*
Qun, Will, Joy, and Max
* - Wei Sun*
Yaming, Noah, and Nolan
* - Junjian Qi*

Contents

About the Companion Website   *xiii*

1   **Introduction**   *1*
1.1   Importance of Modeling and Understanding Cascading Failures   *1*
1.1.1   Cascading Failures   *1*
1.1.2   Challenges in Modeling and Understanding Cascading Failures   *4*
1.2   Importance of Controlled System Separation   *6*
1.2.1   Mitigation of Cascading Failures   *6*
1.2.2   Uncontrolled and Controlled System Separations   *7*
1.3   Constructing Restoration Strategies   *9*
1.3.1   Importance of System Restoration   *9*
1.3.2   Classification of System Restoration Strategies   *10*
1.3.3   Challenges of System Restoration   *13*
1.4   Overview of the Book   *15*
        References   *18*

2   **Modeling of Cascading Failures**   *23*
2.1   General Cascading Failure Models   *23*
2.1.1   Bak–Tang–Wiesenfeld Sandpile Model   *23*
2.1.2   Failure-Tolerance Sandpile Model   *24*
2.1.3   Motter–Lai Model   *30*
2.1.4   Influence Model   *30*
2.1.5   Binary-Decision Model   *33*
2.1.6   Coupled Map Lattice Model   *34*
2.1.7   CASCADE Model   *35*
2.1.8   Interdependent Failure Model   *37*
2.2   Power System Cascading Failure Models   *39*
2.2.1   Hidden Failure Model   *39*
2.2.2   Manchester Model   *40*
2.2.3   OPA Model   *42*
2.2.4   Improved OPA Model   *46*
2.2.5   OPA Model with Slow Process   *49*

2.2.6 AC OPA Model *58*

2.2.7 Cascading Failure Models Considering Dynamics and Detailed Protections *62*

References *64*

3 **Understanding Cascading Failures** *69*

3.1 Self-Organized Criticality *70*

3.1.1 SOC Theory *70*

3.1.2 Evidence of SOC in Blackout Data *71*

3.2 Branching Processes *72*

3.2.1 Definition of the Galton–Watson Branching Process *74*

3.2.2 Estimation of Mean of the Offspring Distribution *74*

3.2.3 Estimation of Variance of the Offspring Distribution *75*

3.2.4 Processing and Discretization of Continuous Data *78*

3.2.5 Estimation of Distribution of Total Outages *81*

3.2.6 Statistical Insight of Branching Process Parameters *81*

3.2.7 Branching Processes Applied to Line Outage Data *82*

3.2.8 Branching Processes Applied to Load Shed Data *84*

3.2.9 Cross-Validation for Branching Processes *85*

3.2.10 Efficiency Improvement by Branching Processes *85*

3.3 Multitype Branching Processes *87*

3.3.1 Estimation of Multitype Branching Process Parameters *88*

3.3.2 Estimation of Joint Probability Distribution of Total Outages *90*

3.3.3 An Example for a Two-Type Branching Process *91*

3.3.4 Validation of Estimated Joint Distribution *92*

3.3.5 Number of Cascades Needed for Multitype Branching Processes *94*

3.3.6 Estimated Parameters of Branching Processes *96*

3.3.7 Estimated Joint Distribution of Total Outages *98*

3.3.8 Cross-Validation for Multitype Branching Processes *100*

3.3.9 Predicting Joint Distribution from One Type of Outage *102*

3.3.10 Estimating Failure Propagation of Three Types of Outages *104*

3.4 Failure Interaction Analysis *105*

3.4.1 Estimation of Interactions between Component Failures *106*

3.4.2 Identification of Key Links and Key Components *108*

3.4.3 Interaction Model *111*

3.4.4 Validation of Interaction Model *113*

3.4.5 Number of Cascades Needed for Failure Interaction Analysis *115*

3.4.6 Estimated Interaction Matrix and Interaction Network *119*

3.4.7 Identified Key Links and Key Components *121*

3.4.8 Interaction Model Validation *125*

3.4.9 Cascading Failure Mitigation *129*

3.4.10 Efficiency Improvement by Interaction Model *134*

References *137*

4   **Strategies for Controlled System Separation**   *141*
4.1   Questions to Answer   *141*
4.2   Literature Review   *142*
4.3   Constraints on Separation Points   *144*
4.4   Graph Models of a Power Network   *148*
4.4.1   Undirected Node-Weighted Graph   *149*
4.4.2   Directed Edge-Weighted Graph   *152*
4.5   Generator Grouping   *153*
4.5.1   Slow Coherency Analysis   *154*
4.5.2   Elementary Coherent Groups   *158*
4.6   Finding Separation Points   *160*
4.6.1   Formulations of the Problem   *160*
4.6.2   Computational Complexity   *164*
4.6.3   Network Reduction   *167*
4.6.4   Network Decomposition for Parallel Processing   *173*
4.6.5   Application of the Ordered Binary Decision Diagram   *175*
4.6.6   Checking the Transmission Capacity and Small
         Disruption Constraints   *185*
4.6.7   Checking All Constraints in Three Steps   *190*
         References   *192*

5   **Online Decision Support for Controlled System Separation**   *197*
5.1   Online Decision on the Separation Strategy   *197*
5.1.1   Spectral Analysis-Based Method   *198*
5.1.2   Frequency-Amplitude Characteristics of Electromechanical
         Oscillation   *199*
5.1.3   Phase-Locked Loop-Based Method   *204*
5.1.4   Timing of Controlled Separation   *210*
5.2   WAMS-Based Unified Framework for Controlled
         System Separation   *212*
5.2.1   WAMS-Based Three-Stage CSS Scheme   *212*
5.2.2   Offline Analysis Stage   *214*
5.2.3   Online Monitoring Stage   *216*
5.2.4   Real-Time Control Stage   *221*
         References   *223*

6   **Constraints of System Restoration**   *225*
6.1   Physical Constraints During Restoration   *225*
6.1.1   Generating Unit Start-Up   *225*
6.1.2   System Sectionalizing and Reconfiguration   *230*
6.1.3   Load Restoration   *233*
6.2   Electromagnetic Transients During System Restoration   *235*
6.2.1   Generator Self-Excitation   *237*

6.2.2   Switching Overvoltage   *237*

6.2.3   Resonant Overvoltage in the Case of Energizing No-Load Transformer   *242*

6.2.4   Impact of Magnetizing Inrush Current on Transformer   *245*

6.2.5   Voltage and Frequency Analysis in Picking up Load   *247*
      References   *251*

7   **Restoration Methodology and Implementation Algorithms**   *255*

7.1   Algorithms for Generating Unit Start-Up   *255*

7.1.1   A General Bilevel Framework   *255*

7.1.2   Algorithms for the Primary Problem   *260*

7.1.3   Algorithms for the Second Problem   *265*

7.2   Algorithms for Load Restoration   *269*

7.2.1   Estimate Operational Region Bound   *271*

7.2.2   Formulate MINLR Model to Maximize Load Pickup   *272*

7.2.3   Branch-and-Cut Solver: Design and Justification   *275*

7.2.4   Selection of Branching Methods   *278*

7.3   Case Studies   *278*

7.3.1   Illustrative Example for Restoring Generating Units   *278*

7.3.2   Optimal Load Restoration Strategies for RTS 24-Bus System   *283*

7.3.3   Optimal Load Restoration Strategies for IEEE 118-Bus System   *287*
      References   *291*

8   **Renewable and Energy Storage in System Restoration**   *295*

8.1   Planning of Renewable Generators in System Restoration   *295*

8.1.1   Renewables for System Restoration   *295*

8.1.2   The Offline Restoration Tool Using Renewable Energy Resources   *296*

8.1.3   System Restoration with Renewables' Participation   *298*

8.2   Operation and Control of Renewable Generators in System Restoration   *305*

8.2.1   Prerequisites of Type 3 WTs for System Restoration   *307*

8.2.2   Problem Setup of Type 3 WTs for System Restoration   *308*

8.2.3   Black-Starting Control and Sequence of Type 3 WTs   *314*

8.2.4   Autonomous Frequency Mechanism of a Type 3 WT-Based Stand-Alone System   *317*

8.2.5   Simulation Study   *320*

8.3   Energy Storage in System Restoration   *323*

8.3.1   Pumped-Storage Hydro Units in Restoration   *323*

8.3.2   Batteries for System Restoration   *332*

8.3.3   Electric Vehicles in System Restoration   *340*
      References   *351*

9   Emerging Technologies in System Restoration *357*
9.1   Applications of FACTS and HVDC   *357*
9.1.1   LCC-HVDC Technology for System Restoration   *357*
9.1.2   VSC-HVDC Technology for System Restoration   *363*
9.1.3   FACTS Technology for System Restoration   *370*
9.2   Applications of PMUs   *376*
9.2.1   Review of PMU   *376*
9.2.2   System Restoration with PMU Measurements   *378*
9.3   Microgrid in System Restoration   *385*
9.3.1   Microgrid-Based Restoration   *385*
9.3.2   Demonstration and Practice   *388*
        References   *393*

10   Black-Start Capability Assessment and Optimization *399*
10.1   Background of Black Start   *399*
10.1.1   Definition of Black Start   *399*
10.1.2   Constraints During BS   *400*
10.1.3   BS Service Procurement   *401*
10.1.4   Power System Restoration Procedure   *403*
10.2   BS Capability Assessment   *404*
10.2.1   Installation Criteria of New BS Generators   *404*
10.2.2   Optimal Installation Strategy of BS Capability   *407*
10.2.3   Examples   *408*
10.3   Optimal BS Capability   *411*
10.3.1   Problem Formulation   *411*
10.3.2   Solution Algorithm   *418*
10.3.3   Examples   *421*
         References   *431*

Index *433*

About the Companion Website

The companion website for this book is at

www.wiley.com/go/sun/cascade

The website includes:

- A toolbox on cascading failure models (related to chapters 2-3), which include a simple example of influence model, and code for branching process and interaction model
- A toolbox on system separation strategies (related to chapters 4-5)
- A toolbox on power system restoration (related to chapters 6-10)

Some PowerPoint slides related to the content of the book.

Scan this QR code to visit the companion website.

1

Introduction

1.1 Importance of Modeling and Understanding Cascading Failures

1.1.1 Cascading Failures

Cascading failures can happen in many different systems, such as in electric power systems [1–7], the Internet [8], the road system [9], and in social and economic systems [10]. These low-probability high-impact events can produce significant economic and social losses.

In electric power grids, cascading blackouts are complicated sequences of dependent outages that could bring about tremendous economic and social losses. Large-scale cascading blackouts have substantial risk and pose great challenges in simulation, analysis, and mitigation. It is important to study the mechanisms of cascading failures so that the risk of large-scale blackouts may be better quantified and mitigated. Cascading blackouts are usually considered rare events, but they are not that uncommon. The frequency of these high-impact events is not as low as expected. The following is a subset of the very famous large-scale blackouts around the world.

- 1965 Northeast blackout: There was a significant disruption in the supply of electricity on November 9, 1965, affecting parts of Ontario in Canada and Connecticut, Massachusetts, New Hampshire, New Jersey, New York, Rhode Island, Pennsylvania, and Vermont in the United States. Over 30 million people and 80,000 square miles were left without electricity for up to 13 hours [11].
- 1996 Western North America blackouts: A disturbance occurred on July 2, 1996, which ultimately resulted in the Western Systems Coordinating Council (WSCC) system separating into five islands and in electric service interruptions to over two million customers. Electric service was restored to most customers within 30 minutes, except on the Idaho Power Company (IPC)

Power System Control Under Cascading Failures: Understanding, Mitigation, and System Restoration, First Edition. Kai Sun, Yunhe Hou, Wei Sun and Junjian Qi.
© 2019 John Wiley & Sons Ltd. Published 2019 by John Wiley & Sons Ltd.
Companion website: www.wiley.com/go/sun/cascade

system, a portion of the Public Service Company of Colorado (PSC), and the Platte River Power Authority (PRPA) systems in Colorado, where some customers were out of service for up to 6 hours [12]. The first significant event was a single phase-to-ground fault on the 345-kV Jim Bridger–Kinport line due to a flashover (arc) when the conductor sagged close to a tree. On July 3, a similar blackout occurred, also initiated by the tree flashover of the 345 kV Jim Bridger–Kinport line.

- 2003 U.S.-Canadian blackout: A widespread power outage occurred throughout parts of the northeastern and midwestern United States and the Canadian province of Ontario on August 14, 2003, affecting an estimated 10 million people in Ontario and 45 million people in eight U.S. states [13]. The initiating events were the out-of-service of a generating plant in Eastlake, Ohio, and the following tripping of several transmission lines due to tree flashover. Key factors include inoperative state estimator due to incorrect telemetry data and the failure of the alarm system at FirstEnergy's control room.

- 2003 Italy blackout: There was a serious power outage that affected all of Italy – except the islands of Sardinia and Elba – for 12 hours and part of Switzerland near Geneva for 3 hours on September 28, 2003. It was the largest blackout in the series of blackouts in 2003, affecting a total of 56 million people [14]. The initiating event was the tripping of a major tie line from Switzerland to Italy due to tree flashover. Then a second 380-kV line also tripped on the same border (Italy–Switzerland) due to tree contact. The resulting power deficit in Italy caused Italy to lose synchronism with the rest of Europe, and the lines on the interface between France and Italy were tripped by distance relays. The same happened for the 220-kV interconnection between Italy and Austria. Subsequently, the final 380-kV corridor between Italy and Slovenia became overloaded and it too was tripped. Due to a significant amount of power shortage, the frequency in the Italian system started to fall. The frequency decay was not controlled adequately to stop generation from tripping due to underfrequency. Thus, over the course of several minutes, the entire Italian system collapsed, causing a nationwide blackout [15].

- 2012 Indian blackout: On July 30 and 31, 2012, there was a major blackout in India that affected over 600 million people. On July 30, nearly the entire north region covering eight states was affected, with a loss of 38 000 MW of load. On July 31, 48 000 MW of load was shed, affecting 21 states. These major failures in the synchronously operating North-East-Northeast-West grid were initiated by overloadin of an interregional tie line on both days [16–18].

- 2015 Ukrainian blackout: On December 23, 2015, the Ukrainian Kyivoblenergo, a regional electricity distribution company, reported service outages to customers [19]. The outages were due to a third party's illegal entry into the company's computer and supervisory control and data acquisition (SCADA) systems: Starting at approximately 3:35 p.m. local time, seven 110-kV and 23 35-kV substations were disconnected for 3 hours. Later statements indicated that the cyber-attack impacted additional portions of the distribution grid and forced operators to switch to manual mode. The event was elaborated on by the Ukrainian news media, who conducted interviews and determined that a foreign attacker remotely controlled the SCADA distribution management system. The outages were originally thought to have affected approximately 80,000 customers, based on the Kyivoblenergo's update to customers. However, later it was revealed that three different distribution companies were attacked, resulting in several outages that caused approximately 225,000 customers to lose power across various areas.

- 2016 Southern California disturbance: On August 16, 2016, the Blue Cut fire began in the Cajon Pass and quickly moved toward an important transmission corridor that is composed of three 500-kV lines owned by Southern California Edison (SCE) and two 287-kV lines owned by Los Angeles Department of Water and Power (LADWP) [20]. The SCE transmission system experienced 13 500-kV line faults, and the LADWP system experienced two 287-kV faults because of the fire. Four of these fault events resulted in the loss of a significant amount of solar photovoltaic (PV) generation. The most significant event related to the solar PV generation loss occurred at 11:45 a.m. Pacific Time and resulted in the loss of nearly 1200 MW.

- 2016 South Australia (SA) blackout: On September 28, 2016 there was a widespread power outage in SA power grid which caused around 850,000 customers to lose their power supply [21]. Before the blackout the total load including loss in the SA power grid was 1826 MW, among which around 883 MW was supplied by wind generation, corresponding to a very high renewable penetration [22]. Late in the afternoon a severe storm hit SA and damaged several remote transmission towers. The SA grid subsequently lost around 52% of wind generation within a few minutes. This deficit had to be compensated by the power import from the neighboring state, Victoria, through the Heywood AC interconnection. The significantly increased power flow was beyond the capability of the interconnection. Ultimately the SA system was separated from the rest of the system before it collapsed [22].

Some of the past cascading blackouts share similarities. For example, the two significant outages in the western North America in 1996 [12], the U.S.-Canadian blackout on August 14, 2003 [13], and the outage in Italy on September 28, 2003 [14], all had tree contact with transmission lines [23]. Modeling and understanding these common features will help prevent future cascading blackouts that might be initiated by the same reason. At the same time, each blackout has its own unique features due to the characteristics of the particular system, which makes the modeling and understanding of cascading failures challenging.

1.1.2 Challenges in Modeling and Understanding Cascading Failures

The modeling and understanding of cascading failures, or in particular cascading blackouts, can be very challenging in the following aspects:

1) Size of the system: The size of the interconnected power system can be very large. For example, in the United States utility companies build power system models, which are then used to create the North American Electric Reliability Corporation (NERC) interconnection-wide models, with over 50 000 buses. Modeling and understanding the possible ways that such a big system fails can be really challenging.

2) Limited computational power: The computational power is constantly improving as technologies for both hardware and software advance. However, it is still very limited. Although $N-1$ contingency analysis is usually achievable, even only $N-2$ contingency analysis for a system with thousands of components can lead to formidable computational burden [24].

3) Mechanisms in cascading blackouts: There can be many mechanisms during a cascading blackout, which can include thermal dynamics of the transmission line and tree contact, human error, power flow redistribution, protection misoperation, voltage collapse, transient instability, oscillation, and so on [25].

4) Complexities of the system: The power system is not only large but also very complex. The components in the power system can have very complex, tight, and even poorly understood interactions. As mentioned in Perrow [26], from the perspective of complex systems, the system-level failures are not caused by any specific event but by the property that the components in the system are tightly coupled and interdependent of each other. Therefore, complex and difficult to understand component interactions make it difficult to capture the failure propagation patterns [5, 27, 28].

5) Evolving system: The power system is evolving all the time. As the economy and the population grow, the load is also constantly increasing. With heavier loads, the margins of the transmission lines will decrease and the stress level of the system will increase, thus increasing the risk of cascading blackouts. To lower the corresponding risk, there are various engineering responses to blackouts, either through upgrading and maintenance of the equipment such as transmission lines or by operational strategies such as improved dispatch and control. It has been conjectured in Carreras et al. and Dobson et al. [29, 30] that these opposing processes lead to a dynamic equilibrium that is self-organized critical.

6) External factors: Many external factors outside the power grid can also contribute to the initiation and propagation of cascading blackouts. If we only pay attention to the power grid itself, we may neglect important risks. For example, if the trees under some lines are not pruned or cleared

properly, it is possible that the lines can be tripped due to tree contact even when the lines are not overloaded, especially in very hot days with low wind [4]. Tree-caused outages are important on system reliability and can cause a large portion of preventable power outages in regions with trees [31–33]. The growing or falling into overhead lines of trees is generally regarded as the single largest cause of electric power outages [23]. Two significant outages in the western North America in 1996 [12], the U.S.-Canadian blackout on August 14, 2003 [13], and the outage in Italy on September 28, 2003 [14], can all be attributed to trees to some extent [23]. More generally, extreme weather can significantly increase outage rates and interacts with cascading effects [34]. Extreme space weather due to coronal mass ejections has the potential to damage the transformers and cause blackouts [35, 36]. Between 2003 and 2012, roughly 679 power outages, each affecting atleast 50000 customers, occurred due to weather events [37]. The number of outages caused by severe weather is expected to rise as climate change increases the frequency and intensity of hurricanes, blizzards, floods, and other extreme weather conditions [38].

7) Emerging problems: The traditional power grid is undergoing a massive transformation across its entire spectrum – generation, transmission, and distribution – through the various smart grid initiatives. For example, it is increasingly dependent on its cyber infrastructure to support the numerous power system applications necessary to provide improved grid monitoring, protection, and control capabilities. Besides, high penetration of renewable generation can bring more risks due to the interactions between the grid and the renewables [20]. With the heavily integrated distributed energy resources, the power grid architecture is evolving fast from a utility-centric structure to a distributed smart grid [39]. These significant changes bring emerging problems and challenges to the modeling and understanding of cascading blackouts.

8) Difficulty in benchmarking and validation: In the last 20 years, many models, software tools, and analytical tools have been developed to study cascading blackouts. Benchmarking and validation are necessary to understand how closely a model corresponds to reality, what engineering conclusions may be drawn from a particular tool, and what improvements need to be made to the tool in order to reach valid conclusions [40]. However, because cascading blackouts are rare events and are usually very complex with several different mechanisms happening at the same time, it is very difficult to verify and validate each of these models or tools.

In Chapters 2 and 3 of this book, we will discuss various models that can be used to model cascading failures and to help understand why and how cascading blackouts happen, based on which effective mitigation measures can be developed to reduce the cascading risks.

1.2 Importance of Controlled System Separation

1.2.1 Mitigation of Cascading Failures

Catastrophic power blackouts can cause tremendous losses and influence up to tens of millions of people. The Northeast Blackout of 1965 was one of the first significant, widespread power outages of the world, causing the disruption of electricity supplies on November 9, 1965, in a huge area covering parts of eight northeast states in the United States and Ontario in Canada [11]. Since then, many efforts have been made to avoid blackouts in North America and other countries all over the world. However, large blackouts continue to happen. Some recent blackout events include the Northeast blackout on August 14, 2003 [13], and the Southwest blackout on September 8, 2011, in North America [41], the European blackout event on November 4, 2006 [42], and the Indian blackout events on July 30 and 31, 2012 [17].

In general, each historical blackout was the consequence of a long chain of cascading failures of individual components triggered by a variety of events such as concurrent multiple equipment failures, natural disasters, mistakes in operations, and so on. Once cascading failures initiate, they successively weaken the transmission network until inevitable system collapse and power outage over a large area [43]. Especially in a late stage of cascading failures, the time allowed for the grid control center to take system-level corrective actions is often very limited, a matter of a few minutes or tens of seconds. Therefore, automatic systemwide protection and control schemes are critical in mitigating cascading failures once failures start to propagate from a local area towards a wide area. At present, most of the existing protection systems lack systemwide coordination. They are either prone to trip equipment under potentially insecure conditions or designed in an event-based control mechanism to react when detecting any planned contingency. Thus, uncoordinated protective actions may not stop cascading failures; rather, they may even trip more equipment to exit operations, cause overload of some remaining equipment due to shifting of load, weaken the transmission network, and as a result, accelerate propagation of failures.

To mitigate cascading failures in a smart grid, it is necessary to develop more adaptive, systemwide control and protection schemes that can coordinate local protective relays to automatically execute an optimized control strategy in real time using wide-area measurements. In many power grids, widely dispersed synchrophasors such as phasor measurement units (PMUs) are important elements of smart grid technology applied to transmission systems. Synchrophasors provide GPS-synchronized real-time voltage and current phasor data at a high sampling rate (e.g., 30–60 Hz) comparable to or exceeding the AC system frequency and can be networked to construct a wide-area measurement system (WAMS) by means of advanced communication infrastructures. A WAMS enables the real-time monitoring of global dynamics of the whole power grid under disturbances and can help coordinate protection and control actions at the system level. A framework for developing a WAMS-based protection and control scheme is suggested here for mitigation of cascading failures:

1) A comprehensive table of control strategies is designed based on adequate offline studies for a wide range of contingency scenarios and operating conditions.
2) The strategy table is maintained online and updated in a timely manner, for example, every 1–15 minutes, to cover any new appearing scenario or operating condition not addressed by offline studies.
3) Once the WAMS detects or reliably predicts any major instability, a control strategy that matches best the current situation from the table with adaptive refining should be performed in time.

1.2.2 Uncontrolled and Controlled System Separations

Consider an interconnected power system having multiple control areas. In general, tie lines connecting the control areas are more critical in power system operations than the other transmission lines because tie lines have relatively high voltage levels and transmission capacities and are key components constituting the backbone of the transmission network. The outage of any tie line causes the areas it connects to be more vulnerable to changes in operating conditions and disturbances. On one hand, the power flow originally carried by the opened tie line will be shifted to the lines on other transmission paths connecting the same two areas. These lines will have reduced power transfer margins. As a result, the transmission network will decrease its robustness and increase its risk to overload other lines. On the other hand, the trip of any tie line can energize power swings on electromechanical oscillation modes and increase the risk of rotor angle instability of the generators in the areas near the tripped tie line. Either line overloading or power swings may trigger protective actions. Local protective relays are seldom coordinated at the system level and are prone to be responsive to power swings, either stable or unstable, and as a result, that may cause unwanted trips of lines, transformers, or generators and even disconnect the transmission network into electrical islands. This is referred to as "uncontrolled system separation."

Some unintentionally formed islands may have a high generation-load imbalance. Thus, maintaining the frequency and voltage levels of such islands must either shed a large amount of load in load-rich islands or trip many generators in generation-rich islands. Moreover, the group of generators isolated into an island may not easily keep synchronism and can lead to an out-of-step condition. Without effective control to synchronize and stabilize generators of the group, some of the generators will have to be tripped by out-of-step protection relays. Also, formation of islands can cause significant power-flow redispatch in some islands and overloads of some lines or transformers, which will be tripped by further protective actions. Thus, line or transformer failures will continue to cause propagation of failures in those islands.

An effective protection and control measure called controlled system separation (CSS) is considered the final resort against a power blackout under cascading failures [44, 45]. It is also called controlled or intentional system islanding in some literature to distinguish it from uncontrolled system separation or islanding [46, 47].

When loss of integrity of the transmission network becomes unavoidable under cascading failures, rather than waiting for uncontrolled system separation to occur, the control center may proactively perform CSS by separating the power transmission system into electrical islands in a controlled manner so that each island is strategically formed with matched generation to stably support its load through a subsystem within the island. Thereafter, although the network loses its integrity, electricity continues to be supplied to most of customers of the system in parallel islands and thus a power blackout is effectively prevented. Once all failures are fixed, the whole system can be restored by resynchronization of all electrical islands. Such a resynchronization requires lower efforts and costs than a blackstart process since most of loads are saved. The latter has to restart the system in the blackout area following a long procedure including starting blackstart generation units to crank other nonblackstart units, energizing transmission lines and picking up loads, which usually takes several hours.

At present, CSS is not widely implemented by power industry. The power grids having schemes of CSS deployed are those experiencing a number of uncontrolled separation events in the past or considering uncontrolled separation as a major threat to grid operations under extreme weather conditions like storms or hurricanes. Typically, CSS schemes are designed based on out-of-step protection schemes with both tripping and blocking functions, which are traditionally applied to protect generators and power systems during unstable conditions [44]. The out-of-step tripping function can isolate unstable generators or control the separation of two areas going out of step, while the blocking function can block unnecessary actions of distance relay elements that are prone to operate during unstable power swings. Out-of-step protection schemes are mainly applied to individual generators or transmission corridors whose conditions are simple enough to detect from local measurements.

If highly reliable, fast, wide-area measuring and communication infrastructures are available in a smart grid, the existing out-of-step protection schemes can be upgraded and coordinated at multiple locations to implement more adaptive and effective CSS. The increasing deployment of synchrophoasor-based WAMS in many countries enables the development and deployment of a practical WAMS-based CSS scheme in the near future based on upgraded out-of-step protection schemes. The WAMS can play critical roles in performing CSS. For example, the WAMS can contribute the following to a CSS scheme:

- It helps operators monitor real-time electromechanical oscillations between control areas and identify the most vulnerable tie lines as potential separation points.
- The system dynamics authentically captured by the WAMS may help detect or predict out-of-step conditions with generators to identify the right timing to conduct CSS.
- Once electrical islands are formed after CSS, the WAMS can continuously monitor frequency and voltage excursions in each island and trigger necessary remedial actions to stabilize the subsystem in the island.
- All electrical islands can be resynchronized more easily to restore the whole system with the aid of synchrophasors placed in different islands.

Chapters 4 and 5 of this book will elaborate key questions about CSS and what useful techniques can be used to answer the questions. Specifically, an important problem of finding separation points for a power system is formulated based on the constraints on islands formed after CSS. Techniques that support WAMS-based online CSS will be introduced and integrated to establish a WAMS-based CSS scheme.

1.3 Constructing Restoration Strategies

1.3.1 Importance of System Restoration

Electric power systems, like transportation, communication, and gas systems, are among the most critical infrastructures of modern societies. These infrastructures share some common characteristics, such as wide geographical coverage, interconnection of numerous components, high reliability requirements, and so forth.

Power systems do have some defining features that contribute to the exceptional complexity of this type of artificial system. First, the generation and consumption of large amounts of electricity have to be implemented simultaneously. The imbalance between generation and consumption instantly causes disturbances between various components spanning a large geographical area. Second, the diversification of the interconnected components in power systems results in various nonlinear properties and dynamics under hybrid time scales. Third, uncertainties from the generation side (such as renewable energy sources), the customer side, and the exogenous factors (for example, extreme weather conditions) can constantly create disturbances that impact the security and reliability of a power supply. As such, power system operation and control are highly complicated. Some significant unforeseen disturbances may cause power systems collapse.

The history of major failures in power systems, namely total blackouts or widespread outages, is almost as long as that of the power industry itself [48]. Some major failures from the late 1970s to the year 2003 with high adverse impacts can be found in Adibi et al. [49, 50], among which the most severe one is the US-Canada blackout of August 14, 2003. In this catastrophic event, a greater than 60 GW load was interrupted for 50 hours. These outages have a huge adverse impact on the economy of modern societies. For example, the Electric Power Research Institute of the United States (EPRI, US) has estimated that US economic loss from outages has reached US$104–164 billion per year [51]. Therefore, minimizing the duration of outages and their impact on the public has become a major concern of system operators.

After a power system collapses, power system restoration is a time-consuming and complicated process involving a series of control actions, taken by the system operators with rigorous temporal dependency, to rebuild the generation and transmission of the power supply and eventually to restore the interrupted electric supply to the affected customers. The studies in power system restoration aim to reduce the possibility and the impact of major outage events by providing solutions for restoration planning, real-time situation awareness, preventive controls, corrective controls, and restorative controls, in the context of restoration processes. After a major outage happens, system operators attempt to restore the power system back to a secure normal operating condition. The control actions may involve ascertaining the extent of outage, maintaining the stable operation of generating units with blackstart capability, cranking nonblackstart units, energizing transmission lines and transformers to establish cranking paths, picking up loads (i.e., restoring the electric supply to customers), and performing voltage and frequency control.

1.3.2 Classification of System Restoration Strategies

1.3.2.1 Diversification of Restoration Strategies

The effectiveness and efficiency of a restoration strategy after an outage depend primarily on the match of this strategy with both the characteristics and the post-fault state of the target power system.

On one hand, the portfolio of generating units will affect the restoration process substantially. According to the proportion of various types of generating units, the power systems can be categorized into thermal systems, hydrothermal systems, and primarily hydro systems [48]. For the thermal systems, the restoration process can be divided into four phases; namely, restarting steam units, reconnecting generation stations, picking up loads, and interconnecting islands. For the hydrothermal systems, the restoration process differs from thermal systems, in that the entire transmission network can be energized in one step. For primarily hydro systems, the restart of generating units is not a major concern. Instead, the switching of long-distance high voltage transmission lines becomes the critical emphasis.

On the other hand, the restoration process should consider the actual outage scenarios. The restoration strategies considering diverse postdisturbance conditions can be grouped into the following five general philosophies [52]: build-upward, build-downward, build-inward, build-outward, and build-together. For example, for a total blackout, the build-upward can be applied to sectionalize the power system into islands, then to restore and firm up these islands, and eventually synchronize all of them. In situations where tie lines are available, the build-inward can be employed, in which some transmission lines are established with the aid of tie lines. Important stations can thus be restored and the restoration process can proceed.

1.3.2.2 General Restoration Phases

The diversification of restoration strategies causes difficulty in refining the general technical challenges and proposing systematic solutions to facilitate the restoration process. By surveying the comprehensive industry practice, associated manuals and guidelines, general restoration phases are proposed in Fink et al. [52]. These phases are termed preparation, system restoration (or network reconfiguration; to avoid ambiguity between system restoration and power system restoration, network reconfiguration will be used hereafter in this book to refer to the second phase), and load restoration, identifying the common features underlying various restoration processes. The major concern and measures taken in each phase are different from those in other phases.

In the preparation phase, the priority is to identify the system status and take action as quickly as possible. The initial energy sources – surviving generating units and those with blackstart capacity (hydro units, gas turbines, etc.) – are identified and are used to crank nonblackstart units. The critical loads, such as the offline power demand of nuclear generating units [53], are also to be restored. Necessary transmission switches will be conducted to establish paths from the surviving generating units to the nonblackstart units and critical loads. This phase typically lasts 1–2 hours.

In the network reconfiguration phase, the major objective is to reestablish the skeleton of the transmission network. Tasks that will be carried out include (i) energize key transmission lines and important substations, (ii) firm up islands and synchronize them as appropriate, and (iii) restore a small number of loads to stabilize generation and for voltage control. This phase typically lasts 1–3 hours.

In the load restoration phase, the goal is to restore service as quickly as reasonable to the interrupted customers based on the load importance/priority. In this final phase, the load pickup can be carried out in a larger increment since the generation and the transmission network have been firmed up in the previous phases. However, caution should be taken to limit the total load pickup subject to the total available frequency response capacity. In addition, load pickup should not be done in a hasty fashion due to the uncertainty of the complicated load behaviors.

1.3.2.3 Reliability Guidelines and Standards for Restoration
To enhance reliability and impose critical general guidelines in the context of power system restoration, both North America and Europe have established formal documentation for the planning, drilling, and operation of restoration processes.

The European Network of Transmission System Operators for electricity (ENTSO-E) has drafted the network code on emergency control and restorative control for regional transmission networks [54], such as Baltic, Continental Europe, Britain, Nordic, and so on. This draft collects the existing rules and practice in various areas and identifies common practice and well-recognized critical issues. This draft also highlights the importance of wide-area system state assessment, information exchange among transmission system operators, and coordination among participants during the restoration process.

As a comparison, North America has established more rigorous restoration standards for utilities to follow. The Federal Energy Regulatory Commission (FERC) established the NERC for developing and enforcing reliability standards for power systems, in the context of planning, operation, emergency, and restoration [55–60]. The NERC standards related to power system restoration are as follows.

1) EOP-005-1: System Restoration Plans
2) EOP-005-2: System Restoration from Blackstart Resources
3) EOP-006-1: Reliability Coordination – System Restoration
4) EOP-006-2: System Restoration Coordination
5) EOP-007-0: Establish, Maintain, and Document a Regional Blackstart Capability Plan

The key information carried in these standards is outlined as follows.

1) The plans, procedures, and resources should be available to restore the power system on the occurrences of actual outage events.
2) The transmission operator's system should have adequate blackstart resources and reliable paths that reach the nonblackstart units. Personnel should regularly drill the procedures to start up the blackstart resources.
3) The control actions taken by various restoration participants must be coordinated to ensure reliability in each phase of restoration.
4) The regional blackstart capacity plan plays a central role in enabling sufficient blackstart capacity to function as expected, and therefore should be maintained and tested on a periodical basis.

1.3.2.4   Emerging Challenges and New Research Opportunities

Although the standards mentioned provide general guidelines during the restoration process, a wide spectrum of challenges needs to be solved. Besides the technical challenges summarized in Adibi et al. [48, 49], some new challenges, resulting from both the deregulation and growth of concern in modern resilient power systems, have created profound theoretical and practical questions. Moreover, new types of components, such as high-voltage direct-current (HVDC) transmission systems, provide new options in conducting the restoration process. These emerging requirements and new components have added new research dimensions to power system restoration. Furthermore, state-of-the-art optimization theory and high-performance computing technologies have permeated almost all engineering disciplines, including power system engineering. These advancements present powerful tools for modeling and solving mathematical problems arising from power system restoration. However, applying these theories and technologies is not a straightforward task.

1.3.3   Challenges of System Restoration

1.3.3.1   Restoration for Power Systems Under Market Environments

Whether the restructuring of the power industry contributes to the frequency of the major outage events is still an issue with diverse options. There seems to exist no strong evidence to relate the power market to the blackout risk [61]. However, under market environments, the operation of both the power system and the power plant is quite close to the reliability margin and therefore impacts the security of the power system [50].

Restoration in a market-based power system is far more complicated than in a vertically structured one. This is because the various independent restoration participants, such as transmission owners, distribution owners, and generation owners, have their own concerns and obligations, whereas the control actions taken by each party should be coordinated [62]. How to implement a smooth coordination during restoration is still an open question.

In addition to technical challenges, regulatory and economic issues in the restoration should also be taken into consideration [62, 63]. The blackstart service is commonly regarded as a type of ancillary service. The pricing of blackstart service and the blackstart cost are difficult to calculate, since the benefits are embedded in the entire restoration process and are distributed among collaborative yet independent entities. This difficulty adds barriers for market participants to invest in the blackstart capacity or associated technological innovation.

1.3.3.2   Resilience Requirements for Future Power Grids

Power systems have been experiencing a paradigm shift toward more reliable, sustainable, environmentally friendly, economic, and efficient modern power grids [64]. The self-healing capability is the first defining feature of modern power grids [64, 65]. This capability is in essence the immune system of modern power grids that enables problematic elements of a power grid to be identified, isolated, and restored, with little or no manual intervention, such that the interruption of electric supply can be minimized [66].

To fulfill the goal of enabling a self-healing resilient power grid, the concept of "smart restoration" [51] has been proposed, with the key idea being that power grids are able to automatically perform self-assessment, response to disturbances, and conduct preventive/corrective/restorative control as necessary.

Designing and deploying such modern power grids is not an easy task. The following fundamental challenges must be met. First, the traditional restoration methodology based on offline planning together with operators' experience must be changed. Emerging requirements include, but are not limited to, an advanced measurement system for situation awareness, efficient and reliable control methods based on rigorous computation to enable the online close-loop operation, and a reliable communication system that enables smooth coordinated controls. Second, owing to the growth of environmental concerns, renewable energy sources (RES) are playing an increasingly important role in the generation side. The percentage of RES capacity in the generation portfolio has reached such a point that RES cannot be neglected in any operation scenario. The uncertainty and variability of RES, however, will hinder the participation of RES in the restoration process, where security and reliability are crucial issues.

On the bright side, new types of power system components and advancement of control, optimization, and computation technologies provide powerful new tools to overcome these fundamental challenges.

1.3.3.3 Emerging Components and Novel Measurement Methods

There are ample new types of components and technological innovations that benefit the development of modern resilient power grids.

HVDC has been widely used in power systems [67], with the appealing feature of transferring a large amount power over a long distance in an economic and flexible fashion. Depending on the topologies and control methodologies, HVDC can be categorized into two types, the line commutated converter (LCC-HVDC) and the voltage source converter (VSC-HVDC).

LCC-HVDC is generally recognized as incapable of supplying a passive load; therefore, it has no blackstart capability. Yet recent research shows that this type of HVDC could serve as a blackstart source if novel control strategies are employed [68, 69]. By contrast, VSC-HVDC inherently has the capability as a blackstart source [70]. How to efficiently utilize HVDC links in different phases of restoration needs further studies.

On the generation side, thermal generating units are commonly regarded as lacking blackstart capability. Fortunately, the fast cut back function (FCB) may enable thermal generating units to have some blackstart capability within a considerable period after a major outage happens. FCB function enables a unit in normal loading level to instantaneously reduce its output down to the house load level in a stabilized operation mode after it is tripped from the power system [71]. This is done by rapidly cutting back the fuel, feed water, and air in response to the turbine-generator output. By triggering this function, the running unit is always ready for parallel operation back to the power system. Thus, FCB capability will benefit power system restoration as a new option.

1.3.3.4 Advancement of Optimization and Computation Technologies
It is common practice that decision-making during power system restoration is modeled into various optimization models. The optimization models bridge the objective with the large spectrum of constraints that power system restoration must take into account.

One fundamental challenge in power system optimization is the convexification of power flow equations [72], such that the state-of-the-art global optimization algorithms can be applied to obtain the solution within polynomial time. This challenge is also important because the power system restoration must consider the power flow as constraints. If one applies numerical optimization algorithms to solve power-flow-constrained optimization models in power system restoration, the convexification of power flow equations is the critical step. However, it is still an open question, providing ample research opportunities. The other challenge is how to solve optimization models with integral and differential equations, which in nature reflect the power system dynamics. Generally, there is no analytical solution to this type of dynamic optimization model. The construction of an efficient algorithm is important to implement an online control for power system restoration. Both challenges will be addressed in this book.

As for high-performance computing technology, graphics processing units (GPUs) have been successfully applied in many engineering disciplines, including power systems [73, 74]. When applying GPUs in power system computations involving power flow, it is still unknown how to facilitate the sparse feature of the power system in the data-parallel computation paradigm, since GPUs are not designed to solve the sparse linear system with significant data dependency. This question will also be addressed in this book, particularly aiming to improve the efficiency of time-domain simulation for the load restoration strategies.

1.4 Overview of the Book

This book will, for the first time, provide a comprehensive introduction to power system control under cascading failures that covers all three major topics related to cascading failures in power transmission grids: (i) modeling and understanding cascading failures (Chapters 2 and 3); (ii) mitigation of cascading failures by controlled system separation (Chapters 4 and 5); and (iii) power system restoration from cascading failures (Chapters 6–10). Related state-of-the-art technologies will be introduced and illustrated in detail with hands-on examples for the readers to learn how to use them to address specific problems. The following is a brief summary of the rest of the book.

Chapter 2 introduces typical models for the simulation of cascading failures, categorized into two classes. The first class of models is general cascading failure models, which are abstract and applicable to many complex systems, including power systems. They include the Bak–Tang–Wiesenfeld sandpile model, failure-tolerance sand-pile model, Motter–Lai model, influence model, binary-decision model, coupled map lattice model, CASCADE model, and interdependent failure model. The second class of models is specifically power system cascading failure models, developed for power system cascading blackouts. They include the hidden failure model, Manchester model, OPA model, and its variants and cascading failure models considering dynamics and detailed protections.

To understand cascading failures, Chapter 3 discusses several theories and models that can extract useful and actionable information from simulated or historical cascading failure data, answering why and how cascading failures happen in electric power grids. Specifically, this chapter focuses on analyses using the self-organized criticality theory, branching process model, multitype branching process model, and failure interaction models considering dynamics and detailed protections.

Chapter 4 first introduces three important questions on CSS (controlled system separation): Where to separate? When to separate? How to separate? Then the chapter focuses on finding separation points for CSS. Constraints on separation points are presented in detail. The mathematical problems on separation points are formulated based on graph theory. Useful techniques for solving separation points, such as slow-coherency-based generator grouping, graph-theory-based power network reduction, and the ordered binary decision diagram method, are introduced. The three-step approach is suggested to integrate these techniques for checking constraints on separation points.

Chapter 5 first presents several techniques that enable an online CSS scheme using synchrophasor-based wide-area measurements. These techniques include measurement-based spectral analysis to determine a two-cut separation boundary by mode shapes, the "frequency-amplitude curve" to predict angular instability from drifting oscillation frequencies, a phase-locked loop-based method for accurate online identification of generation grouping, and an algorithm for real-time out-of-step prediction across a determined separation boundary. Then a unified framework for practical implementation of WAMS-based CSS is suggested to address the questions "where," "when," and "how."

Chapter 6, as the first chapter on power system restoration, introduces the constraints to be addressed during a practical process of system restoration. The first set of constraints are general physical constraints on the startup procedures of generating units, system sectionalizing, and reconfiguration and load restoration, which concern most power systems. The second set of constraints are specifically about electromagnetic transient behaviors of a power system during restoration, such as the constraints on generator self-excitation, switching overvoltage issues with lines, resonant overvoltage issues and magnetizing currents with transformers, and voltage and frequency during the load picking up process.

Chapter 7 presents the methodology and implementation algorithms for optimization of a restoration strategy satisfying the constraints on both generation starting up and load restoration. The optimization of generation starting up is achieved under a bi-level framework by solving a primary problem minimize the duration of restarting generating units and establishing the network and a secondary problem determining the outputs of restarted generating units and picking up dispatchable loads. The optimization for load restoration is to formulate and solve a mixed-integer nonlinear load restoration problem.

Chapter 8 presents the optimal strategy of harnessing renewable and energy storage in power systems restoration. The optimization models and solution algorithms are introduced to address the uncertainty and variability of renewable energy resources in restoration planning. The operations and control of wind turbines are presented for blackstart and load restoration. The roles of energy storage (including pumped-storage hydro units, batteries, and electric vehicles) in alleviating uncertainties of renewables are discussed with the optimal coordination strategies.

Chapter 9 introduces emerging issues and related technologies for power system restoration. The emerging issues include considerations and applications of flexible alternating current transmission system (FACTS) and HVDC technologies during system restoration, applications of PMUs to facilitate the process of restoration, frequency deviations with load restoration and other reliability concerns during system restoration, and the dispatch of microgrids to speed up system restoration.

Finally, Chapter 10, from a system planning perspective, presents the methodology for assessment and optimization of the blackstart capability with a power system. A decision support tool is introduced to assist utility companies in planning studies on their blackstart capabilities.

This book is written as a reference for postgraduate students and researchers who work in the field of power system control under cascading failures. The book can also be used as a reference by electrical or system engineers to improve industry practices in power system planning and operations against cascading failures and blackouts.

References

1 Dobson, I., Carreras, B.A., Lynch, V.E., and Newman, D.E. (2007). Complex systems analysis of series of blackouts: cascading failure, critical points, and self-organization. *Chaos* 17 (2): 026–103.

2 Carreras, B.A., Lynch, V.E., Dobson, I., and Newman, D.E. (2002). Critical points and transitions in an electric power transmission model for cascading failure blackouts. *Chaos* 12 (4): 985–994.

3 Dobson, I., Carreras, B., Lynch, V., and Newman, D. (2001). An initial model for complex dynamics in electric power system blackouts. In *Proceedings of the 34th Annual Hawaii International Conference on System Sciences, February 2001 (HICCS)* IEEE, pp. 710–718.

4 Qi, J., Mei, S., and Liu, F. (2013). Blackout model considering slow process. *IEEE Trans. Power Syst.* 28 (3): 3274–3282.

5 Qi, J., Sun, K., and Mei, S. (2015). An interaction model for simulation and mitigation of cascading failures. *IEEE Trans. Power Syst.* 30 (2): 804–819.

6 Qi, J., Dobson, I., and Mci, S. (2013). Towards estimating the statistics of simulated cascades of outages with branching processes. *IEEE Trans. Power Syst.* 28 (3): 3410–3419.

7 Qi, J., Ju, W., and Sun, K. (2017). Estimating the propagation of interdependent cascading outages with multi-type branching processes. *IEEE Trans. Power Syst.* 32 (2): 1212–1223.

8 Strogatz, S.H. (2001). Exploring complex networks. *Nature* 410 (6825): 268–276.

9 Dorogovtsev, S.N. and Mendes, J.F. (2002). Evolution of networks. *Adv. Phys.* 51 (4): 1079–1187.

10 Mantegna, R.N. and Stanley, H.E. (2000). *An Introduction to Econophysics: Correlation and Complexity in Finance.* Cambridge University Press.

11 Vassell, G.S. (1990). The northeast blackout of 1965. Public Utilities Fortnightly (United States), 126 (8).

12 North American Electric Reliability Council (2002). 1996 system disturbances.

13 U.S.-Canada Power System Outage Task Force (2004). Final report on the August 14th blackout in the United States and Canada. Department of Energy and National Resources Canada.

14 UCTE Investigation Committee (2003). Interim report of the investigation committee on the 28 September 2003 blackout in Italy. UCTE Report, October, 27.

15 Andersson, G., Donalek, P., Farmer, R. et al. (2005). Causes of the 2003 major grid blackouts in North America and Europe, and recommended means to improve system dynamic performance. *IEEE Trans. Power Syst.* 20 (4): 1922–1928.

16 Romero, J.J. (2012). Blackouts illuminate India's power problems. *IEEE Spectr.* 49 (10): 11–12.

17 Gailwad, A. (2013). Indian blackouts – July 30 & 31 2012 recommendations and further actions, in IEEE PES General Meeting.

18 Rampurkar, V., Pentayya, P., Mangalvedekar, H.A., and Kazi, F. (2016). Cascading failure analysis for Indian power grid. *IEEE Trans. Smart Grid* 7 (4): 1951–1960.

19 Lee, R.M., Assante, M.J., and Conway, T. (2016). Analysis of the cyber attack on the Ukrainian power grid. SANS Industrial Control Systems.

20 North American Electric Reliability Corporation (2017). 1,200 MW fault induced solar photovoltaic resource interruption disturbance report.

21 Australian Energy Market Operator (AEMO) (2017). Black System South Australia 28 September 2016.

22 Yan R., Al Masood N., Saha T.K. et al. (2018). The anatomy of the 2016 South Australia blackout: A catastrophic event in a high renewable network. *IEEE Transactions on Power Systems*.

23 Cieslewicz, S. and Novembri, R. (2004). Utility vegetation management final report. US Federal Energy Regulatory Commission.

24 Kaplunovich, P. and Turitsyn, K. (2016). Fast and reliable screening of N-2 contingencies. *IEEE Trans. Power Syst.* 31 (6): 4243–4252.

25 Baldick, R., Chowdhury, B., Dobson, I., Dong, Z., Gou, B., Hawkins, D., et al. (2008). Initial review of methods for cascading failure analysis in electric power transmission systems in Power and Energy Society General Meeting – Conversion and Delivery of Electrical Energy in the 21st Century, IEEE, 1–8.

26 Perrow, C. (2011). *Normal Accidents: Living with High Risk Technologies*. Princeton University Press.

27 Ju, W., Qi, J., and Sun, K. (2015). Simulation and analysis of cascading failures on an NPCC power system test bed, in *2015 IEEE Power Energy Society General Meeting*, 1–5, doi:10.1109/PESGM.2015.7286478.

28 Ju, W., Sun, K., and Qi, J. (2017). Multi-layer interaction graph for analysis and mitigation of cascading outages. *IEEE J. Emerging Sel. Top. Circuits Syst.* 7 (2): 239–249. doi: 10.1109/JETCAS.2017.2703948.

29 Carreras, B.A., Newman, D.E., Dobson, I., and Poole, A.B. (2000). Initial evidence for self-organized criticality in electric power system blackouts, in *Proceedings of the 33rd Annual Hawaii International Conference on System Sciences*, p. 6.

30 Dobson, I., Carreras, B.A., Lynch, V.E., and Newman, D.E. (2001). An initial model for complex dynamics in electric power system blackouts, in *Proceedings of the 34th Annual Hawaii International Conference on System Sciences*, 710–718.

31 Simpson, P. and Van Bossuyt, R. (1996). Tree-caused electric outages. *J. Arboric.* 22: 117–121.

32 Radmer, D.T., Kuntz, P.A., Christie, R.D. et al. (2002). Predicting vegetation-related failure rates for overhead distribution feeders. *IEEE Trans. Power Delivery* 17 (4): 1170–1175.

33 Guikema, S.D., Davidson, R.A., and Liu, H. (2006). Statistical models of the effects of tree trimming on power system outages. *IEEE Trans. Power Delivery* 21 (3): 1549–1557.

34 Dobson, I., Zhou, K., Carrington, N.K. (2018). Exploring cascading outages and weather via processing historic data. In: 51st Hawaii International Conference on System Sciences (HICSS), IEEE.

35 North American Electric Reliability Corporation (2012). Effects of Geomagnetic Disturbances on the Bulk Power System., North American Electric Reliability Corporation, Atlanta, GA.

36 Oughton, E.J., Skelton, A., Horne, R.B. et al. (2017). Quantifying the daily economic impact of extreme space weather due to failure in electricity transmission infrastructure. *Space Weather*, 15: 65–83.

37 U.S. Department of Energy, Office of Electricity Delivery and Energy Reliability. Electric Disturbance Events (OE-417) Annual Summaries.

38 Executive Office of the President. Council of Economic Advisers (2013). Economic Benefits of Increasing Electric Grid Resilience to Weather Outages.

39 Qi, J., Hahn, A., Lu, X. et al. (2016). Cybersecurity for distributed energy resources and smart inverters. *IET Cyber-Phys. Syst. Theory Appl.* 1 (1): 28–39.

40 Bialek, J., Ciapessoni, E., Cirio, D. et al. (2016). Benchmarking and validation of cascading failure analysis tools. *IEEE Trans. Power Syst.* 31 (6): 4887–4900.

41 FERC/NERC, Arizona-Southern California Outages on September 8, 2011: Causes and Recommendations, April 2012.

42 Union for the Co-ordination of Transmission of Electricity. Final Report – System Disturbance on 4 November 2006.

43 IEEE (2008). PES CAMS Task Force on Understanding Prediction, Mitigation and Restoration of Cascading Failures, Initial review of methods for cascading failure analysis in electric power transmission systems, *IEEE PES General Meeting, Pittsburgh*, June 2008.

44 Adibi, M., Kafka, R.J., Maram, S., and Mili, L.M. (2006). On power system controlled separation. *IEEE Trans. Power Syst.* 21 ((4): 1894–1902.

45 Sun, K., Hur, K., and Zhang, P. (2011). A new unified scheme for controlled power system separation using synchronized phasor measurements. *IEEE Trans. Power Syst.* 26 (3): 1544–1554.

46 You, H., Vittal, V., and Yang, Z. (Feb. 2003). Self-healing in power systems: an approach using islanding and rate of frequency decline-based load shedding. *IEEE Trans. Power Syst.* 18: 174–181.

47 Sun, K., Zheng, D., and Lu, Q. (May 2003). Splitting strategies for islanding operation of large-scale power systems using OBDD-based methods. *IEEE Trans. Power Syst.* 18 (2): 912–923.

48 Adibi, M., Clelland, P., Fink, L. et al. (May 1987). Power system restoration – a task force report. *IEEE Trans. Power Syst.* 2 (2): 271–277.

49 Adibi, M.M., Borkoski, J.N., and Kafka, R.J. (Nov. 1987). Power system restoration - the second task force report. *IEEE Trans. Power Syst.* 2 (4): 927–932.

50 Adibi, M.M. and Fink, L.H. (Sep.-Oct. 2006). Overcoming restoration challenges associated with major power system disturbances – restoration from cascading failures. *IEEE Power Energ. Mag.* 4 (5): 68–77.

51 Liu, S., Hou, Y., Liu, C.-C., and Podmore, R. (Jan.–Feb. 2014). The healing touch: tools and challenges for smart grid restoration. *IEEE Power Energ. Mag.* 12 (1): 54–63.

52 Fink, L.H., Liou, K.-L., and Liu, C.-C. (May 1995). From generic restoration actions to specific restoration strategies. *IEEE Trans. Power Syst.* 10 (2): 745–752.

53 Adibi, M.M., Adsunski, G., Jenkins, R., and Gill, P. (Aug. 1995). Nuclear plant requirements during power system restoration. *IEEE Trans. Power Syst.* 10 (3): 1486–1491.

54 ENTSO-E, (2014). Current practices in Europe on emergency and restoration, May 2014 [Online]. Available: https://www.entsoe.eu/Documents/ Network%20codes%-20documents/NC%20ER/140527_NC_ER_Current_ practices_on_Emergency_and_Restoration.pdf.

55 Kafka, R. J. (2008). Review of PJM restoration practices and NERC restoration standards, in *Proceedings of 2008 IEEE Power and Energy Society General Meeting*, 2008.

56 North American Electric Reliability Corporation (NERC), (2005). EOP-007-0: establish, maintain, and document a regional blackstart capability plan," Apr. 1, 2005 [Online]. Available: http://www.nerc.com/ files/EOP-007-0.pdf.

57 North American Electric Reliability Corporation (NERC), (2007). EOP-006-1: reliability coordination-system restoration, Jan. 1, 2007 [Online]. Available: http://www.nerc.com/files/EOP-006-1.pdf.

58 North American Electric Reliability Corporation (NERC), (2013). EOP-006-2: system restoration coordination, Jan. 1, 2013 [Online]. Available: http://www. nerc.com/files/EOP-006-2.pdf.

59 North American Electric Reliability Corporation (NERC), (2006). EOP-005-1: system restoration plans, May 2, 2006 [Online]. Available: http://www.nerc. com/files/EOP-005-1.pdf.

60 North American Electric Reliability Corporation (NERC) (2013). EOP-005-2: system restoration from blackstart resources, Nov. 21, 2013 [Online]. Available: http://www.nerc.com/files/EOP-005-2.pdf.

61 Hines, P., Apt, J., and Talukdar, S., (2008). Trends in the history of large blackouts in the United States. In *Proceedings of the 2008 IEEE Power and Energy Society General Meeting, 2008*.

62 Feltes, J. and Grande-Moran, C. (Jan.–Feb. 2014). Down, but not out: a brief overview of restoration issues. *IEEE Power Energ. Mag.* 12 (1): 34–43.

63 Lin Z. and Wen F. (2007). Power system restoration in restructured power industry. In *Proceedings of the 2007 IEEE Power Engineering Society General Meeting, June 2007*.

64 US. Department of Energy (DOE), (2007). A systems view of the modern grid, Jan. 2007 [online]. Available: https://www.smartgrid.gov/sites/default/files/pdfs/a_systems_view_of_the_modern_grid.pdf.

65 US. Department of Energy (DOE), (2010). Understanding the benefits of the smart grid, Jun. 2010 [online]. Available: https://www.smartgrid.gov/sites/default/files/doc/files/Understanding_Benefits_Smart_Grid_201003.pdf.

66 US. Department of Energy (DOE), (2010). Anticipates and responds to system disturbances (self-heals), Sep. 2, 2010 [online]. Available: https://www.smartgrid.gov/sites/-default/files/pdfs/self_heals_final 08312010.pdf.

67 Povh, D. (Feb. 2000). Use of HVDC and FACTS. *Proc. IEEE* 88 (2): 235–245.

68 Kotb, O. and Sood, V. K. (2010). A hybrid HVDC transmission system supplying a passive load. In *Proceedings of the 2010 IEEE Electric Power and Energy Conference (EPEC)*, 2010.

69 Andersen, B.R. and Lie, X. (Oct. 2004). Hybrid HVDC system for power transmission to island networks. *IEEE Trans. Power Delivery* 19 (4): 1884–1890.

70 Ying, J.-H., Duchen, H., Karlsson, M., Ronstrom, L. and Abrahamsson, B. (2008). HVDC with voltage source converters – a powerful standby black start facility. In *Proceedings of the 2008 IEEE/PES Transmission and Distribution Conference and Exposition, 2008*.

71 Huang, W., Zhang, X., and Zhang, Z. (2011). Research and study of FCB test based on conventional configuration. In *Proceedings of the 2011 International Conference on Electronics, Communications and Control (ICECC), 2011*, pp. 4448–4450.

72 Low, S., Gayme, D., and Topcu, U. (2013). Convexifying optimal power flow: recent advances in OPF solution methods. In *Proceedings of the 2013 American Control Conference (ACC)*, 2013, pp. 5245–5245.

73 Jalili-Marandi, V. and Dinavahi, V. (Aug. 2010). SIMD-based large-scale transient stability simulation on the graphics processing unit. *IEEE Trans. Power Syst.* 25 (3): 1589–1599.

74 Garcia, N. (2010). Parallel power flow solutions using a biconjugate gradient algorithm and a Newton method: A GPU-based approach. In *Proceedings of the 2010 IEEE Power and Energy Society General Meeting, 2010*.

2

Modeling of Cascading Failures

In this chapter, typical models for the simulation of cascading failures will be introduced. These models are divided into two main classes: general cascading failure models and power system cascading failure models. The first class of models may be applicable to different types of systems if they are properly adjusted to consider the specific properties of the system. The second class of models is specially developed for simulating power system cascading blackouts.

2.1 General Cascading Failure Models

2.1.1 Bak–Tang–Wiesenfeld Sandpile Model

The Bak–Tang–Wiesenfeld (BTW) sandpile model [1–3] is a prototypical theoretical model that exhibits avalanche behavior. It was originally introduced on the one- or higher-dimensional lattices [1, 2]. The main feature of the BTW sandpile model on Euclidean space is the emergence of a power law with an exponential cutoff in the avalanche size distribution [1, 2]:

$$p_a(s) \sim s^{-\tau} e^{-s/s_c}, \tag{2.1}$$

where s is the avalanche size (the number of toppling events) and s_c is its characteristic size.

It was then generalized to networks [4–6] and has been widely used to study cascading failures and self-organizing dynamics on complex networks.

Consider a network with n nodes that hold grains of sand. The topology of the network is fixed, but the number of grains of sand on each node may change in time. Let $k(i)$ and $\mathcal{N}(i)$, respectively, be the degree and the set of neighbors of node i. Each node i holds a certain number z_i of grains of sand. A node is called s-sand if it holds s grains of sand. The maximum amount of sand that a node can hold is called the *capacity* of that node, which is denoted by $K(i)$ for

Power System Control Under Cascading Failures: Understanding, Mitigation, and System Restoration, First Edition. Kai Sun, Yunhe Hou, Wei Sun and Junjian Qi.
© 2019 John Wiley & Sons Ltd. Published 2019 by John Wiley & Sons Ltd.
Companion website: www.wiley.com/go/sun/cascade

node i. $K(i)$ is usually set to be $k(i) - 1$. A node over capacity topples by shedding one grain of sand to each of its neighbors. We begin a new cascade by dropping a grain on a randomly chosen root node. After a transient event, the system reaches a steady state in which the input and output of energy is balanced. On the network, cascades occur as follows:

1) Initially each node has a random initial load not exceeding its capacity.
2) Drop a grain of sand on a randomly chosen node i so that $z_i \rightarrow z_i + 1$. This node is called the *root* of the cascade.
3) If this addition of sand does not bring the root over capacity, the cascade is finished. Otherwise, the root topples by shedding one grain to each of its neighbors as $z_i \rightarrow z_i - k(i), z_j \rightarrow z_j + 1, j \in \mathcal{N}(i)$, where $\mathcal{N}(i)$ is the neighbors of i.
4) Any other node that now exceeds its capacity topples in the same way until all nodes are under or at capacity. When a grain of sand moves from one node to another, it can dissipate (disappear) with a small probability ϵ.

Bonabeau [4] studied the BTW sandpile model on the Erdös-Rényi (ER) random network and found that the avalanche size distribution follows a power law with the exponent $\tau \approx 1.5$, which is consistent with the mean-field solution in Euclidean space [7].

In [5] the BTW sandpile model was studied on scale-free networks [8] where the capacity of the node is distributed heterogeneously. A scalefree network is a network whose degree distribution follows a power law as $p_d(k) \sim k^{-\gamma}$ with the degree exponent γ [8]. It was found that the avalanche size distribution also follows a power law with an exponent τ. When $2 < \gamma < 3$, τ is a function of the degree exponent γ of the scale-free network as $\tau = \gamma/(\gamma - 1)$. For $\gamma > 3$ the mean-field value of τ is 1.5.

2.1.2 Failure-Tolerance Sandpile Model

Different controls have been proposed in order to reduce the size of the avalanche (cascade). In Cajuero and Andrade and Noël et al. [6, 9–11], the control adjusts where a cascade begins. However, in real systems the cascade's origin is usually uncontrollable. In D'Agostino et al. [12] the self-organizing dynamics of a sandpile model are controlled by introducing immunization to some nodes, assuming they can absorb an infinite amount of sand. In Qi and Pfenninger [13], a more realistic control is proposed by exploiting the failure tolerance of the nodes, which are usually designed to be able to sustain abnormal operating conditions for some time.

For example, the electric power systems can be considered a dual network that maps the transmission lines into nodes. Because the transmission lines are generally designed to have some margin over the normal power flow operating conditions, they can actually operate at an overloading power flow condition for a short time. Although allowing some components to work at an abnormal operating condition has a potential risk of damage, it may secure time for the operators to perform proper control, such as generation redispatch and load shedding. In some cases this may help prevent the propagation of cascading failures.

The basic sandpile model described in Section 2.1.1 is modified in Qi and Pfenninger [13] to suppress cascades by failure tolerance. Different from the basic model, if the root node exceeds its capacity it will not immediately topple but will operate for a short time above its capacity. This creates a time window in which some control strategy can be used to eliminate the overcapacity before the node topples and causes a cascading failure. Specifically, the following mechanisms are added.

1) The node that operates above its capacity is damaged with a probability ϵ_{dam}. If a node is damaged, its grains of sand will be redistributed to its neighbors. At the same time, the edges connecting it to its neighbors will be removed and it will not be able to hold any sand. Furthermore, the degree and the capacity of its neighbors will decrease by 1 as $k(j) \to k(j) - 1, K(j) \to K(j) - 1, j \in \mathcal{N}(i)$, thus decreasing the system's total capacity of holding sand. If all the edges of a node i are removed, all the sand it holds will be removed as $z_i \to 0$ if $k(i) = 0$.
2) If the node over capacity is not damaged, each grain of sand above capacity will dissipate with a probability ϵ_{act}. This can be considered as an active shedding in response to the overcapacity, leading to decreased system stress and reduced likelihood of cascading failures. If this active dissipation removes all extra $z_i - K(i)$ grains of sand, the node will now be at its capacity and thus the cascade stops; otherwise, it topples in a similar way to the basic BTW sandpile model.
3) If any additional node is over capacity after receiving additional load due to a damaged or toppled neighbor, the same process applies again until all nodes are at or under capacity. Note that the topology of the network can change during a cascade due to node damage, which is very different from the basic BTW sandpile model. Consequently, it is possible that $z_i > K(i) + 1 = k(i)$. This occurs when the neighbor of a node i damages and the sand on this neighbor is redistributed to node i. In this case the grains of sand on node i will be $z_i \to 0$. For its neighbors, the grains of sand on node i will first be evenly distributed as $z_j \to z_j + z_i / k(i), j \in \mathcal{N}(i)$ and the remaining $z_i \bmod k(i)$ grains of sand will be redistributed to the same number of randomly chosen neighboring nodes.
4) When the cascade ends, the network topology is recovered and the grains of sand are reset to what they would be in the basic model. In this way the control is only considered a temporary measure so that the dynamics of the system will not be significantly changed.

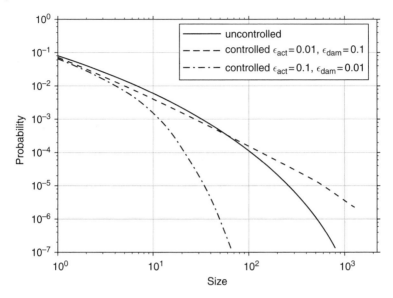

Figure 2.1 Probability distribution of cascade size $n = 10^5$, $n = 10^7$, and $\epsilon = 0.05$.

This control strategy introduces a benefit through active dissipation, but also an additional risk from the possible damage of overloaded nodes and the consequent degradation of the network's overall capacity. The probability distributions of cascade sizes of a random 4-regular network of size $n = 10^5$ under different ϵ_{act} and ϵ_{dam} are shown in Figure 2.1. Here a r-regular graph is a graph where each vertex has the same number of r neighbors, that is, every vertex has the same degree of r. When the risk for node damage outweighs the active dissipation, the probability for large cascades significantly increases. Otherwise, the probability for large cascades significantly decreases.

The total cost C_t for one cascade consists of three parts: the cost of the size of cascades C_{cas}, the cost of active sand dissipation C_{act}, and the cost of node damage C_{dam}. When cascade size is greater than zero, $C_{cas} = c\,[\text{size}]^\alpha$, where $\alpha > 1$. When cascade size is zero, there is a benefit of 1 (i.e., $C_{cas}[\text{size} = 0] = -1$) [6]. This benefit defines the scale of costs and can represent profits on uneventful days for infrastructures and investment portfolios.

By setting the cost in this way, the cost for one node damage is greater than that for each sand dissipation, which is greater than that for cascade size one. This is reasonable because the damage of one node can greatly degrade the capacity of the network to hold sand. The control will cause intentional extra sand dissipation while the cascade mainly transfers grains of sand from one node to another, and the only possible loss is the very weak sand dissipation with probability ϵ.

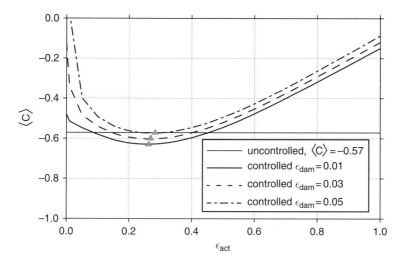

Figure 2.2 Control effect under different parameters. The triangles represent optimal control values (ϵ_{act}^*).

Here c and α are chosen as 0.005 and 1.5. For C_{act} and C_{dam}, the cost for each sand dissipation and each node damage are, respectively, set as 0.6 and 1.0, which correspond to the costs of cascade sizes of 24 and 34. Equivalently $C_{act} = n_{act}c\,[24]^\alpha$ and $C_{dam} = n_{dam}c\,[34]^\alpha$, where n_{act} and n_{dam} are, respectively, the grains of actively dissipated sand and the number of damaged nodes in one cascade.

This is realistic in real systems since the damage of a component can cause great economic loss due to loss of functionality of that component and the cost of repair after a cascade ends. The control strategy that actively dissipates load will intentionally make some load disappear. This dissipated load corresponds to an economic loss, while a cascade itself does not directly cause economic loss. For example, in power systems, if a transmission line is damaged during a cascading blackout it usually cannot be recovered until the cascade ends. It therefore cannot transmit power at the time when it is most needed to do so. Furthermore, it then has to undergo costly repairs or even replacement. By contrast, the control can be implemented by actively shedding some load, preventing further overloading of transmission lines. Although shedding load will cause the disruption of some consumers and will surely cause economic loss, it does not cause component damage or influence the functioning of the transmission network. Furthermore, during a cascade caused by the tripping of one transmission line, the power generated by the generators can still be transmitted to the consumers through other lines, thus not directly causing economic loss.

Given a fixed ϵ_{dam} we can find an optimal control parameter ϵ_{act}^* that minimizes the average total cost $\langle C_t \rangle$. Figure 2.2 shows the cost under different ϵ_{dam}

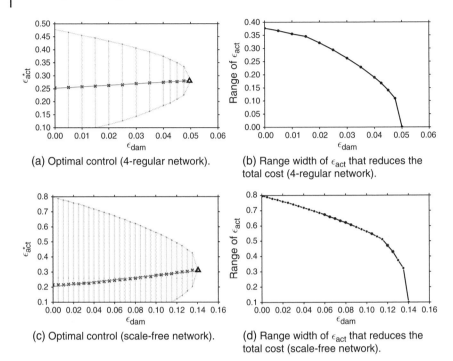

(a) Optimal control (4-regular network).

(b) Range width of ϵ_{act} that reduces the total cost (4-regular network).

(c) Optimal control (scale-free network).

(d) Range width of ϵ_{act} that reduces the total cost (scale-free network).

Figure 2.3 Optimal control and range width of ϵ_{act} that reduces the total cost for both the 4-regular and scale-free case. (a) and (c): The "×" indicates the optimal control values (ϵ_{act}^*), and the vertical range indicates those values of ϵ_{act} for which the controlled case has lower cost than the uncontrolled case. The triangle is the critical ϵ_{dam} where the range becomes zero. (a) Optimal control (4-regular network). (b) Range width of ϵ_{act} that reduces the total cost (4-regular network). (c) Optimal control (scale-free network). (d) Range width of ϵ_{act} that reduces the total cost (scale-free network).

and ϵ_{act}, obtained by sampling over a discrete set of values for ϵ_{dam} and ϵ_{act}. For each curve, ϵ_{dam} is fixed and there is a nontrivial optimal ϵ_{act} for which the total cost is minimized (to determine them, a third-degree polynomial is fitted to the sampled cost curves to obtain smooth curves). The cost for the basic model without control is also plotted for comparison. When ϵ_{act} is increased, the control becomes more successful and it is more easily possible to stop the propagation of cascades, thus decreasing the cost of cascades. However, the control itself has a cost, and thus the increased ϵ_{act} will increase the cost of control. Therefore, there is an optimal ϵ_{act} that will guarantee minimum total cost. Figure 2.2 also shows an increased risk for performing the control, that is, higher values for ϵ_{dam} will require higher optimal ϵ_{act}.

For any fixed ϵ_{dam}, the ϵ_{act} for optimal control is shown in Figure 2.3a. The optimal ϵ_{act} slightly increases when ϵ_{dam} increases. This is because the increased risk of damage requires more successful control in order to limit the total cost. Furthermore, with the increase of ϵ_{dam} the range of ϵ_{act}, denoted by $[\epsilon_{\text{act}}^{\text{min}}(\epsilon_{\text{dam}}), \epsilon_{\text{act}}^{\text{max}}(\epsilon_{\text{dam}})]$, in which the total cost decreases, will shrink and finally disappear, indicating that no matter how one adjusts ϵ_{act} (how successful the active dissipation control is), the total cost will surely increase after the control is added. Here $\epsilon_{\text{act}}^{\text{min}}(\epsilon_{\text{dam}})$ and $\epsilon_{\text{act}}^{\text{max}}(\epsilon_{\text{dam}})$, respectively, denote the minimum and maximum ϵ_{act} under ϵ_{dam} that can guarantee a decreased total cost compared with the uncontrolled case. The critical ϵ_{dam} that corresponds to zero range width $(\epsilon_{\text{act}}^{\text{max}}(\epsilon_{\text{dam}}) = \epsilon_{\text{act}}^{\text{min}}(\epsilon_{\text{dam}}))$ for ϵ_{act} can be used as an indicator of the robustness of the network. It is denoted by $\epsilon_{\text{dam}}^{*}$ and the bigger it is the more robust the network is.

The range width $W = \epsilon_{\text{act}}^{\text{max}}(\epsilon_{\text{dam}}) - \epsilon_{\text{act}}^{\text{min}}(\epsilon_{\text{dam}})$ is shown in Figure 2.3b. W first decreases approximately linearly and then drops rapidly after the ϵ_{dam} exceeds some value (0.04 here). The critical $\epsilon_{\text{dam}}^{*}$ corresponding to zero W is around 0.05, which is a very small value and indicates that the 4-regular network we are considering is not very robust. When ϵ_{dam} is greater than 0.05, no matter how successful the active dissipation control is, the total cost of the system will have to increase.

We examine how the random network described earlier compares to a scale-free network. For a scale-free network, for the same ϵ_{act}, high-degree nodes are easier to control, but the effect of damaging a high-degree node can also be significant. Therefore, it is not obvious whether it is easier or harder to control a scale-free network. For this comparison, a scale-free network is generated with the Barabási–Albert preferential attachment model [8] as implemented in NetworkX [14], with $m = 2$ (where m is the number of edges to attach from a new node to existing nodes), resulting in a mean degree of 4 to match the 4-regular graph. Again, the initial load on each node is randomly assigned (lesser or equal to its capacity). As in the 4-regular case, $n = 10^5$ and $N = 10^7$. The control proceeds in exactly the same manner as described above for the 4-regular case.

The optimal control and range width of ϵ_{act} that reduce the total cost are shown in Figure 2.3c and d. For the same ϵ_{dam} the range of ϵ_{act} reducing the total cost for the scale-free network is larger than that for the random network. The critical damage probability for the scale-free network is around 0.14, which is also much higher than the random network. From these results, it appears that the scale-free network is more robust to random failures, which is consistent with the conclusion in Albert et al. [15]. The robustness of scale-free networks to random failures is due to their extremely inhomogeneous connectivity distribution. Power-law distribution implies that the majority of nodes have only a few edges; nodes with small connectivity will be influenced with much higher probability. Besides, being easier to control for high-degree nodes seems to play a more important role than the more significant effect of their damage.

2.1.3 Motter–Lai Model

The Motter–Lai model is proposed to study cascading failures on networks [16–18]. On a network, at each time step each pair of nodes exchanges one unit of the relevant quantity, such as information or energy, through the shortest path connecting the two nodes. The *load* of a node is the total number of shortest paths passing through it. The *capacity* of a node is the maximum load it can handle. It is assumed that the capacity of node i, denoted by C_i, is proportional to the initial load L_i as $C_i = (1 + \alpha)L_i$, $i = 1, 2, \ldots, n$, where $\alpha \geq 0$ is the tolerance parameter and n is the initial number of nodes.

The removal of nodes can change the distribution of shortest paths and further the load of nodes. If the load of one node becomes larger than its capacity, the node fails. This can lead to a new redistribution of loads and may cause subsequent failures. The potential cascading failure may stop after a few steps but can also propagate to a great extent and shut down a large fraction of the network.

It is found in Motter and Lai [16] that large-scale global cascading failures occur if the distribution of loads is highly heterogeneous and the nodes with heavy loads are removed. Heterogeneous networks, such as those with scale-free distribution of links, are robust to random breakdown but not resistant to intentional attacks. This is mainly because the trigger of random breakdown is probably a node with small load while that of intentional attack is a node with heavy load. By contrast, homogeneous networks, such as those with uniform, exponential, and Poisson degree distributions, do not experience cascading failures under random breakdown or intentional attacks.

For the western U.S. power grid that has 4941 nodes and an average degree 2.67, the degree distribution is approximately exponential and is thus relatively homogeneous. The distribution of loads, however, is more skewed than that displayed by semirandom networks. On this network global cascades can be triggered by load-based intentional attacks but not by random or degree-based removal of nodes [16], as shown in Figure 2.4. G is the relative size of the largest connected component, the ratio between the numbers of nodes in the largest component before and after the cascade.

After a long time of evolution, real networks tend to be very resistant to random failure of nodes. However, on these networks there usually exist a few nodes with exceptionally large load. The attack on a single important node with heavy load may trigger a cascade of overload failures capable of disabling a large part of the network.

2.1.4 Influence Model

The influence model [19, 20] comprises a network of n interacting nodes. Each node is represented by a Markov chain. It is assumed that node i can be in m_i possible states. The state of node i at time k is represented by a vector

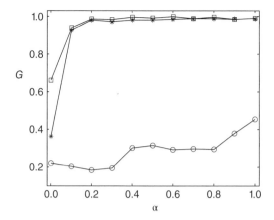

Figure 2.4 Cascading failure in the western U.S. power grid as triggered by the removal of a single node chosen at random (squares), or among those with largest degrees (asterisks) or highest loads (circles) [16].

s_i^k which has m_i elements. This vector has a single 1 at the entry corresponding to the present state and 0 everywhere else. The state of node i is updated as follows.

1) Node i randomly chooses one of its neighboring nodes or itself to be its determining node. Node j is selected with probability $d_{ij} \geq 0$ and there is $\sum_j d_{ij} = 1$.

2) The current state s_j^k of the determining node j fixes the probability vector $p_i^{k+1} = \sum_{j=1}^{n} d_{ij} s_j^k A_{ji}$. Here A_{ji} is a fixed row-stochastic $m_j \times m_i$ matrix, that is, a matrix whose rows are probability vectors, with nonnegative entries that sum to 1. p_i^{k+1} is a vector with nonnegative entries that sum to 1. It will be used to randomly choose the next state of node i.

3) The next state s_i^{k+1} is realized according to p_i^{k+1}.

The probabilities d_{ij} with which determining neighbors are selected define the network associated with the influence model. These probabilities can be put in a $n \times n$ stochastic matrix D whose ith row jth column element is d_{ij}. This matrix is called the *network influence matrix*.

The influence model introduced here can be written in a compact form. First let

$$s^k = \begin{bmatrix} s_1^k \\ \vdots \\ s_n^k \end{bmatrix}, \quad p^{k+1} = \begin{bmatrix} p_1^{k+1} \\ \vdots \\ p_n^{k+1} \end{bmatrix}.$$

Then the influence model can be summarized in the following way:

1) Calculate the probability vector $\boldsymbol{p}^{k+1}$ as

$$\boldsymbol{p}^{k+1} = \left(\left(\boldsymbol{s}^k \right)^{\mathrm{T}} \boldsymbol{H} \right)^{\mathrm{T}}, \tag{2.2}$$

where $(.)^{\mathrm{T}}$ takes the transpose of a matrix or vector and

$$\boldsymbol{H} = \begin{bmatrix} d_{11}A_{11} & \cdots & d_{n1}A_{1n} \\ \vdots & & \vdots \\ d_{1n}A_{n1} & \cdots & d_{nn}A_{nn} \end{bmatrix}.$$

2) Update the states of the nodes by generating a realization based on the probabilities in $\boldsymbol{p}^{k+1}$.

Consider a network with three nodes. Each node can have two possible states: normal (N) or failed (F). Initially all nodes are in normal state. The network influence matrix is set to be

$$\boldsymbol{D} = \begin{bmatrix} \dfrac{2}{3} & \dfrac{1}{6} & \dfrac{1}{6} \\ \dfrac{1}{6} & \dfrac{2}{3} & \dfrac{1}{6} \\ \dfrac{1}{6} & \dfrac{1}{6} & \dfrac{2}{3} \end{bmatrix},$$

and the stochastic matrix for all i, j is set to be

$$\boldsymbol{A} = \begin{bmatrix} \dfrac{5}{9} & \dfrac{4}{9} \\ \dfrac{1}{11} & \dfrac{10}{11} \end{bmatrix}.$$

If the determining node of node i is Normal, node i will be updated to Normal (Failed) with probability 5/9 (4/9); if the determining node of node i is Failed, node i will be updated to Normal (Failed) with probability 1/11 (10/11). Then one realization of the influence model is:

$$\begin{pmatrix} N \\ N \\ N \end{pmatrix} \rightarrow \begin{pmatrix} N \\ N \\ F \end{pmatrix} \rightarrow \begin{pmatrix} N \\ F \\ F \end{pmatrix} \rightarrow \begin{pmatrix} F \\ F \\ F \end{pmatrix} \rightarrow \begin{pmatrix} F \\ F \\ F \end{pmatrix} \rightarrow \begin{pmatrix} F \\ F \\ F \end{pmatrix}.$$

Note that since A_{22} is much greater than A_{21} while A_{11} is only slightly greater than A_{12}, it is more possible for a node to become failed (F).

2.1.5 Binary-Decision Model

The binary influence model [21] presents an explanation about the large-scale cascading failures triggered by trivial initial perturbations.

Consider a network with n nodes. Each node is connected to k other nodes by edges with probability p_k. The nodes directly connected to a node i are called the *neighbors* of node i. The average number of neighbors is $\langle k \rangle = z$. Each node has two possible states, 0 or 1. Here 0 can represent *normal* and 1 *failed*. Each node is assigned a threshold ϕ that is drawn from a probability distribution $f(\phi)$. The $f(\phi)$ is defined on $[0, 1]$ and satisfies $\int_0^1 f(\phi) d\phi = 1$ to make sure it is indeed a probability distribution. The cascades on the binary influence model occur as follows.

1) Initially all nodes have state 0. A perturbation is added by changing the states of a small fraction $\Psi_0 \ll 1$ of nodes to 1.
2) The state of any node i is updated to 1 if at least ϕ of the nodes that it is directly connected to are in state 1, and is updated to 0 otherwise. If the state of a node becomes 1, its state will remain 1.
3) The states of the nodes will evolve in the following time steps according to the rule in Step 2.

In an infinite network with finite z, assume the initial perturbation is only on one node i, which is called the *seed*. The seed i can grow only if at least one of its neighbors, such as node j, has a threshold $\phi \leq 1/k$ where node j has degree k. Such a node that satisfies this condition is called being *vulnerable*.

The probability that a node j has degree k and is vulnerable is $\rho_k \, p_k$ with $\rho_k = P[\phi \leq 1/k]$. The generating function of vulnerable node degree is:

$$G_0(x) = \sum_k \rho_k p_k x^k, \tag{2.3}$$

where

$$\rho_k = \begin{cases} 1, & k = 0 \\ F\left(\dfrac{1}{k}\right), & k > 0, \end{cases} \tag{2.4}$$

and $F(\phi) = \int_0^\phi f(\psi) d\psi$. $P_v = G_0(1) = \sum_k \rho_k p_k$ is the fraction of vulnerable nodes and $z_v = G_0'(1) = \sum_k k \rho_k p_k$ is the average degree of the vulnerable vertices.

The normalized generating function is:

$$G_1(x) = \frac{\sum_k k \rho_k p_k x^{k-1}}{\sum_k k p_k} = \frac{G_0'(x)}{z}.$$

(2.5)

The average vulnerable cluster size is:

$$<n> = G_0(1) + \frac{\left(G_0'(1)\right)^2}{z - G_0''(1)} = P_v + \frac{z_v^2}{z - G_0''(1)}.$$

(2.6)

It diverges when

$$G_0''(1) = \sum_k k(k-1)\rho_k p_k = z.$$

(2.7)

Eq. 2.7 is called the *cascade condition*. When $G_0''(1) < z$, all vulnerable clusters are small so that the early adopters are isolated and are not able to cause global cascades. When $G_0''(1) > z$, the typical size of vulnerable clusters is infinite and a random initial perturbation can trigger global cascades with finite probability.

In [21] a power law distribution of cascade sizes is observed when cascade propagation is limited by the global connectivity of the network. In this case the most connected nodes are far more likely than average nodes to trigger cascades. However, when the network is highly connected, cascade propagation is limited by the local stability of individual nodes, and the size distribution of cascades is bimodal, implying a more extreme instability that is harder to anticipate.

Besides, the heterogeneity plays an ambiguous role in determining a system's stability: increasingly heterogeneous thresholds make the system more vulnerable to global cascades; but an increasingly heterogeneous degree distribution makes it less vulnerable [21].

2.1.6 Coupled Map Lattice Model

The coupled map lattices have been applied to study cascading failures [22, 23]. Define an adjacency matrix A as

$$A_{ij} = \begin{cases} 1, & \text{if there is an edge between } i \text{ and } j \\ 0, & \text{otherwise.} \end{cases}$$

(2.8)

It is assumed that there is up to one edge between any two nodes and no node has an edge with itself.

The coupled map lattices of n nodes can be described as

$$x_i(t+1) = \left| (1-\epsilon) f(x_i(t)) + \epsilon \sum_{j=1, j \neq i}^{n} \frac{a_{ij} f(x_j(t))}{k(i)} \right|, i = 1, \ldots, n, \qquad (2.9)$$

where $x_i(t)$ is the state of node i at time step t, $k(i)$ is the degree of node i, and $0 \leq \epsilon \leq 1$ is the coupling strength. The f function defines local dynamics and is chosen as $f(x) = 4x(1-x)$ in [22], which is a chaotic logistic map.

If $0 < x_i(t) < 1$, node i is in a *normal* state. If $0 < x_i(t) < 1$ for $t \leq m$ and $x_i(t) \geq 1$ at $t = m$, node i is *failed* at the mth time step and will be kept in the failed state by setting $x_i(t) \equiv 0$ for $t > m$.

The cascading failures on the coupled map lattices occur as follows.

1) Initially for any node i there is $0 < x_i < 1$ and all of the nodes are in normal state.
2) At time step m a perturbation $R \geq 1$ is added to node c. This will lead to the failure of node c and for $t \geq m$ there is $x_c(t) \equiv 0$.
3) The states of the nodes directly connected with node c will be affected as in Eq. (2.9). If the state of any node is greater than 1, that node will fail and may lead to further failures of other nodes.

It is found in [22] that a sufficiently large perturbation on a single node can lead to cascading failure of all the other nodes in the network. Cascading failures are much easier to occur in small-world and scale-free coupled map lattices than in globally coupled map lattices. For globally coupled map lattices the perturbation threshold for the occurrence of cascading failure approaches infinity as the number of nodes approaches infinity. By contrast, a small perturbation may trigger large-scale cascading failure in a small-world or scale-free coupled map lattice in a few steps.

2.1.7 CASCADE Model

The CASCADE model is proposed in Dobson et al. [24]. Consider a system with n identical components. For each component, the minimum and maximum initial loads are, respectively, $L_{\min}$ and $L_{\max}$. The initial load L_i of component i is a random variable uniformly distributed in $[L_{\min}, L_{\max}]$. Components fail when their load exceeds L_{fail}. When a component fails, a fixed amount of load P is transferred to each of the other components. In particular, the model produces failures in generations $g = 0, 1, \ldots$ in the following manner.

1) All n components are initially working and have initial loads $L_1, L_2, \ldots, L_n$ that are independent random variables uniformly distributed in $[L_{\min}, L_{\max}]$. Set the generation serial number $g = 0$.
2) Add the initial disturbance D to the load of each component.

3) If the load of a component is greater than L_{fail}, this component fails. A failed component will remain in a failed state. Suppose that M_g components fail in this step.

4) Increase the loads of the components that do not fail according to the number of failures M_g. Specifically, add $M_g P$ to the load of each component where M_g is the number of components that fail in generation g. Increase g by 1 and go back to Step 3.

5) The cascade ends if there are no component failures in a generation.

The loads and model parameters are normalized so that the initial loads lie in $[0, 1]$ and $L_{\text{fail}} = 1$ while preserving the sequence of component failures. The normalized modified initial disturbance and the normalized load increase when a component fails are

$$d = \frac{D + L_{\text{max}} - L_{\text{fail}}}{L_{\text{max}} - L_{\text{min}}} \tag{2.10}$$

$$p = \frac{P}{L_{\text{max}} - L_{\text{min}}}. \tag{2.11}$$

The distribution of the total number of component failures can be written as

$$P(S = r) = \begin{cases} \binom{n}{r} \phi(d) (d + rp)^{r-1} \left(\phi(1 - d - rp) \right)^{n-r}, & r = 0, 1, \ldots, n-1 \\ 1 - \sum_{s=0}^{n-1} P(S = s), & r = n, \end{cases} \tag{2.12}$$

where $p \geq 0$ and the saturation function is

$$\phi = \begin{cases} 0, & x < 0 \\ x, & 0 \leq x \leq 1 \\ 1, & x > 1. \end{cases} \tag{2.13}$$

More details about the derivation of Eq. (2.12) can be found in Dobson et al. [24].

If $d \geq 0$ and $d + np \leq 1$, there is no saturation and Eq. (2.12) can be reduced to the quasibinomial distribution

$$P(S = r) = \binom{n}{r} d(d + rp)^{r-1} (1 - d - rp)^{n-r}, \tag{2.14}$$

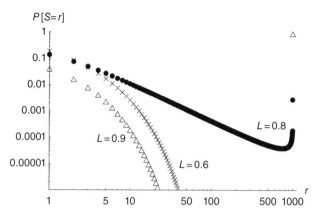

Figure 2.5 Log-log plot of the distribution of the number of components failed for three values of average initial load L [24]. L = 0.9 has an isolated point at (1000,0.8), indicating probability 0.8 of all 1000 components failed. The probability of no failures is 0.61 for L = 0.6, 0.37 for L = 0.8, and 0.14 for L = 0.9.

which can be approximated by letting $n \to \infty$, $p \to 0$, and $d \to 0$ in such a way that $\lambda = np$ and $\theta = nd$ are fixed to give the generalized Poisson distribution [25]:

$$P(S = r) \approx \theta(r\lambda + \theta)^{r-1} \frac{e^{-r\lambda-\theta}}{r!}. \qquad (2.15)$$

The critical case corresponds to $\lambda = np = 1$. It has been proven in [26] that this distribution has a power tail with exponent −1.5.

Assume the system has $n = 1000$ components. The initial component loadings vary from L^{min} to L^{max}. Set $L^{max} = L^{fail} = 1$. The average initial component loading $L = (L^{min} + 1)/2$. The initial disturbance $D = 0.0004$ and the load transfer amount $P = 0.0004$. Then the parameters p and d are actually $0.0004/(2 - 2L)$, $0.5 \le L < 1$.

The probability distribution of the number of components failed as L increases from 0.6 is shown in Figure 2.5. $L = 0.6$ corresponds to a nonsaturating case for which the distribution is approximately exponential. As L increases, the tail becomes heavier. The distribution for the critical case $L = 0.8$, $np = 1$ has approximate power-law region with an exponent of around −1.4, which compares to the analytically obtained exponent of −1.5 in [26].

2.1.8 Interdependent Failure Model

Most cascading failure models study only one system or one network. An interdependent failure model is proposed in Buldyrev et al. [27], in which the failure of nodes in one network may lead to failure of dependent nodes in other

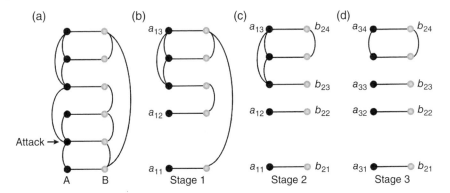

Figure 2.6 A process of a cascade of failures [27].

systems or networks. Between the pairs of interdependent nodes there are also interconnecting edges.

Consider two networks, A and B, both of which have n nodes. It is assumed that nodes A_i and B_i depend on each other. If A_i (B_i) fails, B_i (A_i) will also fail.

1) Initially all nodes are in normal status. A perturbation is added by removing $1 - p$ fraction of nodes of network A. All edges that are connected to the removed nodes are also removed (Figure 2.6a). a_1 clusters are the clusters of network A remaining after a fraction $1 - p$ of A-node are removed.

2) The nodes in network B that are connected with the removed nodes in network A and the interconnecting edges are also removed (Figure 2.6b). The edges in network B that are connected to the removed nodes are removed.

3) In the second stage, b_1 sets are defined as the sets of B-nodes that are connected to a_1 clusters by interconnecting links. All the B-links connecting different b_1-sets are removed. Because the two networks are connected differently, each b_1-set may split into several clusters, which is defined as b_2 clusters (Figure 2.6c). The b_1-sets that do not split and hence coincide with a_1-clusters are mutually connected.

4) In the third stage, a_3-clusters are determined in a similar way (Figure 2.6d) and in the fourth stage the b_4 clusters are determined. This process continues until no further splitting and link removal can occur.

A process of a cascade of failures is shown in Figure 2.6. This process leads to a percolation phase transition for two interdependent networks at a critical threshold $p = p_c$. This critical threshold is much greater than the threshold for a single network. There is no giant mutually connected component when p is smaller than p_c. Therefore, for the interdependent networks, there is no giant mutually connected component when more than a much smaller $1 - p_c$ fraction of nodes are removed, compared with a single network.

2.2   Power System Cascading Failure Models

2.2.1   Hidden Failure Model

The hidden failure model [28, 29] focuses on the hidden failure of the protection systems in power grids. A hidden failure of a protection system is a permanent defect that can cause a relay to incorrectly remove circuit elements after another switching event happens. A hidden failure can be difficult to detect during normal operations but may be exposed as a direct consequence of other system disturbances. It has been shown in Thorp et al. [30] that if one line trips, all the lines connected to its ends are exposed to the incorrect tripping. This may lead to cascading misoperations of the relays and even major system disturbances.

The simulation procedure of the hidden failure model begins from a base load flow and follows these steps:

1) A randomly selected transmission line is tripped as the initial triggering event and DC power flow is calculated.
2) The lines that violate line flow constraints are tripped. If there is no violation, the currently exposed lines, which are the lines connected to the last tripped line, are determined and tripped with a small probability p of incorrect tripping.
3) If the system breaks into multiple islands, the following redispatch is performed to minimize the load shedding:

$$\min \sum_{i \in L} C_i \tag{2.16}$$

$$\text{s.t. } \sum_{i \in G} P_i + \sum_{i \in L} C_i - \sum_{i \in L} D_i = 0 \tag{2.17}$$

$$P_i^{\min} \le P_i \le P_i^{\max}, i \in \mathcal{S}_{\mathrm{G}} \tag{2.18}$$

$$-F_j^{\min} \le F_j \le F_j^{\max}, j \in \mathcal{S}_{\mathrm{Line}} \tag{2.19}$$

$$0 \le C_k \le D_k, k \in \mathcal{S}_{\mathrm{L}}, \tag{2.20}$$

where P_i is the real power output of the generator i, C_k is the load shed at bus k, D_k is the initial load at bus k, F_j is the line flow of line j, $P_i^{\min}$ and $P_i^{\max}$ are the minimum and maximum real power outputs of generator i, the absolute value of the line flow of line j cannot exceed $F_j^{\max}$, and $\mathcal{S}_{\mathrm{G}}$, $\mathcal{S}_{\mathrm{L}}$, and $\mathcal{S}_{\mathrm{Line}}$ are, respectively, the set of generators, loads, and lines.

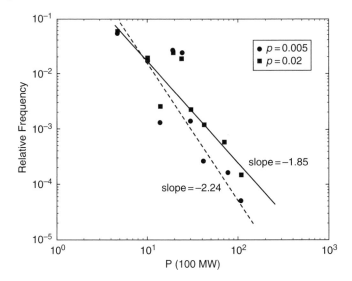

Figure 2.7 Distribution of the blackout size under different hidden failure probabilities [24].

4) If no further lines are tripped, record the total load loss, the sequence of line outages and its associated path probability, and then stop. Otherwise, go back to Step 2.

The simulation is repeated over an ensemble of randomly selected transmission lines as the initiating triggering event. In Figure 2.7 the impact of the hidden failure probability p on the blackout size distribution of the WSCC 179-bus system is shown [29]. It is seen that the probability for large-scale blackouts decreases when the hidden failure probability decreases. Upgrading the protection system or performing consistent maintenance can reduce the cascading risk.

2.2.2 Manchester Model

The Manchester model [31, 32] as shown in Figure 2.8 considers the following time-dependent phenomena:

1) Cascading tripping of transmission lines due to thermal overload: a fault and the subsequent tripping of one heavily loaded line can lead to overloads of the other lines. It is possible that these overloaded lines may be tripped by overload protection relays or faults due to sagging of transmission lines. Therefore, the overloaded lines are tripped at probability r.

2) Sympathetic tripping of power system components following a fault: If a failure occurs in one element's vulnerability region, this element will be

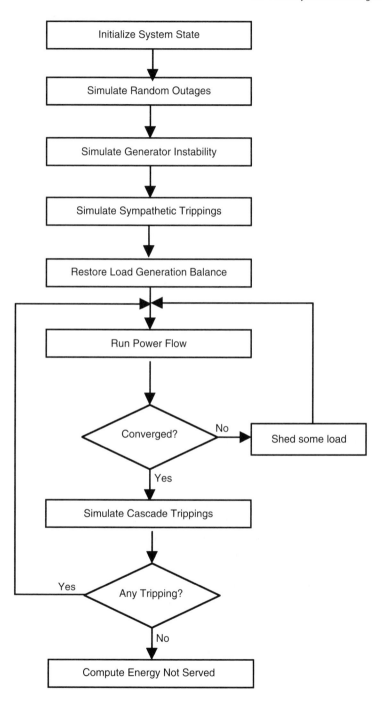

Figure 2.8 Flow chart of the Manchester model [32].

tripped at probability p. Here vulnerability region defines the portion of the system where a fault may provoke the tripping of the element.

3) Loss of stability of generators: The probability of stability due to a fault on line k (PST_k) is evaluated as [33]:

$$PST_k = \sum_{j=1}^{N_j}\sum_{i=1}^{4} P_k\left(S|ij\right) \times P_i \times P_j, \tag{2.21}$$

where $P_k(S|ij)$ is the probability of stability for a given fault of type i at location j of line k, P_i is the probability of a fault of type i, and P_j is the probability of a fault at location j. Offline computations are necessary to obtain the set of probabilities, $P_k(S|ij)$, P_i, P_j, and PST_k.

The time-dependent phenomena are incorporated into a Monte Carlo simulation. The random outages, sympathetic trippings of transmission lines due to protection malfunctions and by the disconnection of generators because of transient instability are simulated in a probabilistic manner. When the convergence of the power flow has been achieved, a series of cascade tripping events may occur because some lines may be overloaded. In this case, a new power flow computation is required. If these cascade outages are severe, the power flow may diverge and load is again shed until a new equilibrium point is reached.

2.2.3 OPA Model

The OPA model [34–37] simulates the patterns of cascading blackouts of a power system under the complex dynamics of a growing demand and the engineering responses to failure. OPA stands for **O**ak Ridge National Laboratory, **P**ower Systems Engineering Research Center at the University of Wisconsin, University of **A**laska to indicate the institutions collaborating to devise the simulation.

The OPA model represents transmission lines, loads, and generators and computes the network power flows with a DC load flow. Each simulation run starts from a solved base case solution for the power flows, generation, and loads that satisfy circuit laws and constraints. To obtain diversity in the runs, the system loads at the start of each run are varied randomly around their mean values by multiplying by a factor uniformly distributed in $[2-\gamma, \gamma]$. Initial line outages are generated randomly by assuming that each line can fail independently. Whenever a line fails, the generation and load is redispatched to satisfy the transmission line and generation constraints using standard linear programming methods. The optimization cost function is weighted to ensure that load shedding is avoided wherever possible. If any lines were overloaded during the optimization, then these lines are those that are likely to have experienced high stress, and each of these lines fails independently with a specified probability. The process of redispatch and testing for line outages is iterated until there are no more outages.

The OPA model has a two-loop structure including the fast dynamics and slow dynamics. The fast dynamics in the inner loop simulates power flow and cascading failure that occur in a fast time scale of minutes to hours. The slow dynamics in the outer loop simulates the evolution of a power grid. The slow dynamics are indexed by days so that load growth and responses to blackouts are updated daily. After the simulation of the current day the load will be increased and the system will be upgraded, based on which the simulation of the next day will be performed.

2.2.3.1 Fast Dynamics of Cascading Events in OPA Model

- Step 1: Consider day k. The generator capacities, outputs of generators, and the load are determined by the slow dynamics (see slow dynamics for details).
- Step 2: Initial line outages are randomly generated by assuming that each line can fail independently with probability p_0.
- Step 3: If any limits are violated or if a line has been outaged in the previous step, redispatch the power injections according to the following DC optimal power flow (OPF) problem:

$$\min \sum_{i \in \mathcal{S}_G} c_i \left| p_i - P_i^k \right| + \sum_{i \in \mathcal{S}_L} W_i \left(p_i - P_i^k \right)$$

$$\text{s.t.} \sum_{i \in \mathcal{S}_G \cup \mathcal{S}_L} \left(p_i - P_i^k \right) = 0$$

$$\boldsymbol{F} = \boldsymbol{A}\boldsymbol{p}$$

$$P_i^k \leq p_i \leq 0, \; i \in \mathcal{S}_L$$

$$0 \leq p_i \leq P_i^{\max,k}, \; i \in \mathcal{S}_G$$

$$-F_l^{\max,k} \leq F_l \leq F_l^{\max,k}, \; l \in \mathcal{S}_{\text{Line}},$$

$$(2.22)$$

where $\mathcal{S}_G$, $\mathcal{S}_L$, and $\mathcal{S}_{\text{Line}}$ are the sets of generators, loads, and lines. P_i^k is the real power output of a generator bus or the real power demand of a load bus on day k. p_i is the power injection at bus i. For generator buses, $p_i > 0$ while for load buses $p_i \leq 0$. Since only $-p_i$ is supplied at the load bus i, $(p_i - P_i^k)$ amount of the load is shed. c_i is the unit generation redispatch cost for generator i. W_i is the economic loss coefficient for the load i, which is set as a number that is much larger than c_i since load shedding is much more expensive than generation redispatch. $P_i^{\max,k}$ is the maximum real power output of generator i and $F_l^{\max,k}$ is the maximum power flow of line l on day k. $F = A p$ is used to calculate the real power flow of all lines from the real power injections of all buses except the slack bus, which can be obtained from DC power flow equations. Specifically, in DC power flow the relationship between the real power injections p and the phase angles θ of all buses except the slack bus (note that the phase angle of the slack bus is zero) can be expressed as

$$p = -B\theta, \tag{2.23}$$

where B is the imaginary component of the bus admittance matrix calculated neglecting line resistance and excepting the slack bus row and column. Then the phase angles can be obtained as

$$\theta = -B^{-1}p. \tag{2.24}$$

Since the power flow on a line l from bus i to bus j is

$$F_l = \frac{\theta_i - \theta_j}{X_{ij}}, \tag{2.25}$$

putting all line flows in a vector F and writing Eq. (2.25) in matrix form, we have

$$F = N\theta. \tag{2.26}$$

Substituting Eq. (2.24) into Eq. (2.26), we have

$$F = -NB^{-1}p, \tag{2.27}$$

and in Eq. (2.22) $A = -NB^{-1}$.

If the optimization is infeasible, stop the iteration and produce a list of lines that are overloaded during the optimization.

- Step 4: For each line that is overloaded in Step 3, it is tripped by probability β.
- Step 5: If lines are outaged in Step 4, then go to Step 3. If no lines are outaged in Step 4, then stop the iteration.

2.2.3.2 Slow Dynamics of System Evolution in OPA Model

There are also some distinctions between the two models. In the sandpile, the avalanches are coincident with the relaxation of high gradients. In OPA, each blackout occurs on fast time scale (less than one day), but the capacities of the lines that fail during the blackout are upgraded after the blackout. Besides, the OPA model has inherent inhomogeneity due to the inhomogeneity of the transmission network and also the inhomogeneous distribution of loads and generators [34].

The competing forces in the OPA model, which are the increase of the load and the upgrading of the power system as a response, may lead to complex

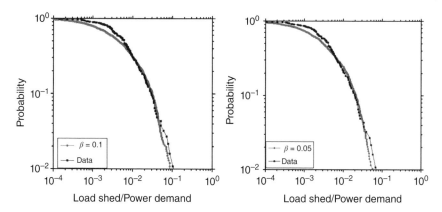

Figure 2.9 CCDs of the normalized load shed from OPA for the WECC 1553-bus system compared with the data for the western interconnection [37].

dynamical behaviors indicated by the power tail of the probability distribution of the blackout size near the critical transitions, either due to the limit on total generator capacity or the limit of power flow [35]. Interestingly, analyses of 15 years of North American blackout data also show a probability distribution of blackout size with a power tail [38], which suggests that the North American power system may be operated close to these critical transitions.

In [37] the OPA model is validated on the Western Electricity Coordinating Council (WECC) 1553-bus system model. The parameters are chosen as $\gamma = 1.15$, $\lambda = 1.00005$, $\mu = 1.07$, $p_0 = 0.0001$ based on real data.

Define the complementary cumulative distribution (CCD) of a random variable x as

$$P(y) = \int_y^\infty p(x)dx, \tag{2.28}$$

where $p(x)$ is the probability density function of the random variable x.

Figure 2.9 shows a comparison of the CCDs of the normalized blackout size (ratio between the load shed and the total power demand) obtained from the OPA model simulation with the NERC historical outage data. When β is chosen either as 0.1 or 0.05, the agreement between the data and the OPA results is reasonably good.

Note that when the slow dynamics of the system evolution is not considered and there is no load increase or system upgrade, the OPA model is called open-loop OPA.

- Step 1: In day $k+1$, all loads are increased by multiplying a fixed parameter λ as

$$P_i^{k+1} = \lambda P_i^k, i \in \mathcal{S}_\mathrm{L}. \tag{2.29}$$

Table 2.1 Analogy between OPA model and sandpile model [34].

	OPA Model	Sandpile Model
System state	Fractional overloads	Gradient profile
Driving force	Load increase	Addition of sand
Relaxing force	Line improvements	Gravity
Event	Line limit or outage	Sand topples
Cascade	Cascading line outages	Avalanche

- Step 2: In day $k+1$, the generator capacity is increased as

$$P_i^{\max,k+1} = \lambda P_i^{\max,k}, i \in \mathcal{S}_G. \tag{2.30}$$

- Step 3: In day $k+1$, the overloaded transmission lines are updated as

$$F_l^{\max,k+1} = \mu F_l^{\max,k}, \tag{2.31}$$

where μ is the rate of increase in transmission capacity and l belongs to the set of overloaded lines on day k.

The analogy between the main quantities in the OPA model and the sandpile model [1–3] is summarized in Table 2.1.

2.2.4 Improved OPA Model

Compared with basic OPA, the improved OPA [39] considers the unwanted operation of protective relays and the failure of the energy management system (EMS) or communications. The unwanted operation of relays is simulated by tripping lines that are not overloaded with probability $\xi \times |F/F^{\max}|^a$. ξ is the base probability of unwanted operation of relays and $|F/F^{\max}|$ is the load ratio of the transmission line. Besides, the control center may be disrupted due to contingencies, such as communication interruption or the breakdown of the EMS. The failure rate of the dispatching center is denoted by $1 - \eta$ and the optimization problem used to simulate redispatch is calculated by probability η.

Another problem with the basic OPA model is that it does not sufficiently consider the role of system planning in upgrading the system and preventing future outages. The main idea of the basic OPA model is to increase the transmission capacity of the lines that are tripped in the inner iteration. In other words, the line will not be updated until they are tripped. In practical systems, however, the transmission lines are updated based on the load forecasting, that is, the future slow dynamics, which is considered in the improved OPA model. The flow chart of the improved OPA model is shown in Figure 2.10.

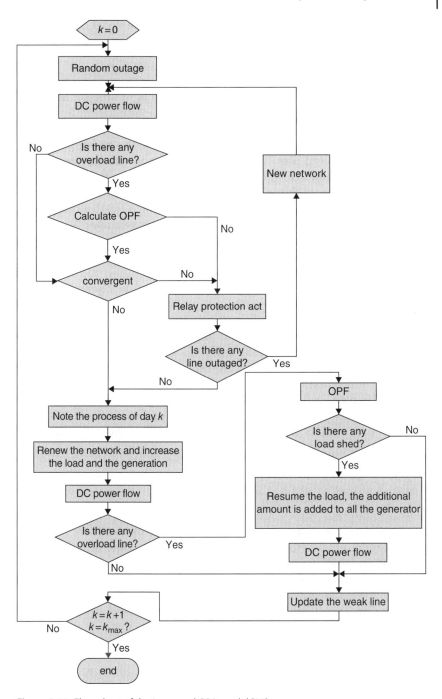

Figure 2.10 Flow chart of the improved OPA model [39].

2.2.4.1 Fast Dynamics of Cascading Events in Improved OPA Model

- Step 1: In day k, the capacities and outputs of generators and the load are determined by slow dynamics.
- Step 2: Accidental faults of lines
  Each transmission line is tripped with probability p_0 to simulate accidental faults.
- Step 3: DC power flow
  DC power flow is calculated. Due to line outages, the power grid might separate into several islands. For some islands it is necessary to regulate generation and load to eliminate unbalanced power. Specifically, if the total generation is greater than the total load, the output of each generator is decreased proportionally to their initial output in Step 1; if the total generation is less than the total load, the output of every generator is increased according to their reserve capacities; if the total generation capacity is less than the total load, load shedding will be implemented in every load bus proportional to its load amount.
- Step 4: If there are some lines whose line flows exceed their limits, then go to step 5. If not, exit the fast dynamics.
- Step 5: Response of the control center
  Calculate the optimization problem in Section 2.2.3.1 with probability η (note that $1 - \eta$ is the failure rate of the control center). If the optimization program is not implemented or is not convergent, go to Step 6. Otherwise, exit the fast dynamics process.
- Step 6: Tripping overloaded or normal lines
  The lines whose real power P exceeds the transmission capacity (denoted by $P_l^{\max}$) are tripped with probability β. The normal lines are tripped by probability $\xi \times \left| P/P_l^{\max} \right|^n$ where ξ is the base probability of the unwanted operation of relays. In Mei et al. [39], n is chosen as 10. If there is any line tripped, go back to Step 3; otherwise, exit the fast dynamics.

2.2.4.2 Slow Dynamics of System Evolution in Improved OPA Model

In the slow dynamics of the improved OPA model, the power demand and the generator capacities are upgraded in the same way as in Steps 1 and 2 of the slow dynamics of the basic OPA model. Then power flow is calculated based on the upgraded power demand, and the generator capacities and the overloaded transmission lines are updated. Compared with the basic OPA model, the effects of the planning and operation are considered to have a more practical upgrade strategy. Specifically, the transmission line upgrade is implemented in the following steps.

- Step 1: Calculate the DC power flow. If there are not any overloaded lines, go to Step 5, otherwise go to Step 2.

- Step 2: Calculate the DC OPF (the same as in basic OPA model). Some transmission lines may not be adequate, and some loads need to be shed. If loads are shed, go to Step 3, otherwise, go to Step 5.
- Step 3: Restore the shed loads. Each generator increases its output proportional to its generation, so the power could be balanced.
- Step 4: Calculate the DC power flow.
- Step 5: Rebuild the lines whose $|P/P_l^{max}|$ is larger than ϵ, which is the load rate of weak lines. This step mimics the power system planning. The identified weak lines are updated in the same way in Step 3 of the slow dynamics of the basic OPA model.

The basic and improved OPA models are applied to the Northeastern Power Grid of China (NPGC). The NPGC system consists of Heilongjiang, Jilin, Liaoning, and the northern part of Inner Mongolia of China. The system covers an area of more than 1.2 million square meters and serves more than 100 million people. Most of the hydropower plants are located in the east and most of the thermal power plants are located in the west and Heilongjiang province. The major consumers are in the middle and south of Liaoning province. Hence, the power is transmitted from the west and the east to the middle and from the north to the south.

A total of k_{max} = 2000 simulations corresponding to 2000 days are performed. The parameters of the improved OPA model are reasonably chosen as p_0 = 0.0007, η = 0.95, β = 0.999, ξ = 0.001, n = 10, λ = 1.00041, ε = 0.9, μ = 1.005 based on the operation practices of the NPGC system.

In the first 600 days, there are 22 blackouts in the OPA model and 14 in the improved model. In the last 1400 days, there are 463 blackouts in the OPA model and 56 in the improved model. The number of blackouts by the basic OPA model is much larger than the one by the improved OPA model. This is due to the differences between these two models on the outage simulation and upgrade criteria of transmission lines. In NPGC usually no more than 10 disturbances happen within 1 year, which is much closer to the outcome by the improved model.

The complementary cumulative distributions of the load shed from basic and improved OPA models are shown in Figure 2.11. The blackout size distribution of the improved model agrees approximately with a power law with exponent of about −0.8288, and this value agrees better with the statistical analysis of the historical data −0.8641 [40]. Therefore, the improved OPA model produces more realistic results compared to the basic OPA model.

2.2.5 OPA Model with Slow Process

Tree-caused outages are important in system reliability and can cause a large portion of preventable power outages in regions with trees [41–43]. The growing or falling into overhead lines of trees is generally regarded as the single largest cause of electric power outages [44]. Both distribution and

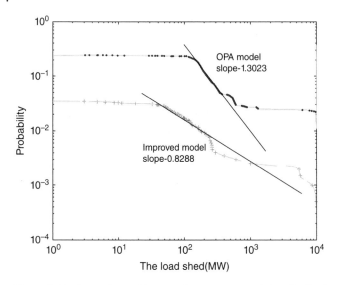

Figure 2.11 Comparison of the complementary cumulative distributions (CCDs) of the load shed from basic and improved OPA models [39].

transmission tree-related outages are experienced by utilities on a regular basis. Two significant outages in the western North America in 1996 [45], the U.S.-Canadian blackout on August 14, 2003 [46], and the outage in Italy on September 28, 2003 [47], can all be attributed to trees to some extent [44].

Many factors outside the power grid can greatly influence the power system operation. For example, if the trees under some lines are not pruned or cleared properly, it is possible that the lines can be tripped for tree contact even when the lines are not overloaded. In the U.S.-Canadian blackout on August 14, 2003, at 15:05, FirstEnergy Corporation's Chamberlin-Harding 345-kV line tripped and locked out while loaded at less than 45% of its rating, because it contacted overgrown trees within its right-of-way [48]. If we only pay attention to the power grid itself we may neglect this risk, which, under some circumstances, may lead to large-scale blackouts.

In order to address this problem, in Qi et al. [49] a model is proposed in which the tree contact and failure of transmission lines due to the temperature evolution are modeled. Because the model can be considered as an extension of the basic OPA model and because the line tripping due to the temperature evolution is usually slow, the model is called the OPA model with slow process. After Qi et al. [49], the slow process of the cascading blackouts was also discussed in Henneaux et al. [50].

Similar to OPA [34–37, 51, 52] and improved OPA [39], the OPA model with slow process also has a two-loop structure, fast dynamics and slow dynamics. The flow chart is shown in Figure 2.12, in which the upper and lower dotted-line boxes, respectively, denote fast dynamics and slow dynamics.

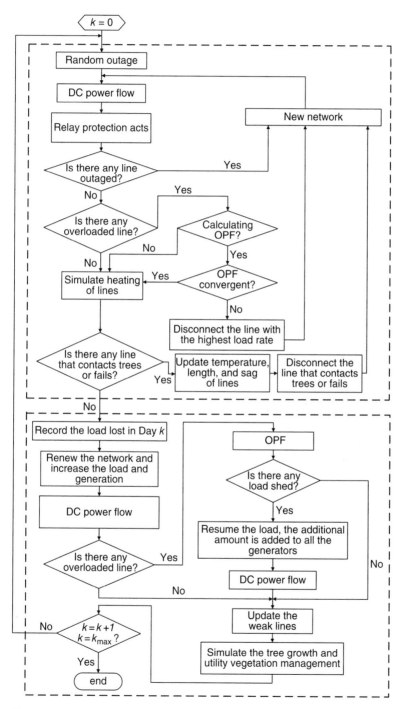

Figure 2.12 Flow chart of the OPA model with slow process.

2.2.5.1 Fast Dynamics of Cascading Events in OPA Model with Slow Process

The fast dynamics in the inner loop simulates power flow and cascading failure. Tree contact and failure of lines are considered in order to simulate the slow process. The simulation of overcurrent relays and the dispatching center is also improved. Specifically, the model is implemented in the following six steps.

- Step 1: In day k, the capacities and outputs of generators and the load are determined by slow dynamics.
- Step 2: Accidental faults of lines
  Each transmission line is tripped with probability p_0 to simulate accidental faults.
- Step 3: DC power flow
  DC power flow is calculated. Due to line outages, the power grid might separate into several islands. For some islands it is necessary to regulate generation and load to eliminate unbalanced power. The regulation strategy is the same as the improved OPA model.
- Step 4: Operation and maloperation of overcurrent relays
  The lines whose real power P exceeds the operation value of overcurrent relays (denoted by P_{set}) are tripped with probability β. The operation current of overcurrent relays is generally set as [53]

$$I_{\text{set}} = \frac{K_{\text{rel}} K_{\text{Ms}}}{K_{\text{re}}} I_{\text{L.max}}, \tag{2.32}$$

where K_{rel} is the reliable coefficient and is generally $1.25 - 1.5$, K_{Ms} is the motor coefficient, K_{re} is the return coefficient that is usually $0.85 - 0.95$, and $I_{\text{L.max}}$ is the maximum load current under normal conditions. We choose $K_{\text{rel}} = 1.2$, $K_{\text{Ms}} = 1.8$, and $K_{\text{re}} = 0.9$. Since the operation value of overcurrent relays is denoted by real power, P_{set} is set to be 2.4 times the line capacity.

The other lines are tripped by probability $\xi \times |P/P_{\text{set}}|^n$, which is the same as the improved OPA model. If there is any line tripped, go back to Step 3; otherwise, go to Step 5.
- Step 5: Response of the control center
  If there are overloaded lines, the dispatching center will calculate the DC OPF in Section 2.2.3.1 with probability η ($1 - \eta$ is the failure rate of the control center). If the optimization program is convergent, go to Step 6; otherwise, disconnect the line or transformer with the highest load rate (the ratio between line flow P and the line capacity P_l^{max}) to avoid its failure for heating and go back to Step 3.
- Step 6: Tree contact or failure of lines
  If some line contacts trees or fails due to heating, disconnect it and go back to Step 3; otherwise, exit the fast dynamics. The details of this step are discussed as follows.

A simple equation is used to describe the time evolution of line temperature by neglecting the spatial variation in temperature can be modeled as [54]:

$$\frac{dT(t)}{dt} = \alpha I^2 - v\left(T(t) - T_0\right),$$

(2.33)

where $T(t)$ is the line temperature at time t, $I = P/V$ is the current in the line measured in amperes, $\alpha = 0.239/(\rho C \omega^2 \kappa)$, $v = Gp/(\rho C \omega)$, ρ is the density, C is the specific heat, ω is the area of cross-section, κ is the electrical conductivity, G is the surface conductance, p is the perimeter, and T_0 is the temperature of the medium.

If the initial line temperature is $T(0)$ and the real power through the line is constant value P, the evolution of line temperature can be modeled as [54]:

$$T(t) = e^{-vt}\left(T(0) - T_e(P)\right) + T_e(P),$$

(2.34)

where

$$T_e(P) = \frac{\alpha}{v}\frac{P^2}{V^2} + T_0$$

(2.35)

is the equilibrium temperature that the line will reach when $t \rightarrow \infty$. The equilibrium temperature under the transmission capacity $P_l^{\max}$ is $T_{cl} = T_e\left(P_l^{\max}\right)$, which is the maximum temperature the line can endure.

Note that the "line" here is actually the line between a typical horizontal span of a transmission line rather than the whole transmission line. Correspondingly, the line length is the length of the line between the considered horizontal span. The deviation of line length ΔL caused by temperature change is [55]

$$\Delta L = \Delta T \cdot \alpha_L \cdot L_0,$$

(2.36)

where ΔT is the line temperature change, α_L is the linear elongation coefficient of the line, and L_0 is the initial line length.

Assuming the suspended points of the line have equal heights, the relationship between line length L and sag f with known horizontal span l is [56]

$$L = l + \frac{8f^2}{3l}.$$

(2.37)

The initial line length L_0 can be obtained from initial sag f_0 with Eq. (2.37). Let d_0 denote the initial distance between line and trees. When the line contacts trees the line sag will be $f = f_0 + d_0$. After we get the temperature change ΔT with Eqs. (2.36) and (2.37), the tree contact time will be

$$t_{\text{tree}} = -\frac{1}{v}\ln\frac{T(0) - T_e(P) + \Delta T}{T(0) - T_e(P)}.$$

(2.38)

For overloaded lines, their temperature can reach T_{cl} within limited time t_{cl} as

$$t_{cl} = -\frac{1}{v} \ln \frac{T_{cl} - T_e(P_l)}{T(0) - T_e(P_l)}. \tag{2.39}$$

Let f_{cl} denote the line sag of an overloaded line at t_{cl}. If $d_0 > f_{cl} - f_0$, the line will fail by heating at t_{cl}; otherwise, it will contact trees within a shorter time t_{tree}.

For the rest of the lines, their temperatures cannot reach T_{cl}. However, they might contact trees for the increase of line sag caused by heating. Let f_e denote the line sag when the line temperature is $T_e(P)$. If $d_0 \geq f_e - f_0$, the line cannot contact trees; otherwise, it will contact trees within t_{tree}.

For overloaded lines, record t_{tree} if they can contact trees; otherwise, record t_{cl}. For other lines, record t_{tree} if they can contact trees. The line with the shortest recorded time $t_{\min}$ will be tripped. The initial temperature, the initial line length, and the initial line sag of all the lines except the tripped line can be updated to be the values at $t_{\min}$.

For fast dynamics, the differences between the OPA model with slow process and the improved OPA model [39] are summarized as follows.

1) The OPA model with slow process adds the simulation of tree contact and failure of lines, which helps describe the cascading failure process more realistically.
2) Since in practical power grid the operation of a relay is much faster than the response of the dispatching center, in the OPA model with slow process the relays operate before the response of the control center.
3) The operation value of overcurrent relays is not the transmission capacity but a greater value, which agrees with the practical conditions.
4) If the optimization problem is calculated but is not convergent, the dispatching center will disconnect the line with the highest load rate to avoid its failure due to heating.

2.2.5.2 Slow Dynamics of System Evolution in OPA Model with Slow Process

The slow dynamics in the outer loop simulates the evolution of the power grid. Compared with the improved OPA model [39], it includes the simulation of tree growth and utility vegetation management (UVM), which can be considered as a factor driving the system toward self-organized criticality as the increase of load, generation, and line capacity.

The tree height is actually the height of the tallest tree under the line and correspondingly the distance between the line and trees is that between the line and the tallest tree. We simulate the tree growth with the Chapman–Richards model [57, 58]:

$$H(A) = a\left(1 - e^{-bA}\right)^c, \tag{2.40}$$

where $H(A)$ is the tree height at age A, and a, b, and c are constant parameters. Gaussian noise with zero mean is added to the tree growth parameters a, b, c of Mongolian Scots Pine, a typical tree species in northeast China [59]. The unit of A is usually year, but we use day and divide it by 365 because the tree growth in different days needs to be simulated.

For transmission lines of different voltage levels, there exist specific smallest allowable distances between the line and the trees [60], which are called "safe distances" and are denoted by d_s.

Because it is difficult to get the detailed information of initial tree heights from real power grids, the initial heights are determined randomly rather than deterministically. We consider that the tree growth under most lines can be reasonably maintained because the utilities perform UVM to meet the requirements for safe distance. Specifically, the initial tree height under a specific line is set to be $H_0 = \tilde{H} + r$, where $\tilde{H}$ is the tree height that enables a $1.2d_s$ distance, r is Gaussian noise with zero mean and standard deviation of $0.2d_s$. The distance between line and trees will range from $0.8d_s$ to $1.6d_s$ with probability 95.5%. After the tree height is determined, the initial distance between the line and trees, d_0, can be determined.

Standard values for transmission lines are chosen as the distance between the line and ground h_0, the horizontal span l, and the initial sag f_0.

The UVM that aims at preventing trees from growing unreasonably is simulated in the following manner.

1) UVM after outages caused by tree contact
   If there are some lines that contact trees, the trees under them will be removed by setting the tree height to zero.
2) Routine UVM
   UVM cycle is a term used by utility arborists to generally describe the time to complete identified pruning or removal of certain trees on their entire electric system [44]. Assume the cycle is Y years and UVM is performed once every month, the work to be finished in each month will be:

$$W = \frac{N}{12Y}, \tag{2.41}$$

where N is the total number of lines to be dealt with. The actual spread for UVM cycles can array from 1 to more than 10 years [44].

We can calculate the ratio μ between the work needed to be done in the routine UVM in a month and that left undone in the cycle:

$$\mu = \frac{W - n_{\text{tree}}}{N - n_{\text{done}}}, \tag{2.42}$$

where n_{tree} is the number of lines that contact trees in the month and n_{done} is the number of lines whose trees have been dealt with in the cycle. If $\mu > 1$, let $\mu = 1$; if $\mu < 0$, let $\mu = 0$. For a specific line, the trees under it are treated with UVM with probability μ.

Since line patrol and UVM may not be performed perfectly, in real systems it is practical that not all lines are performed with ideal UVM. We assume that the lines to be dealt with are performed with ideal UVM with probability γ, which is called UVM effective execution rate. Since UVM can be effectively executed in most cases, we set γ to be 0.95. Here ideal UVM means the trees will be pruned or removed in an expected way. Specifically, if the distance between the line and trees is greater than $1.5\,d_s$, no UVM is performed; Otherwise, according to Cieslewicz and Novembri [44], we assume that the trees under two-thirds of the lines are removed and the rest are pruned by setting their height to enable a $1.5\,d_s$ distance. For those lines that are not trimmed, only the natural growth of trees is simulated.

The OPA model with slow process is applied to NPGC. The parameters that also appear in the improved OPA model remain the same except η is increased to 0.98 ($1 - \eta$ is the failure rate of the control center) since the probability that the control center cannot obtain the system state due to communication interruption or breakdown of EMS is usually very low. In order to simulate typical blackouts in which a few tree contacts trigger the cascading failure and finally lead to many line outages and a large amount of load shed, we choose the UVM cycle Y as 6 years and the UVM effective execution rate γ as 0.8.

We simulate 4000 days on NPGC and, among those 4000 days, 67 lines contact trees in 54 days and 10 lines fail due to heatng in 10 days. The largest load shed is 18 759 MW. Tree contact of a line is an important cause of this blackout, and there is also a line that fails due to heating. A total of 89 lines or transformers are disconnected and 21.98% of loads are shed. The process of this blackout can be divided into the following four stages. The tripped lines and transformers are shown in Figure 2.13, in which i–j denotes the lines or transformers tripped in stage i, j denotes the tripping order in the stage, and the number in the parenthesis is the number of tripped transformers. When relays operate, it is possible for more than one line or transformer to be tripped. However, when lines are disconnected by the dispatching center or tripped for tree contact or failure for heating, only one line will be tripped.

1) Cause of blackout: The trigger is the accidental outage of a 220-kV line LYS-QGA in Liaoning province. This leads to the overload of neighboring lines. The dispatching center calculates DC OPF and sheds part of the load in QGA. After that there is no overloaded line.

   If the tree contact or failure of lines is not considered, the system will operate in a normal state. However, if this is considered things will be totally different. The load rate of a 220-kV line JDS-FMP, which is in Jilin province and is quite far away from LYS-QGA, is as high as 0.927 and at the same time the distance between this line and trees is only 2.24 m. The sag of this line increases with heating, and after 61.1 minutes this line contacts trees and is disconnected by relay, which further leads to the overload of another seven lines.

2) Gradual development: The influenced area gradually expands in this stage. The dispatching center calculates OPF to eliminate overload, but it is difficult for OPF to converge since the system already operates in an emergency state. To avoid line fail from heating, the dispatching center disconnects the line with the highest load rate. In this stage, blackout mainly occurs in the region near JDS-FMP in Jilin province and small areas in Heilongjiang and Liaoning provinces.

Although some relays act, the operation of relays does not cause other relays to act. The influenced area is denoted by dashed lines in Figure 2.13.

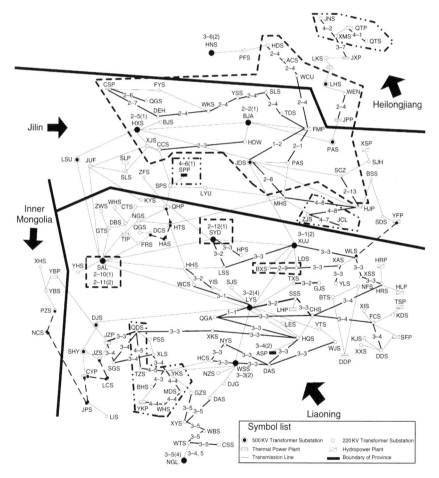

Figure 2.13 Process of blackout in the Northeast Power Grid of China simulated by the OPA model with slow process.

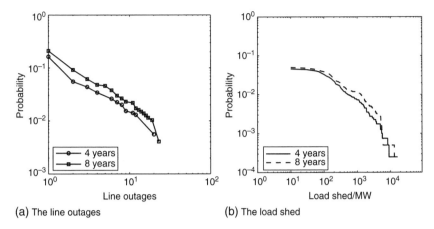

(a) The line outages   (b) The load shed

Figure 2.14 CCDs of the line outages and the load shed under different Y. (a) The line outages. (b) The load shed.

3) Relays operate successively: In this stage some lines are tripped by relays, and this further causes other lines to be tripped by relays. The operation of relays successively disconnects 3, 8, 22, 9, 9, 2, and 1 line. The blackout extends from Jilin province to a large area of Liaoning province. Heilongjiang province is also slightly influenced.

4) Blackout gradually stops: In this stage the relays stop acting successively. Only a small number of lines and power plants are influenced. After the dispatching center disconnects some lines and sheds part of the loads at 25 substations, there is no overloaded line and the blackout finally ends. The influenced area is denoted by dot-dash lines in Figure 2.13.

The CCDs of the number of line outages and the amount of load shed under different UVM cycles are shown in Figure 2.14. When the UVM cycle is increased to 8 years, the number of lines contacting trees, the probability for small-scale and large-scale blackouts, and the blackout risk all increase.

The CCDs of the number of line outages and the amount of load shed under UVM effective execution rates are shown in Figure 2.15. When γ is decreased, it is more likely that the tree growth cannot be controlled effectively. Thus the number of lines contacting trees and the blackout risk increase.

2.2.6 AC OPA Model

The basic OPA, the improved OPA, and the OPA model with slow process all use DC power flow, in which only real power is considered while the bus voltages are considered constant. In contrast, the AC OPA model [61, 62] uses AC power flow and thus can consider reactive power and voltage. The operation mode of the system is first determined by AC OPF and load shedding and will be readjusted by AC OPF until there are no further outages.

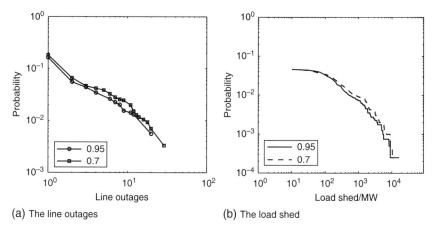

(a) The line outages (b) The load shed

Figure 2.15 CCDs of the line outages and the load shed under different γ. (a) The line outages. (b) The load shed.

2.2.6.1 Fast Dynamics of Cascading Events in AC OPA Model

One AC OPA simulation is implemented in the following steps.

- Step 1: Consider day k. The loads at the start of each run are randomly varied around their mean values by multiplying a factor uniformly distributed in $[2-\gamma, \gamma]$.
- Step 2: Accidental faults of lines
  Each transmission line is tripped with probability p_0 to simulate accidental faults.
- Step 3: Calculate the following AC OPF problem:

$$\min \sum_{i \in \mathcal{S}_B} \left(P_{gi} - P_{di}^k \right)$$

$$\text{s.t. } P_{gi} - P_{di}^k = V_i \sum_{j \in i} V_j \left(G_{ij} \cos\theta_{ij} + B_{ij} \sin\theta_{ij} \right), \ i \in \mathcal{S}_B$$

$$Q_{gi} - Q_{di}^k = V_i \sum_{j \in i} V_j \left(G_{ij} \sin\theta_{ij} - B_{ij} \cos\theta_{ij} \right), \ i \in \mathcal{S}_B$$

$$P_{gi}^{\min,k} \le P_{gi} \le P_{gi}^{\max,k}, \ i \in \mathcal{S}_G$$

$$Q_{gi}^{\min,k} \le Q_{gi} \le Q_{gi}^{\max,k}, \ i \in \mathcal{S}_G$$

$$V_i^{\min} \le V_i \le V_i^{\max}, \ i \in \mathcal{S}_B$$

$$-F_l^{\max,k} \le F_l \le F_l^{\max,k}, \ l \in \mathcal{S}_{\text{Line}},$$

where $\mathcal{S}_G$, $\mathcal{S}_L$, $\mathcal{S}_B$, and $\mathcal{S}_{\text{Line}}$ are the sets of generators, loads, buses, and lines. P_{gi} and Q_{gi} are the real and reactive power outputs of generator i. P_{di}^k and Q_{di}^k are the real and reactive loads of load bus i on day k. $P_{gi}^{\min,k}$ and $P_{gi}^{\max,k}$ are the lower and upper limits of the real power outputs of generator i on day k. $Q_{gi}^{\min,k}$ and $Q_{gi}^{\max,k}$ are the lower and upper limits of the reactive power outputs of generator i on day k. V_i is the voltage magnitude of bus i and $V_i^{\min}$ and $V_i^{\max}$ are its lower and upper limits. $\theta_{ij} = \theta_i - \theta_j$ is the phase angle difference between bus i and bus j, $j \in i$ denotes the buses that are connected to bus i including bus i itself, F_l is the power flow of line l, and $F_l^{\max,k}$ is the line flow limit on day k. G_{ij} and B_{ij} are the real and imaginary elements of the bus admittance matrix.

If the OPF converges, go to Step 4; otherwise, shed load until the OPF converges and then go to Step 4.

- Step 4: If there are lines that violate their line limits, go to Step 5; otherwise, stop the simulation.
- Step 5: The overloaded lines in Step 4 are independently tripped with probability β. If there are line outages, go to Step 6; otherwise, stop the simulation.
- Step 6: If the system is separated into islands, balance the generation and load in each island by shedding load or adjusting the generators' outputs and go to Step 3; otherwise, directly go to Step 3.

2.2.6.2 Slow Dynamics of System Evolution in AC OPA Model
For slow dynamics, there are

$$P_{di}^{k+1} = \lambda P_{di}^{k}, \ i \in \mathcal{S}_{\mathrm{L}} \tag{2.43}$$

$$Q_{di}^{k+1} = \lambda Q_{di}^{k}, \ i \in \mathcal{S}_{\mathrm{L}} \tag{2.44}$$

$$P_{gi}^{\max,k+1} = \lambda P_{gi}^{\max,k}, \ i \in \mathcal{S}_{\mathrm{G}} \tag{2.45}$$

$$Q_{gi}^{\max,k+1} = \lambda Q_{gi}^{\max,k}, \ i \in \mathcal{S}_{\mathrm{G}} \tag{2.46}$$

$$F_l^{\max,k+1} = \mu F_l^{\max,k}, \ \text{for outaged lines on day } k. \tag{2.47}$$

Similar to the basic OPA, when the slow dynamics of the system evolution is not considered, the AC OPA model is called open-loop AC OPA. Here open-loop AC OPA simulation is performed on the standard RTS-96 three-area system model [63], as shown in Figure 2.16. This system has 73 buses and 120 branches, with a total of 8550 MW of load. Compared to the initial model, each reactor of 100 Mvar at buses 106, 206, and 306 is split into two; 50 Mvar at each extremity of the line. These reactors are considered to be automatically disconnected in case of the outage of the corresponding line. The precontingency steady-state is based on a preventive-security-constrained OPF, such that the system is $N-1$ secure [64].

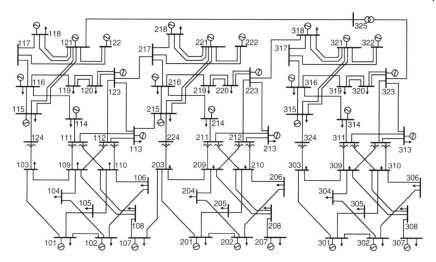

Figure 2.16 RTS-96 three-area system.

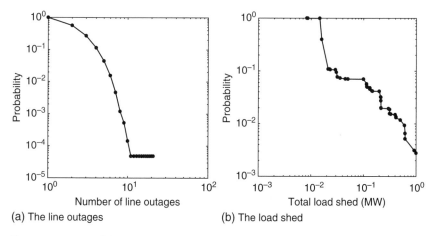

(a) The line outages

(b) The load shed

Figure 2.17 CCDs of the line outages and the load shed from AC OPA simulation. (a) The line outages. (b) The load shed.

The initial failure probability of each line p_0 is assumed to be 0.01. A line is considered to be overloaded when its power flow is larger than 99% of its line limits, and is disconnected by probability $\beta = 1$ when it is overloaded. The open-loop AC OPA simulation is run for 30 000 times, in which the slow dynamics for system evolution is ignored. The CCDs of the total number of line outages and the total load shed are shown in Figure 2.17. Note that the CCD of the line outages is obtained on the condition that there is at least one

Table 2.2 Number of line outages in each generation for AC OPA model.

Generation	Number of Outages
0	36 129
1	5489
2	604
3	31
4	2

Table 2.3 Critical lines identified by AC OPA simulation.

	Critical lines			
Rank	**Initial Outages (Buses)**	**Frequency**	**Subsequent Outages (Buses)**	**Frequency**
1	77 (220, 223)	633	89 (306, 310)	1882
2	38 (120, 123)	629	10 (106, 110)	1600
3	66 (215, 221)	613	51 (206, 210)	1144
4	113 (319, 320)	609	90 (307, 308)	891
5	27 (115, 121)	601	52 (207, 208)	420

line outage and the CCD of the load shed is for the case in which load shedding does occur.

The total number of line outages in each generation or stage is listed in Table 2.2. Similar to the basic OPA and its other variants, the AC OPA model can naturally produce line outages and the load shed in generations or stages; each iteration of the "main loop" of the simulation produces another generation [65, 66]. In Table 2.3 we also list the critical lines that are identified by the AC OPA simulation. Here, a line is considered to be critical if it outages most frequently during the simulation, either as initial outages or subsequent outages. In the parentheses next to the line number we list its two buses.

2.2.7 Cascading Failure Models Considering Dynamics and Detailed Protections

Most existing cascading failure models only perform quasi-steady-state simulations based on DC or AC power flow or optimal power flow, ignoring the dynamics of the system. In Song et al. [67], a dynamic cascading failure model is proposed, in which the dynamics of the power system are simulated by solving the following differential-algebraic equations:

$$\dot{x} = f(x,y,z) \tag{2.48a}$$

$$g(x,y,z) = 0 \tag{2.48b}$$

$$h(x,y,z) < 0, \tag{2.48c}$$

where x is the continuous state vector, y is the vector of algebraic variables, and z is a vector of integer variables. The differential equation (Eq. (2.48a)) represents the machine dynamics. The algebraic constraints (Eq. (2.48b)) are the standard AC power flow equations. When constraint h_i fails, an associated counter function d_i activates. Each z_i changes state if d_i reaches its limit.

During cascading failures, power systems undergo many discrete changes caused by exogenous events (e.g., manual operations, weather) and endogenous events (e.g., automatic protective relay actions). The discrete events will consequently change algebraic equations and the system's dynamic response, which may result in cascading failures, system islanding, and large blackouts.

Five types of protective relays are modeled: overcurrent (OC) relays, distance (DIST) relays, temperature (TEMP) relays, undervoltage load shedding (UVLS) relays, and underfrequency load shedding (UFLS) relays. Specifically,

- OC relays monitor the instantaneous current flow along each branch.
- DIST relays monitor the apparent admittance of the transmission line.
- TEMP relays monitor the line temperature obtained from

$$\dot{T}_i = r_i F_i^2 - k_i T_i, \tag{2.49}$$

  where T_i is the temperature difference relative to the ambient temperature for line i, F_i is the current flow of line i, and r_i and k_i are the heating and time constants for line i. r_i and k_i are chosen so that line i's temperature reaches 75°C if current flow hits the rate-A limit, and its TEMP relay triggers in 60 seconds when current flow jumps from rate-A to rate-C.
- When the voltage magnitude or frequency are lower than the specified thresholds, the UVLS relay or UFLS relay will shed 25% of the initial load to avoid the onset of voltage instability and reduce system stress. Both the UVLS and UFLS relays used a fixed-time delay of 0.5 seconds.

Song et al. [67] and Yan et al. [68] compare DC power flow-based cascading failure simulation with transient stability analysis. It is concluded that when the oscillation or disturbance is confined within a certain range, the power flow-based cascading failure simulator can well approximate the power system behavior. However, when cascading failures continue to develop, the power flow-based cascading failure simulation can fail in capturing the actual power system behavior as the steady-state assumption does not hold any more. In this case, transient stability analysis models are more suitable for the simulation of power system behavior so that proper critical control action can be taken to address severe power grid disturbances.

The TRELSS industry cascading failure model also has groups of protection devices acting together but ignores transient dynamics [29]. Besides, a dynamic contingency analysis tool has been developed to deterministically model the typical behavior of protection models and dynamics with an applicability to cascading in large systems [69].

In [70], an integrated dynamics and protection simulator, TS3ph-CAPE, is used for producing cascading data. TS3ph-CAPE uses a full three-phase representation of the power system to enable analysis of unbalanced operating conditions, such as unbalanced loads and disturbances, and includes a variety of generator, control, and load models. To simulate the role of protection in cascading failures, the simulator interfaces with Electrocon's commercial package CAPE through a socket at the end of each time domain integration step. Once the dierential and algebraic equations describing the generation and transmission systems have been solved, TS3ph passes the three-phase voltages and currents to CAPE which evaluates the relay and noties TS3ph of any future breaker operations.

CAPE has hundreds of manufacturer specific relay models. For example, the following common relay elements can be incorporated into the various substations.

- distance (line protection)
- over-current (backup protection)
- current dierential (bus and transformer protection)
- under-/over-frequency (generator protection)

Note that the models discussed in this chapter are non-exhaustive. A review of general cascading failure models and power system cascading failure models can be found in [71] and [72], respectively.

References

1 Bak, P., Tang, C., and Wiesenfeld, K. (1987). Self-organized criticality: an explanation of the 1/f noise. *Physical Review Letters* 59 (4): 381.

2 Bak, P., Tang, C., and Wiesenfeld, K. (1988). Self-organized criticality. *Physical Review A* 38 (1): 364.

3 Bak, P. and Tang, C. (1989). Earthquakes as a self-organized critical phenomenon. *Journal of Geophysical Research* 94 (15): 635–615.

4 Bonabeau, E. (1995). Sandpile dynamics on random graphs. *Journal of the Physical Society of Japan* 64(1): 327–328.

5 Goh, K.I, Lee, D.S., Kahng, B., and Kim, D. (2003). Sandpile on scale-free networks. *Physical Review Letters* 91 (14): 148701.

6 Noël, P.A., Brummitt, C.D., and D'Souza, R.M. (2013). Controlling self-organizing dynamics on networks using models that self-organize. *Physical Review Letters* 111 (7): 078701.

7 Alstrøm, P. (1988). Mean-field exponents for self-organized critical phenomena. *Physical Review A* 38 (9): 4905.

8 Barabási, A.L. and Albert, R. (1999). Emergence of scaling in random networks. *Science* 286 (5439): 509–512.

9 Cajueiro, D.O. and Andrade, R.F. (2010). Controlling self-organized criticality in sandpile models. *Physical Review E* 81 (1): 015102.

10 Cajueiro, D.O. and Andrade, R.F. (2010). Controlling self-organized criticality in complex networks. *The European Physical Journal B-Condensed Matter and Complex Systems* 77 (2): 291–296.

11 Cajueiro, D.O. and Andrade, R.F. (2010). Dynamical programming approach for controlling the directed Abelian Dhar-Ramaswamy model. *Physical Review E* 82 (3): 031108.

12 D'Agostino, G., Scala, A., Zlatic, V., and Caldarelli, G. (2012). Robustness and assortativity for diffusion-like processes in scale-free networks. *EPL* 97 (6): 68006.

13 Qi, J. and Pfenninger, S. (2015). Controlling the self-organizing dynamics in a sandpile model on complex networks by failure tolerance. *EPL* 111 (3): 38006.

14 Schult, D.A. and Swart, P. (2008). Exploring network structure, dynamics, and function using network. In: *Proceedings of the 7th Python in Science Conferences (SciPy 2008)*, 2008, 11–16.

15 Albert, R., Jeong, H., and Barabási, A.L. (2000). Error and attack tolerance of complex networks. *Nature* 406 (6794): 378–382.

16 Motter, A.E. and Lai, Y.C. (2002). Cascade-based attacks on complex networks. *Physical Review E* 66 (6): 065102.

17 Zhao, L., Park, K., and Lai, Y.C. (2004). Attack vulnerability of scale-free networks due to cascading breakdown. *Physical Review E* 70 (3): 035101.

18 Lai, Y.C., Motter, A.E., and Nishikawa, T. (2004). Attacks and cascades in complex networks. In: *Complex Networks*, 299–310. Springer.

19 Asavathiratham, C. (2000). The influence model: A tractable representation for the dynamics of networked Markov chains, Ph.D. thesis, Citeseer.

20 Asavathiratham, C., Roy, S., Lesieutre, B., and Verghese, G. (2001). The influence model. *IEEE Control Systems* 21 (6): 52–64.

21 Watts, D.J. (2002). A simple model of global cascades on random networks. *Proceedings of the National Academy of Sciences* 99 (9): 5766–5771.

22 Wang, X.F. and Xu, J. (2004). Cascading failures in coupled map lattices. *Physical Review E* 70 (5): 056113.

23 Xu, J. and Wang, X.F. (2005). Cascading failures in scale-free coupled map lattices. *Physica A: Statistical Mechanics and its Applications* 349 (3): 685–692.

24 Dobson, I., Carreras, B.A., and Newman, D.E. (2005). A loading-dependent model of probabilistic cascading failure. *Probability in the Engineering and Informational Sciences* 19 (01): 15–32.

25 Consul, P.C. (1989). Generalized Poisson distributions. New York: Marcel Dekker.

26 Otter, R. (1949). The multiplicative process. *Annals of Mathematical Statistics* 20: 206–224.

27 Buldyrev, S.V., Parshani, R., Paul, G. et al. (2010). Catastrophic cascade of failures in interdependent networks. *Nature* 464 (7291): 1025–1028.

28 Phadke, A. and Thorp, J.S. (1996). Expose hidden failures to prevent cascading outages [in power systems]. *IEEE Computer Applications in Power* 9 (3): 20–23.

29 Chen, J., Thorp, J.S., and Dobson, I. (2005). Cascading dynamics and mitigation assessment in power system disturbances via a hidden failure model. *International Journal of Electrical Power & Energy Systems* 27 (4): 318–326.

30 Thorp, J., Phadke, A., Horowitz, S., and Tamronglak, S. (1998). Anatomy of power system disturbances: importance sampling. *International Journal of Electrical Power & Energy Systems* 20 (2): 147–152.

31 Rios, M.A., Kirschen, D.S., Jayaweera, D. et al. (2002). Value of security: modeling time-dependent phenomena and weather conditions. *IEEE Transactions on Power Systems* 17 (3): 543–548.

32 Kirschen, D.S., Jayaweera, D., Nedic, D.P., and Allan, R.N. (2004). A probabilistic indicator of system stress. *IEEE Transactions on Power Systems* 19 (3): 1650–1657.

33 Billinton, R. and Aboreshaid, S. (1995). Security evaluation of composite power systems. *IEE Proceedings-Generation, Transmission and Distribution* 142 (5): 511–516.

34 Dobson, I., Carreras, B.A., Lynch, V.E., and Newman, D.E. (2001). An initial model for complex dynamics in electric power system blackouts. In: *Proceedings of the Annual Hawaii International Conference on System Sciences*, Citeseer, 51.

35 Carreras, B.A., Lynch, V.E., Dobson, I., and Newman, D.E. (2002). Critical points and transitions in an electric power transmission model for cascading failure blackouts. *Chaos* 12 (4): 985–994.

36 Ren, H., Dobson, I., and Carreras, B.A. (2008). Long-term effect of the n-1 criterion on cascading line outages in an evolving power transmission grid. *IEEE Transactions on Power Systems* 23 (3): 1217–1225.

37 Carreras, B.A., Newman, D.E., Dobson, I., and Degala, N.S. (2013). Validating OPA with WECC data. In: *46th Hawaii International Conference on System Sciences (HICSS)*, IEEE, 2197–2204.

38 Carreras, B.A., Newman, D.E., Dobson, I., and Poole, A.B. (2004). Evidence for selforganized criticality in a time series of electric power system blackouts. *IEEE Transaction on Circuits Systems I: Regular Papers* 51 (9): 1733–1740.

39 Mei, S., He, F., Zhang, X. et al. (2009). An improved OPA model and blackout risk assessment. *IEEE Transactions on Power Systems* 24 (2): 814–823.

40 Weng, X., Hong, Y., Xue, A., and Mei, S. (2006). Failure analysis on China power grid based on power law. *Journal of Control Theory and Applications* 4 (3): 235–238.

41 Simpson, P. and Van Bossuyt, R. (1996). Tree-caused electric outages. *Journal of Arboriculture* 22: 117–121.

42 Radmer, D.T., Kuntz, P.A., Christie, R.D. et al. (2002). Predicting vegetation-related failure rates for overhead distribution feeders. *IEEE Transactions on Power Delivery* 17 (4): 1170–1175.

43 Guikema, S.D., Davidson, R.A., and Liu, H. (2006). Statistical models of the effects of tree trimming on power system outages. *IEEE Transactions on Power Delivery* 21 (3): 1549–1557.

44 Cieslewicz, S. and Novembri, R. (2004). Utility vegetation management final report. Federal Energy Regulatory Commission.

45 North America Electric Reliability Council (2002). 1996 System disturbances.

46 U.S.-Canada Power System Outage Task Force (2004). Final report on the August 14th blackout in the United States and Canada.

47 Union for the Co-ordination of Electricity Transmission (UCTE) (2003). Interim Report of the Investigation Committee on the 28 September 2003 Blackout in Italy.

48 NERC Steering Group (2004). Technical analysis of the August 14, 2003, blackout: What happened, why, and what did we learn. Report to the NERC Board of Trustees.

49 Qi, J., Mei, S., and Liu, F. (2013). Blackout model considering slow process. *IEEE Transactions on Power Systems* 28 (3): 3274–3282.

50 Henneaux, P., Labeau, P.E., and Maun, J.C. (2013). Blackout probabilistic risk assessment and thermal effects: impacts of changes in generation. *IEEE Transactions on Power Systems* 28 (4): 4722–4731.

51 Carreras, B.A., Lynch, V.E., Dobson, I., and Newman, D.E. (2002). Dynamics, criticality and self-organization in a model for blackouts in power transmission systems. In: *Proceedings of 35th Annual Hawaii International Conference on System Sciences*, IEEE, 9.

52 Carreras, B.A., Lynch, V.E., Dobson, I., and Newman, D.E. (2004). Complex dynamics of blackouts in power transmission systems. *Chaos* 14 (3): 643–652.

53 Zhang, B. and Yin, X. (2005). *Power System Relay Protection*. Beijing: China Electric Power Press.

54 Anghel, M., Werley, K.A., and Motter, A.E. (2007). Stochastic model for power grid dynamics. In: *Proceedings of 40th Annual Hawaii International Conference on System Sciences*, IEEE, 113.

55 Anis, H., Mahmoud, S., and Abdallah, M. (2000). Probabilistic modeling of power lines magnetic fields. In: *Power Engineering Society Summer Meeting, 2000*. IEEE, 4, 2383–2387.

56 Guo, S. (2009). *Basics of Design of Overhead Transmission Lines*. Beijing: China Electric Power Press.

57 Richards, F. (1959). A flexible growth function for empirical use. *Journal of Experimental Botany* 10 (2): 290–301.

58 Shvets, V. and Zeide, B. (1996). Investigating parameters of growth equations. *Canadian Journal of Forest Research* 26 (11): 1980–1990.

59 Shen, H., Li, S., Hu, X. et al. (1995). Analysis on the relationship between climatic factors and the growth of Mongolian Scots pine in eastern mountain area of northeast China. *Journal of Northeast Forestry University* 23: 33–39.

60 Cui, J. and Chen, J. (2011). *Design and Building of Overhead Lines*. Beijing: China Water Power Press.

61 Mei, S., Weng, X., Xue, A. et al. (2006). Blackout model based on OPF and its self-organized criticality. In: *2006 Chinese Control Conference*, IEEE, 1673–1678.

62 Mei, S., Ni, Y., Wang, G., and Wu, S. (2008). A study of self-organized criticality of power system under cascading failures based on AC-OPF with voltage stability margin. *IEEE Transactions on Power Systems* 23 (4): 1719–1726.

63 Grigg, C., Wong, P., Albrecht, P. et al. (1999). The IEEE reliability test system-1996. *IEEE Transactions on Power Systems* 14 (3): 1010–1020.

64 Ciapessoni, E., Cirio, D., Cotilla-Sanchez, E. et al. (2018). Benchmarking quasi-steady state cascading outage analysis methodologies. In: *IEEE International Conference on Probabilistic Methods Applied to Power Systems (PMAPS)*.

65 Dobson, I., Kim, J., and Wierzbicki, K.R. (2010). Testing branching process estimators of cascading failure with data from a simulation of transmission line outages. *Risk Analysis* 30 (4): 650–662.

66 Qi, J., Dobson, I., and Mei, S. (2013). Towards estimating the statistics of simulated cascades of outages with branching processes. *IEEE Transactions on Power Systems* 28 (3): 3410–3419.

67 Song, J., Cotilla-Sanchez, E., Ghanavati, G., and Hines, P.D. (2016). Dynamic modeling of cascading failure in power systems. *IEEE Transactions on Power Systems* 31 (3): 2085–2095.

68 Yan, J., Tang, Y., He, H., and Sun, Y. (2015). Cascading failure analysis with DC power flow model and transient stability analysis. *IEEE Transactions on Power Systems* 30 (1): 285–297.

69 Samaan, N.A., Dagle, J.E., Makarov, Y.V. et al. (2015). Dynamic Contingency Analysis Tool-Phase 1.

70 Dobson, I., Flueck, A., Aquiles-Perez, S. et al. (2018). Towards incorporating protection and uncertainty into cascading failure simulation and analysis. *IEEE International Conference Probabilistic Methods Applied to Power Systems (PMAPS)*.

71 Boccaletti, S., Latora, V., Moreno, Y. et al. (2006). Complex networks: Structure and dynamics. *Physics reports* 424(4–5): 175–308.

72 Bialek, J., Ciapessoni, E., Cirio, D. et al. (2016). Benchmarking and validation of cascading failure analysis tools. *IEEE Transactions on Power Systems* 31 (6): 4887–4900.

3

Understanding Cascading Failures

Cascading blackouts of transmission systems are complicated sequences of dependent outages that lead to load shed. Of particular concern are the larger cascading blackouts; these blackouts are rare and high-impact events that have substantial risk, and they pose many challenges in simulation, analysis, and mitigation. In the last chapter, we introduced several cascading failure models that can produce massive amounts of data regarding line outages, generator tripping, and load shedding. However, simulations cannot produce statistical insight or metrics with actionable information without a carefully designed information extraction method. In this chapter, we will discuss several theories and models that can extract useful and actionable information from simulated cascading data to help understand why and how cascading failures happen in electric power grids. Specifically,

- The self-organized criticality (SOC) theory in statistical physics [1–6] provides an explanation about why and how large-scale systemwide failures can happen. There has been evidence that indicates power system blackouts are governed by SOC.
- The branching process model [7, 8] provides an overall statistical description about how cascading failures propagate among the components in the system [9, 10]. The descriptive parameters of the branching process can characterize the system resilience to cascading. For example, the average propagation quantifies the extent to which outages cause further outages. The branching process analysis is a bulk statistical analysis that should be regarded as complementary to detailed causal analysis.
- The multitype branching process model [7, 8] is a generalization of the one-type branching process model that can be used to effectively analyze the interdependencies either between different types of failures or between different infrastructure systems [11].
- The failure interaction analysis [12] can explicitly study the interactions between component failures based on the idea that the system-level

Power System Control Under Cascading Failures: Understanding, Mitigation, and System Restoration, First Edition. Kai Sun, Yunhe Hou, Wei Sun and Junjian Qi.

failures are not caused by any specific event but by the property that the components in the system are tightly coupled and interdependent with each other [13]. Compared with the branching process model, it retains more information about the system behavior and thus can be used to study in more detail how the outages propagate in the system from one component to another.

3.1 Self-Organized Criticality

SOC [1–6] has been seen in many natural and engineering systems, such as earthquakes [3], forest fires [4], and power grids [14–16], which means that a system can be self-organized toward the critical point with power-law-distributed event sizes. SOC has been applied to study and understand the cascading failures in complex systems, which are complicated sequences of dependent outages and can take place in electric power systems [10, 12, 14–16], the Internet [17], the road system [18], and in social and economic systems [19].

3.1.1 SOC Theory

In SOC theory, dynamical systems with extended spatial degrees of freedom naturally evolve into self-organized critical structures of states that are barely stable [1]. SOC is a property of the dynamical system to organize its microscopic behaviors to be spatial (and/or temporal) scale independent. In contrast to usual phase transitions, a system displaying SOC organizes itself into a state with critical behaviors through evolutions, not requiring external tuning of the control parameters. Such a state has a complex but rather general structure, under which there is no single characteristic event size or scale. Despite the complexity, the system exhibits simple statistical properties governed by power laws.

In phase space, an SOC state corresponds to a finite-dimensional attractor, which is neither stable nor unstable but is critically stable [20]. When a system in a critically stable state is perturbed, it will spontaneously evolve to another critically stable state. As a result, the system is always critically stable and does not converge to a stable attractor. In the evolution of complex systems, the completely stable or unstable states can hardly exist: if a system is completely stable, it will not evolve; if it is completely unstable, its current state cannot be maintained in a dynamic process. Therefore, only critically stable systems can both evolve and maintain certain stability [20].

Several mathematical models seem to display SOC behavior, such as the Bak–Tang–Wiesenfeld sandpile model [1–3], Critical Forest Fire model [4], Olami–Feder–Christensen earthquake model [21, 22], and Lattice Gas model [23].

3.1.2 Evidence of SOC in Blackout Data

In Carreras et al. [24], a 15-year time series of North American electric power transmission system blackouts is analyzed. It is shown that the probability distribution functions of various measures of blackout size have a power tail, and rescaled range analysis of the time series shows moderate long-time correlations. Figure 3.1 gives the number of blackouts with more than n customers unserved. The empirical data falls off with a power of approximately −0.8 (all tail points considered) or −0.7 (last seven tail points neglected due to sparse data).

Determining whether power system blackouts are governed by SOC is not straightforward, since there are no unequivocal determining criteria. One approach is to compare the characteristic measures of the power system to those obtained from a known SOC system. In Carreras et al. [24], the blackout size distribution from the empirical data is compared with results from a one-dimensional idealized sandpile, which is a prototypical model of an SOC system. Figure 3.2 shows the probability distribution function (PDF) of the avalanche sizes from the sandpile data together with the rescaled PDF of the energy unserved from the blackout data [24]. The same analysis applied to a time series from a sandpile model known to be self-organized critical gives results of the same form. Thus, the blackout data seem consistent with SOC.

In Qi et al. [16], cascading failures are simulated on the Northeast Power Grid of China by using the OPA model with slow process. The probability of load shed is shown on a log–log plot in Figure 3.3. The blackout size distribution of the blackout model agrees approximately with a power law with exponent of −0.8363, which agrees with the statistical analysis of the historical data,

Figure 3.1 Complementary cumulative frequency of the number of customers unserved [24].

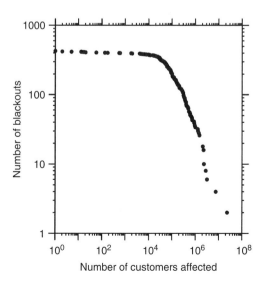

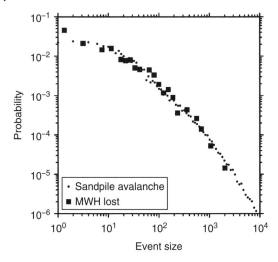

Figure 3.2 Rescaled probability distribution function (PDF) of energy unserved during blackouts superimposed on the PDF of the avalanche size in sandpile [24].

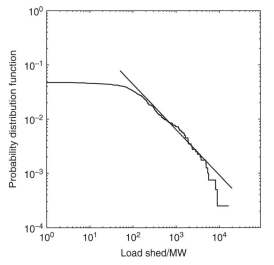

Figure 3.3 Distribution of the load shed from the OPA model with slow process.

−0.8641 [25]. The distribution of the number of line outages also shows power tail nature, as is shown in Figure 3.4. The slope is −1.2742.

3.2 Branching Processes

Simulations of cascading outages or statistical utility outage data can provide many samples of cascades. The simulation data can be generated from cascading failure models, such as OPA model or its variants [16, 26–28], which

Figure 3.4 Distribution of line outages from the OPA model with slow process.

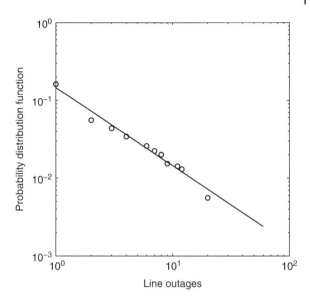

naturally produce line outages and the load shed in generations or stages; each iteration of the "main loop" of the simulation produces another generation [10, 29]. Alternatively, the statistical utility data can be grouped into different cascades and then into different generations within each cascade based on the outages' timing [30].

It is useful to statistically describe these simulated cascades by high-level probabilistic models. One such model is the Galton–Watson branching process [7, 8]. It statistically describes how the number of outages in a blackout propagate. The initial outages propagate randomly to produce subsequent outages in generations. Each outage in each generation (a "parent" outage) independently produces a random number 0, 1, 2, ... of outages ("children" outages) in the next generation. The distribution of the number of children from one parent is called the *offspring distribution*. The children outages then become parents to produce another generation, and so on. If the number of outages in a generation becomes zero, then the cascade stops.

The descriptive parameters of the branching process can characterize the system resilience to cascading. For example, the average propagation quantifies the extent to which outages cause further outages. Moreover, it is much quicker to estimate the parameters of a branching process from a shorter simulation run and then predict the distribution of blackout size using the branching process than it is to run the simulation for a very long time to accumulate enough cascades (especially the rare large cascades) to empirically estimate the distribution of blackout size.

In the branching process modeling, random numbers of outages in each generation are produced to statistically match the outcome of the cascading. When used to track the number of line outages, the branching process does not specify which lines outage, or where they are, or explain why they outaged. It only describes the statistics of the number of line outages in each generation and also the total number of line outages. Similarly, when used to track the load shed, it does not specify which load is shed, or where, or explain why the load is shed. It is worth noting that the underlying cascading processes and interactions are complicated and varied. For example, there are situations in which the load shed tends to inhibit the chance of further cascading, and there are others in which the load shed tends to increase the chance of further cascading. The branching process analysis is a bulk statistical analysis that should be regarded as complementary to detailed causal analysis.

3.2.1 Definition of the Galton–Watson Branching Process

Consider objects that can generate additional objects of the same kind. An initial set of objects is called generation 0 and has children called generation 1, whose children are generation 2, and so on. We denote Z_0, Z_1, Z_2, $\cdots$ the numbers of objects in the zeroth, first, second, … generations, and make the following two assumptions [7, 8].

1) If the size of the n-th generation is known, the probability law governing later generations does not depend on the sizes of generations preceding the n-th. This means Z_0, Z_1, $\cdots$, and so on form a *Markov chain.*
2) Different objects in one generation do not interfere with one another, and the number of children born to an object does not depend on how many other objects are present.

Assume $Z_0 = 1$. If $Z_0 \neq 1$ appropriate adjustments can be easily made due to Assumption 2. The probability distribution of Z_1 is $P(Z_1 = k) = p_k$, $k = 0,1,2, \ldots, \Sigma p_k = 1$, where p_k is the probability that an object existing in the n-th generation has k children in the $(n+1)$-th generation. As mentioned before, this distribution is the *offspring distribution*. It is assumed that p_k does not depend on the generation number n.

3.2.2 Estimation of Mean of the Offspring Distribution

In the case of line outages, the number in each generation are counted. In the case of load shed, the continuously varying load shed amounts in each generation are processed and discretized as described in Section 3.2.4 to produce integer multiples of the chosen discretization unit. In either case, M cascades are collected and arranged as

	generation 0	generation 1	generation 2	$\cdots$
cascade 1	$Z_0^{(1)}$	$Z_1^{(1)}$	$Z_2^{(1)}$	$\cdots$
cascade 2	$Z_0^{(2)}$	$Z_1^{(2)}$	$Z_2^{(2)}$	$\cdots$
$\vdots$	$\vdots$	$\vdots$	$\vdots$	$\vdots$
cascade M	$Z_0^{(M)}$	$Z_1^{(M)}$	$Z_2^{(M)}$	$\cdots$

where $Z_j^{(m)}$ is the number of outages in generation j of cascade m. Each cascade has a nonzero number of outages in generation zero. The shortest cascades stop in generation one by having no further outages, but cascades will continue for several or occasionally many generations before terminating. The assumption of a positive number of outages in generation zero implies that all statistics are conditioned on a cascade starting.

The branching process parameters can be estimated from these cascades. Based on these parameters the distribution of the blackout size (the number of line outages or the amount of the load shed) can be estimated.

Note that all the outages are parent outages, and all the outages in generations one and higher are children outages. It follows that the average propagation λ can be estimated as the total number of children divided by the total number of parents:

$$\hat{\lambda} = \frac{\sum_{m=1}^{M}\left(Z_1^{(m)} + Z_2^{(m)} + \cdots\right)}{\sum_{m=1}^{M}\left(Z_0^{(m)} + Z_1^{(m)} + \cdots\right)}. \tag{3.1}$$

This is the standard Harris estimator of the offspring mean [8, 31].

3.2.3 Estimation of Variance of the Offspring Distribution

The variance of the offspring distribution is another important parameter of the branching process [7, 8]. It is needed for determining the discretization of the load shed data in Section 3.2.4. Although there are estimators of offspring variance in the general branching process study [31, 32], an estimator is proposed in Qi et al. [10] especially for the cascading data in which the number of generations is small and the offspring mean and variance may vary somewhat with generation.

Let $Y_t^{(m)} = Z_0^{(m)} + Z_1^{(m)} + \cdots + Z_t^{(m)}$ be the total outages up to and including generation t of cascade m. Let $Y_\infty^{(m)}$ be the total outages in all generations of cascade m. We group the cascades into K groups so that each group has at least

n initial outages. Let $Y_\infty^{[k]}$ be the sum of the outages in group k and $\hat{\lambda}_\infty^{[k]}$ be the average propagation (offspring mean) estimated by the Harris estimator Eq. (3.1) for the outages in group k.

Given $\hat{\lambda}$ in Eq. (3.1), the variance of the offspring distribution can be estimated as

$$\hat{\sigma}^2 = \frac{1}{K}\sum_{k=1}^{K} Y_\infty^{[k]}\left(\hat{\lambda}_\infty^{[k]} - \hat{\lambda}\right)^2. \tag{3.2}$$

It is desirable to have large n and large K, but there is a tradeoff between n and K. If there are a total of p initial outages in the cascades, $nK \approx p$. To determine values of n and K, we simulated 5000 realizations of an ideal Galton–Watson branching process with Poisson offspring distribution of known mean and variance for a range of values of n. We found that for small n and large K, Eq. (3.2) underestimates the variance, whereas for large n and small K, Eq. (3.2) is a noisy estimate. We chose $n = 20p/5000$ and $K \approx 250$ based on this testing. When load shed data is discretized, the number of initial outages p is inversely proportional to the discretization unit Δ. Choosing n proportional to p ensures that the grouping of the cascades into K groups will remain unchanged when the unit of discretization unit Δ changes.

Now we give justification for Eq. (3.2), mainly by following Janev [33]. Consider a branching process $X_0, X_1, X_2, \ldots$ with $X_0 = n$ and offspring mean $0 < \lambda < 1$ and offspring variance $\sigma^2 < \infty$. It corresponds to one of the K groups of cascades discussed previously. Let $Y_t(n) = X_0 + X_1 + \lambda + X_t$ denote the total outages up to and including generation t. Let $\hat{\lambda}_t(n) = (Y_t(n) - X_0)/Y_{t-1}(n)$ be the Harris estimator for the offspring mean computed from $X_0, X_1, X_2, \ldots, X_t$. Write $X_0^{(i)}, X_1^{(i)}, X_2^{(i)}, \ldots$ for the branching process starting from the ith initial outage only. Let $W_t^{(i)} = \sum_{k=1}^{t}\left(X_k^{(i)} - \lambda X_{k-1}^{(i)}\right)$.

From Janev [33, Eq. (8)], there is

$$Y_{t-1}(n)\left(\hat{\lambda}_t(n) - \lambda\right)^2 = \frac{\frac{1}{n}\left(\sum_{i=1}^{n} W_t^{(i)}\right)^2}{\frac{1}{n}Y_{t-1}(n)}. \tag{3.3}$$

We have

$$\frac{1}{n}\left(\sum_{i=1}^{n} W_t^{(i)}\right)^2 = \frac{1}{n}\sum_{i=1}^{n}\left(W_t^{(i)}\right)^2 + \frac{1}{n}2pt\sum_{\substack{1 \le i,j \le n \\ i \ne j}} W_t^{(i)} W_t^{(j)}.$$

Janev [32] states that $EW_t^2 = \sigma^2 \left(1 + \lambda + \cdots + \lambda^{t-1}\right)$ and $EW_t = 0$. Then the strong law of large numbers implies

$$\frac{1}{n}\left(\sum_{i=1}^{n} W_t^{(i)}\right)^2 \xrightarrow{a.s.} \sigma^2 \left(1 + \lambda + \cdots + \lambda^{t-1}\right).$$

Moreover, $Y_{t-1}(n)/n \xrightarrow{a.s.} EY_{t-1} = 1 + \lambda + \cdots + \lambda^{t-1}$, a nonzero constant, as $n \to \infty$. Hence, from Eq. (3.3),

$$Y_{t-1}(n)\left(\hat{\lambda}_t(n) - \lambda\right)^2 \xrightarrow{a.s.} \sigma^2 \quad as \ n \to \infty.$$

Now letting $t \to \infty$, we get $Y_\infty(n)\left(\hat{\lambda}_\infty(n) - \lambda\right)^2 \xrightarrow{a.s.} \sigma^2$ as $n \to \infty$, and $E\left[Y_\infty(n)\left(\hat{\lambda}_\infty(n) - \lambda\right)^2\right] \to \sigma^2$ as $n \to \infty$.

Write $\hat{\lambda}(p)$ to show the dependence of the Harris estimator $\hat{\lambda}$ for all the cascades on the total number of initial outages p. Define

$$\Delta = Y_\infty(n)\left[\left(\hat{\lambda}_\infty(n) - \lambda\right)^2 - \left(\hat{\lambda}_\infty(n) - \hat{\lambda}(p)\right)^2\right] = \Delta_1 + \Delta_2,$$

where

$$\Delta_1 = Y_\infty(n)\left(\hat{\lambda}_\infty(n) - \lambda\right)\left(\hat{\lambda}(p) - \lambda\right)$$

$$\Delta_2 = Y_\infty(n)\left(\hat{\lambda}_\infty(n) - \hat{\lambda}(p)\right)\left(\hat{\lambda}(p) - \lambda\right).$$

By Cauchy–Schwartz,

$$\left|E\Delta_1\right|^2 \le E\left[\left(Y_\infty(n)\left(\hat{\lambda}_\infty(n) - \lambda\right)\right)^2\right]E\left[\left(\hat{\lambda}(p) - \lambda\right)^2\right]. \tag{3.4}$$

Since $Y_\infty(n)\left(\hat{\lambda}_\infty(n) - \lambda\right) = (1 - \lambda)\left[Y_\infty(n) - \dfrac{n}{1 - \lambda}\right]$, from Janev [33, Eq. (6) and Lemma 1] we obtain

$$E\left[\left(Y_\infty(n)\left(\hat{\lambda}_\infty(n) - \lambda\right)\right)^2\right] = (1 - \lambda)^2 \frac{n\sigma^2}{(1 - \lambda)^3} = \frac{n\sigma^2}{1 - \lambda}.$$

From Janev [32, Theorem 2(ii)],

$$E\left[\left(\hat{\lambda}(p) - \lambda\right)^2\right] \to \sigma^2(1 - \lambda)/p \quad as \ p \to \infty.$$

Choose $p \geq n^{1+\delta}$ for some $\delta > 0$. Then from Eq. (3.4),

$$\left| E \Delta_1 \right|^2 \leq \frac{n\sigma^2}{1-\lambda} \frac{\sigma^2 \left(1 - \lambda \right)}{n^{1+\delta}} = \frac{\sigma^4}{n^\delta} \to 0 \quad as \ n \to \infty,$$

and $E\Delta_1 \to 0$ as $n \to \infty$.

$$\Delta_1 - \Delta_2 = Y_\infty \left(n \right) \left(\hat{\lambda} \left(p \right) - \lambda \right)^2 = \frac{Y_\infty \left(n \right)}{n} n \left(\hat{\lambda} \left(p \right) - \lambda \right)^2$$

For $p \geq n^{1+\delta}$, as $n \to \infty$,

$$nE \left[\left(\hat{\lambda} \left(p \right) - \lambda \right)^2 \right] = n \frac{\sigma^2 \left(1 - \lambda \right)}{p} \leq n \frac{\sigma^2 \left(1 - \lambda \right)}{n^{1+\delta}} \to 0.$$

That is, $E \left[n \left(\hat{\lambda} \left(p \right) - \lambda \right)^2 \right] \overset{L_1}{\to} 0$. Moreover, $Y_\infty \left(n \right) / n \overset{L_1}{\to} 1 / \left(1 - \lambda \right)$, and hence, $\Delta_1 - \Delta_2 \overset{L_1}{\to} 0$ and $E[\Delta_1 - \Delta_2] \to 0$ as $n \to \infty$. Since we have already shown $E\Delta_1 \to 0$, we conclude that $E\Delta = E[\Delta_1 + \Delta_2] \to 0$. Hence for $p \geq n^{1+\delta}$,

$$E \left[Y_\infty \left(n \right) \left(\hat{\lambda}_\infty \left(n \right) - \hat{\lambda} \left(p \right) \right)^2 \right] \to \sigma^2 \quad as \ n \to \infty.$$

Since $\hat{\sigma}^2 \to E \left[Y_\infty \left(n \right) \left(\hat{\lambda}_\infty \left(n \right) - \hat{\lambda} \left(p \right) \right)^2 \right]$ as $K \to \infty$, $\hat{\sigma}^2 \to \sigma^2$ as $K, n \to \infty$.

3.2.4 Processing and Discretization of Continuous Data

Analyzing the load shed data is important because the load shed is a measure of blackout size that is of great significance to both utilities and society, whereas line outages are of direct interest only to utilities. Therefore, the capability to describe load shed cascading data is useful. In order to accommodate the continuously varying load shed data, two approaches are tried in Kim et al. [9]. The approach that worked best discretized the load shed data to integer multiples of a chosen discretization unit. The discretized load shed data can then be processed by a Galton–Watson branching process, which works with nonnegative integers. However, in Kim et al. [9] the choice of the discretization unit is ad hoc without a systematic approach.

Another approach is to directly analyze the load shed by a continuous-state branching process model [9]. Each generation is a continuously varying amount of load shed that propagates according to a continuous offspring probability distribution. One major challenge is how to determine the form of the continuous offspring distribution. In Kim et al. [9] a Gamma distribution is assumed for simplicity and computational convenience. But the problem of a justified choice of continuous offspring distribution for cascading load shed remains open. The calculations for continuous state branching processes are analogous to the calculations for Galton–Watson branching processes but can be more technically difficult.

In Qi et al. [10] a systematic way is proposed to select the unit of discretization. The basic idea is based on the assumption that the offspring distribution is a Poisson distribution. There are general arguments suggesting that the choice of a Poisson offspring distribution is appropriate [9], based on the offspring outages being selected from a large number of possible outages that have small probability and are approximately independent. Since the variance of a Poisson distribution equals the mean, the discretization can be chosen so that the offspring distribution has its variance equal to its mean and is therefore consistent with the Poisson distribution.

Next we discuss the detailed procedure for the processing and discretization of the load shed data to produce integer counts of the discretization unit of load shed.

1) Initial processing

    Very small load shed amounts (less than 0.5% of total load) are considered negligible and are rounded to zero. The cascades with no load shed are discarded. For the cascades without load shed in initial generations but with non-negligible load shed in subsequent generations, the initial generations with no load shed are discarded so that generation zero always starts with a positive amount of load shed. Moreover, when any intermediate generation with zero load shed is encountered, the current cascade ends and a new cascade is started at the next generation with nonzero load shed. Finally, a total of M cascades are obtained, and $X_k^{(m)}$ denotes the load shed at generation k of cascade m.

2) Discretization

    To apply a Galton–Watson branching process to the load shed data, we need to discretize the continuously varying load shed data in MW to integer multiples of a unit of discretization Δ. In particular, we use the following discretization to convert the load shed $X_k^{(m)}$ MW to integer multiples $Z_k^{(m)}$ of Δ MW:

$$Z_k^{(m)} = \mathrm{int}\left[\frac{X_k^{(m)}}{\Delta} + 0.5\right], \qquad (3.5)$$

where int[x] is the integer part of x. We add 0.5 before taking the integer part to ensure that the average values of $Z_k^{(m)}$ and $X_k^{(m)}/\Delta$ are equal. The discretization in Eq. (3.5) is straightforward except that the choice of the discretization unit Δ matters.

In Kim et al. [9] a heuristic and subjective choice of Δ is made. Here we give a justifiable way to choose Δ. We first discuss how the choice of Δ affects the mean and variance of the offspring distribution. Consider a second choice of discretization unit Δ' and the corresponding discretized data $Z'^{(m)}_k$ so that

$$Z'^{(m)}_k = \text{int}\left[\frac{X^{(m)}_k}{\Delta'} + 0.5\right].$$

Neglecting the effects of rounding, we have $Z^{(m)}_k \approx X^{(m)}_k / \Delta$ and $Z'^{(m)}_k \approx X^{(m)}_k / \Delta'$, which further yields

$$Z'^{(m)}_k \approx \frac{\Delta}{\Delta'} Z^{(m)}_k. \tag{3.6}$$

The mean λ of the offspring distribution is a dimensionless ratio that does not depend on the scaling or units of the load shed (see Eq. (3.1)). Recall that the offspring distribution is defined as the distribution of the number of units of load shed in generation $k+1$, assuming one unit of load shed in generation k. When the discretization is changed from Δ to Δ', both $Z^{(m)}_k$ and $Z^{(m)}_{k+1}$ are multiplied by Δ/Δ'. Thus, the mean of the offspring distribution does not change.

Although the variance of $Z^{(m)}_{k+1}$ is multiplied by $(\Delta/\Delta')^2$, the offspring distribution variance σ^2, which is the variance in generation $k+1$ arising from one unit of load shed in generation k, is only multiplied by Δ/Δ'. This is because $Z^{(m)}_k$ is also multiplied by Δ/Δ'.

To explain these scalings with an example, suppose that in the first cascade there are 2 units of load shed in generation k with discretization Δ so that $Z^{(1)}_k = 2$. According to the principles of the branching process, the distribution of the load shed $Z^{(1)}_{k+1}$ in generation $k+1$ is the sum of two independent copies of the offspring distribution for discretization Δ, and therefore has mean 2λ and variance $2\sigma^2$. Now change the discretization to $\Delta' = 2\Delta$ so that $Z'^{(1)}_{k+1} \approx Z^{(1)}_{k+1}/2$ and $Z'^{(1)}_k = Z^{(1)}_k/2 = 1$. Then the distribution of $Z'^{(1)}_{k+1}$ is the offspring distribution for discretization Δ', which has mean $EZ'^{(1)}_{k+1} = EZ^{(1)}_{k+1}/2 = 2\lambda/2 = \lambda$, and variance $E\left[\left(Z'^{(1)}_{k+1} - \lambda\right)^2\right] = E\left[\left(Z^{(1)}_{k+1} - 2\lambda\right)^2\right]/4 = 2\sigma^2/4 = \sigma^2/2$. Thus, changing the discretization unit Δ strongly affects σ^2, and in particular increasing Δ decreases σ^2 proportionally.

When we consider the influence of rounding on these scaling approximations, the mean of the offspring distribution λ is only slightly affected by Δ, while the variance of the offspring distribution σ^2 has an overall strong tendency of decreasing proportionally with the increase of Δ (strict monotonicity for small changes in Δ is not guaranteed). We notate these strong and weak dependencies by writing $\sigma^2(\Delta)$ and $\lambda(\Delta)$, respectively, and can get $\sigma^2(\Delta) \approx \sigma^2(\Delta')\,\Delta'/\Delta$ and $\lambda(\Delta) \approx \lambda(\Delta')$ from this analysis.

Our calculations assume a Poisson offspring distribution, for which the variance is equal to the mean. Therefore, to be consistent with the Poisson distribution, we need to choose a discretization so that the variance of the offspring distribution is equal to its mean. That is, we need to choose Δ so that $\sigma^2(\Delta) = \lambda(\Delta)$.

Specifically, we discretize the data for $\Delta = 1$ MW and estimate $\sigma^2(1)$ and $\lambda(1)$. Then the Δ that satisfies $\sigma^2(\Delta) = \lambda(\Delta)$ is approximately $\sigma^2(1)/\lambda(1)$.

3) Processing after discretization

After discretization, the initial load shed in some cascades may become zero. These cascades are discarded.

3.2.5 Estimation of Distribution of Total Outages

The empirical probability distribution of the number of initial outages Z_0 is

$$P(Z_0 = z_0) = \frac{1}{M_u} \sum_{m=1}^{M_u} I\left[Z_0^{(m)} = z_0 \right], \tag{3.7}$$

where M_u is the number of cascades used for estimation and the notation I[event] is the indicator function that evaluates to one when the event happens and evaluates to zero when the event does not happen.

The average number of initial outages θ is estimated as

$$\hat{\theta} = \frac{1}{M_u} \sum_{m=1}^{M_u} Z_0^{(m)}. \tag{3.8}$$

The offspring distribution is assumed to be a Poisson distribution with mean $\hat{\lambda}$. Given the probability distribution of Z_0 and the average propagation $\hat{\lambda}$, the formula for calculating the probability distribution of the total number of outages Y_∞ is the following mixture of Borel–Tanner distributions [34]:

$$P\left(Y_\infty = r\right) = \sum_{z_0=1}^{r} P\left(Z_0 = z_0\right) z_0 \hat{\lambda} \left(r\hat{\lambda}\right)^{r-z_0-1} \frac{e^{-r\hat{\lambda}}}{\left(r - z_0\right)!}. \tag{3.9}$$

The summation in Eq. (3.9) runs from $z_0 = 1$ to r since these are the possible initial outages for r total outages.

3.2.6 Statistical Insight of Branching Process Parameters

The mean of the offspring distribution is the parameter λ, which is the average propagation (the average number of children outages from each parent outage) quantifying the tendency for the cascade to propagate.

The eventual behavior of the branching process is governed by the propagation parameter λ [7,8]. If $\lambda < 1$ (each parent outage has on average less than one child), the outages will always eventually die out. In the supercritical case of $\lambda > 1$ (each parent outage has on average more than one child), the outages can increase exponentially until the system size or saturation effects are encountered. At the critical case of $\lambda = 1$, the branching process has a power-law distribution of the total number of outages with a heavy tail, as observed in the real blackout data.

The branching process model does not directly represent any of the physics or mechanisms of the outage propagation, but after it is validated, it can be used to predict the total number of outages. The parameters of a branching process model can be estimated from a much smaller data set, and then predictions of the total number of outages can be made based on the estimated parameters. While it is sometimes possible to observe or produce large amounts of data to make an empirical estimate of the total number of outages (indeed, this is the way the branching process prediction is validated), the ability to do this via the branching process model with much less data is a significant advantage that enables practical applications. The simplicity of the branching process model also allows a high-level understanding of the cascading process without getting entangled in the various and complicated mechanisms of cascading. The branching process should be seen as complementary to detailed modeling of cascading outage mechanisms.

3.2.7 Branching Processes Applied to Line Outage Data

The cascading outage data are produced by the improved OPA model [35] on the Northeastern Power Grid of China (NPGC) system and the AC OPA model [27, 28] on the IEEE 118-bus system. The NPGC system consists of Heilongjiang, Jilin, Liaoning, and the northern part of Inner Mongolia. The 500 kV and 220 kV transmission lines and substations are considered and there are 568 buses. The NPGC system data includes line limits. The IEEE 118-bus system data is standard, except that the line limits are determined by running the fast dynamics of the improved OPA and the slow dynamics of basic OPA that selectively upgrades lines in response to their participation in blackouts [14], starting from an initial guess of the line limits. This results in a coordinated set of line limits.

For the improved OPA model, we use the same parameters as those in Mei et al. [35] and $p_1 = 0.999$, $\xi = 0.001$, $n = 10$, and $\eta = 0.95$. For the NPGC system $p_0 = 0.0007$, which is the same as Mei et al. [35]. For the IEEE 118-bus system $p_0 = 0.0001$, which is the same as Dobson et al. [29]. For both models the load variability $\gamma = 1.67$ as in Dobson et al. [29].

The simulation is run so as to produce 5000 cascading outages with a nonzero number of line outages or non-negligible load shed. The branching process

Table 3.1 Estimated parameters for line outage data.

System	Model	Load Level	$\hat{\theta}$	$\hat{\lambda}$	$\hat{\lambda}_{500}$
NPGC	Im OPA	1.15	1.69	0.02	0.02
NPGC	Im OPA	1.3	2.14	0.14	0.14
NPGC	Im OPA	1.6	2.54	0.19	0.20
IEEE 118	AC OPA	1.0	4.38	0.40	0.40
IEEE 118	AC OPA	1.2	7.01	0.52	0.52
IEEE 118	AC OPA	1.4	10.08	0.63	0.63

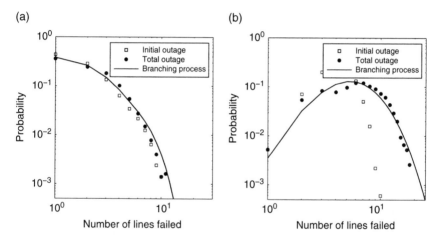

Figure 3.5 Probability distributions of the number of line outages. (a) Improved OPA on NPGC system at load level 1.3 times the base case. (b) AC OPA on IEEE 118-bus system at base case load level.

is tested for estimating the distribution of total line outages on cascading data from the improved OPA simulations on the NPGC system under different load levels.

The estimated branching process parameters $\hat{\theta}$ and $\hat{\lambda}$ are shown in Table 3.1. The distributions of line outages are shown in Figure 3.5. The distributions of total line outages (dots) and initial line outages (squares) are shown, as well as a solid line indicating the total line outages predicted by the branching process. The branching process data are also discrete, but are shown as a line for ease of comparison. It is seen that the branching process prediction of the distribution of total line outages matches the empirical distribution very well.

3.2.8 Branching Processes Applied to Load Shed Data

For the load shed data, the estimated branching process parameters $\hat{\theta}$ and $\hat{\lambda}$ are listed in Table 3.2. In contrast with the line outages, the average propagation $\hat{\lambda}$ does not always increase with load level. When the load level increases from 1.3 times base case to 1.6 times base case for improved OPA simulation on the NPGC system, $\hat{\ }$ decreases from 0.29 to 0.22. The results comparing the distributions of load shed are shown in Figure 3.6. The branching process predictions of the load shed distributions match the empirical distributions well.

Table 3.2 Estimated parameters for load shed data.

System	Model	Load Level	$\hat{\theta}$(MW)	$\hat{\lambda}$	Δ(MW)	$\hat{\lambda}_{500}$
NPGC	Im OPA	1.15	177	0.10	198	0.11
NPGC	Im OPA	1.3	441	0.29	388	0.35
NPGC	Im OPA	1.6	5404	0.22	2465	0.23
IEEE 118	AC OPA	1.0	160	0.42	90	0.44
IEEE 118	AC OPA	1.2	224	0.58	167	0.58
IEEE 118	AC OPA	1.4	407	0.67	425	0.68

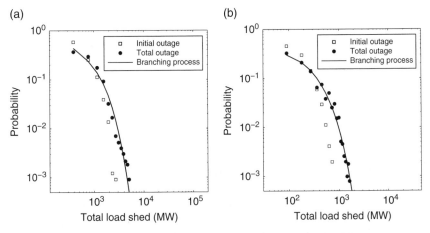

Figure 3.6 Probability distribution of the load shed. (a) Improved OPA on NPGC system at load level 1.3 times the base case. (b) AC OPA on IEEE 118-bus system at base case load level.

3.2.9   Cross-Validation for Branching Processes

A reasonable objection to the comparison in Sections 3.2.7 and 3.2.8 is that the same data are used both to estimate the distribution and to obtain the empirical distribution. To address this objection the data are divided into separate fitting and validation sets. Specifically, we estimate the distribution from the odd numbered cascades by the branching process and compare it with the empirical distribution for the even numbered cascades. The results are shown in Figures 3.7 and 3.8 and the matches are also satisfactory.

3.2.10   Efficiency Improvement by Branching Processes

Tables 3.1 and 3.2 show the propagation $\hat{\lambda}_{500}$ estimated using the first 500 cascades. It is seen that $\hat{\lambda}_{500}$ is close to $\hat{\lambda}$. If the initial load shed distribution is known accurately, then accurately estimating the distribution of the total amount of load shed via discretization and the branching process requires substantially fewer cascades [9].

In particular, let p_{branch} be the probability of shedding total load $S\Delta$ MW, computed via estimating λ from K_{branch} simulated cascades with non-negligible load shed and then using the branching process model. p_{branch} is conditioned on a non-negligible amount of load shed. Let $p_{empiric}$ be the

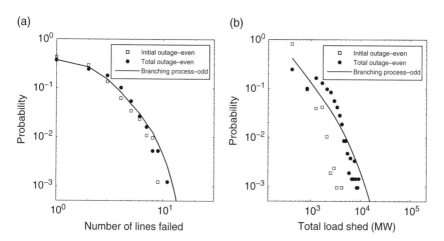

Figure 3.7 Probability distribution of the blackout size (odd numbered cascades used for estimation). (a) The line outages by improved OPA on NPGC system at load level 1.3 times the base case. (b) The load shed by AC OPA on IEEE 118-bus system at load level 1.4 times the base case.

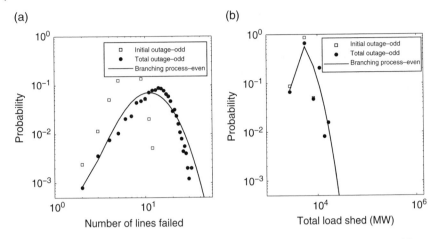

Figure 3.8 Probability distribution of the blackout size (even numbered cascades used for estimation). (a) The line outages by AC OPA on IEEE 118-bus system at load level 1.2 times the base case. (b) The load shed by improved OPA on NPGC system at load level 1.6 times the base case.

probability of shedding total load $S\Delta$ MW, computed empirically by simulating $K_{empiric}$ cascades with non-negligible load shed. If we require the same standard deviation for both methods, then Section IV of Kim et al. [9] derives the following approximation of the ratio R of the required number of simulated cascades as

$$R = \frac{K_{empiric}}{K_{branch}} = \frac{p_{empiric}\left(1 - p_{empiric}\right)\theta}{\lambda\left(1 - \lambda\right)\Delta}\left(\frac{dp_{branch}}{d\lambda}\right)^{-2}. \qquad (3.10)$$

R is a ratio describing the gain in efficiency when using branching process rather than empirical methods. To obtain numerically a rough estimate of R, we evaluate Eq. (3.10) for the almost largest total load shed $S\Delta$ MW for all six load shed cases. Here, we sort the total load shed of $K_{empiric}$ cascades in descending order and choose $S\Delta$ to be the ith one, where $i = \text{int}[5\% K_{empiric} + 0.5]$. Then $dp_{branch}/d\lambda$ is estimated by numerical differencing. The results in Table 3.3 show that $K_{empiric}$ exceeds K_{branch} by an order of magnitude or more.

Similarly, the efficiency gains for line outage data can also be confirmed by evaluating Eq. (3.10) with $\Delta = 1$ for almost largest total line outages S, which can be determined in a similar way as choosing $S\Delta$. The results in Table 3.4 show that $K_{empiric}$ exceeds K_{branch} by an order of magnitude or more. The efficiency gains are not surprising because estimating the parameters of a distribution is generally expected to be more efficient than estimating the distribution empirically.

Table 3.3 Efficiency gain R for load shed data.

System	Model	Load Level	$S\Delta$	R
NPGC	Im OPA	1.15	419	28
NPGC	Im OPA	1.3	1851	27
NPGC	Im OPA	1.6	12759	15
IEEE 118	AC OPA	1.0	830	28
IEEE 118	AC OPA	1.2	1581	111
IEEE 118	AC OPA	1.4	3946	91

Table 3.4 Efficiency gain R for line outage data.

System	Model	Load Level	S	R
NPGC	Im OPA	1.15	4	114
NPGC	Im OPA	1.3	6	68
NPGC	Im OPA	1.6	6	21
IEEE 118	AC OPA	1.0	13	32
IEEE 118	AC OPA	1.2	23	41
IEEE 118	AC OPA	1.4	46	36

3.3 Multitype Branching Processes

In real cascading blackouts, different types of outages such as line outages, load shedding, and isolated buses can exist simultaneously. More importantly, these outages are usually interdependent. If we can describe them jointly, we may gain a better understanding of their propagation. Also, in order to evaluate the time that is needed to restore the system after a cascading event, we need to know how many buses and lines are still in service, as well as the amount of the load shed. However, usually we do not have all these data and have to predict some of them by using the available data. This will require a quantitative analysis of the interdependencies between them.

In addition, it is also useful to be able to study the interactions between different infrastructure systems, such as those between electric power systems and communication networks [36, 37], natural gas networks [38, 39], water systems [40], and transportation networks [41, 42].

As a generalization of the one-type branching process, the multitype branching process [7, 8] can be used to describe these interdependencies. In a multitype branching process, each type i outage in one generation (a type i "parent" outage) independently produces a random nonnegative integer number of outages of the same type (type i "children" outages) and any other type (type k "children" outages where $k \neq i$). All generated outages in different types make up the next generation. The process ends when the number of outages in all types becomes zero.

For an n-type branching process, there are n^2 offspring distributions and correspondingly n^2 offspring means, which can be arranged into a matrix called the *offspring mean matrix* Λ. Recall that the criticality of the branching processes with only one type is directly determined by the offspring mean λ. Differently, the criticality of the multitype branching processes is determined by the largest eigenvalue of Λ, which is denoted by ρ. If $\rho \leq 1$, the multitype branching process will always go extinct. Otherwise, the multitype branching process will go extinct with a probability $0 \leq q < 1$ [8].

3.3.1 Estimation of Multitype Branching Process Parameters

A total of M cascades are arranged as

$$
\begin{array}{cccc}
 & \text{generation 0} & \text{generation 1} & \cdots \\
\text{cascade 1} & \left(Z_0^{1,1},\ldots,Z_0^{1,n}\right) & \left(Z_1^{1,1},\ldots,Z_1^{1,n}\right) & \cdots \\
\text{cascade 2} & \left(Z_0^{2,1},\ldots,Z_0^{2,n}\right) & \left(Z_1^{2,1},\ldots,Z_1^{2,n}\right) & \cdots \\
\vdots & \vdots & \vdots & \vdots \\
\text{cascade } M & \left(Z_0^{M,1},\ldots,Z_0^{M,n}\right) & \left(Z_1^{M,1},\ldots,Z_1^{M,n}\right) & \cdots
\end{array}
$$

where $Z_g^{m,t}$ is the number of type t outages in generation g of cascade m, and n is the number of types of outages. Each cascade has a nonzero number of outages in generation zero for at least one type of outages and each type of outage should have a nonzero number of outages at least for one generation. Note that continuous data such as the load shed need to be first discretized by the method in Section 3.2.4.

Assume M_{u} cascades are used for estimating the parameters. The largest eigenvalue ρ of the mean matrix can be estimated as the total number of all types of children divided by the total number of all types of parents by directly using the simulated cascades and ignoring the types [31].

$$
\hat{\rho} = \frac{\displaystyle\sum_{m=1}^{M_{\mathrm{u}}}\sum_{g=1}^{\infty}\sum_{t=1}^{n} Z_g^{m,t}}{\displaystyle\sum_{m=1}^{M_{\mathrm{u}}}\sum_{g=0}^{\infty}\sum_{t=1}^{n} Z_g^{m,t}}. \tag{3.11}
$$

When the number of type j children to type i parents $S^{(i,j)}$ and the total number of type i parents $S^{(i)}$ are observed, λ_{ij} (the expected number of type j children generated by one type i parent) can be estimated by a maximum likelihood estimator that is the total number of type j children produced by type i parents divided by the total number of type i parents [43].

$$\hat{\lambda}_{ij} = \frac{S^{(i,j)}}{S^{(i)}}, \tag{3.12}$$

where $S^{(i,j)}$ and $S^{(i)}$ can be described by using the available M_u cascades as

$$S^{(i,j)} = \sum_{m=1}^{M_u} \sum_{g=1}^{\infty} Z_g^{m,i \to j} \tag{3.13}$$

$$S^{(i)} = \sum_{m=1}^{M_u} \sum_{g=0}^{\infty} Z_g^{m,i}. \tag{3.14}$$

Here $Z_g^{m,i \to j}$ is the number of type j offspring generated by type i parents in generation g of cascade m.

However, it is usually impossible to have such detailed information. Because there can be many mechanisms in cascading, it is difficult to determine the exact number of type j outages that are produced by type i outages. In other words, $Z_g^{m,i \to j}$ in Eq. (3.13) cannot be determined, and thus $S^{(i,j)}$ cannot be decided and the mean matrix cannot be estimated by Eq. (3.12).

To solve this problem, the Expectation Maximization (EM) algorithm [44] is applied [11]. It fits the problem well as a method for finding maximum likelihood estimates of parameters in statistical models where the model depends on unobserved latent variables. Similar to the last section, we also assume that the offspring distributions of the branching processes are Poisson. The EM algorithm mainly contains two steps, which are the E-step and the M-step. For the estimation of the offspring mean matrix of an n-type branching process, the EM algorithm can be formulated as follows.

1) Initialization: Set initial guess of mean matrix as $\hat{\Lambda}^{(0)}$.
   Since for cascading blackouts the outages will always die out, we have $0 \leq \lambda_{ij} \leq 1$. Based on this, all elements of the initial mean matrix are set to be 0.5, which is the midpoint of the possible range.
2) E-step: Estimate $S^{(i,j)(k+1)}$ based on $\hat{\Lambda}^{(k)}$.
   Under the assumption that the offspring distributions are all Poisson, for generation $g \geq 1$ of cascade m, the number of type j offspring produced by type $t = 1, ..., n$ parents follows Poisson distribution

$$Z_g^{m,t \to j} \sim \text{Pois}\left(Z_{g-1}^{m,t} \hat{\lambda}_{tj}^{(k)}\right). \tag{3.15}$$

Thus, the number of type j offspring in generation $g \geq 1$ of cascade m produced by type i parents in generation $g - 1$ of the same cascade is

$$Z_g^{m,i \to j} = Z_g^{m,j} \frac{Z_{g-1}^{m,i} \hat{\lambda}_{ij}^{(k)}}{\displaystyle\sum_{t=1}^{n} Z_{g-1}^{m,t} \hat{\lambda}_{tj}^{(k)}}. \tag{3.16}$$

After obtaining $Z_g^{m,i \to j}$ for all generations $g \geq 1$ of cascades $m = 1, \ldots, M_u$ we are finally able to calculate $S^{(i,j)}$ by using Eq. (3.13).

3) M-step: Estimate $\hat{\Lambda}^{(k+1)}$ based on $S^{(i,j)(k+1)}$.
   After $S^{(i,j)(k+1)}$ is obtained, the updated mean matrix $\hat{\Lambda}^{(k+1)}$ can be estimated by Eq. (3.12).

4) End: Iterate the E-step and M-step until

$$\max_{i,j \in \{1, \cdots, n\}} \left| \hat{\lambda}_{ij}^{(k+1)} - \hat{\lambda}_{ij}^{(k)} \right| < \epsilon, \tag{3.17}$$

where ϵ is the tolerance that is used to control the accuracy and $\hat{\lambda}_{ij}^{(k+1)}$ is the final estimate of λ_{ij}.

3.3.2 Estimation of Joint Probability Distribution of Total Outages

The probability generating function for the type i individual of an n-type branching process is

$$f_i(s_1, \ldots, s_n) = \sum_{u_1, \ldots, u_n = 0}^{\infty} p_i(u_1, \ldots, u_n) s_1^{u_1} \ldots s_n^{u_n}, \tag{3.18}$$

where $p_i(u_1, \ldots, u_n)$ is the probability that a type i individual generates u_1 type 1, ..., and u_n type n individuals. If we assume that the offspring distributions for various types of outages are all Poisson, (3.18) can be easily written after the offspring mean matrix $\hat{\Lambda}$ is estimated by the method in Section 3.3.1.

According to Harris [8] and Good [45], the probability generating function, $w_i(s_1, \ldots, s_n)$, of the total number of various types of individuals in all generations, starting with one individual of type i, can be given by

$$w_i = s_i f_i(w_1, \ldots, w_n), \qquad i = 1, \ldots, n. \tag{3.19}$$

When the branching process starts with more than one type of individual, the total number of various types can be determined by using the Lagrange–Good inversion [45], in which the following theorem is given.

Theorem 3.1   If the n-type branching process starts with r_1 individuals of type 1, r_2 of type 2, and so on, then the probability that the whole process will have precisely m_1 of type 1, m_2 of type 2, etc., is equal to the coefficient of $s_1^{m_1-r_1} \cdots s_n^{m_n-r_n}$ in

$$f_1^{m_1} \cdots f_n^{m_n} \left\| \delta_\mu^v - \frac{s_\mu}{f_\mu} \frac{\partial f_\mu}{\partial s_v} \right\|, \tag{3.20}$$

where $\left\| a_\mu^v \right\|$ denotes the determinant of the $n \times n$ matrix whose entry is a_μ^v $(\mu, v = 1, \cdots, n)$ and δ_μ^v is Kronecker's delta (=1 if $\mu = v$, otherwise = 0). We denote the coefficient of $s_1^{m_1-r_1} \cdots s_n^{m_n-r_n}$ as $c(r_1, \cdots, r_n; m_1, \cdots, m_n)$.

Given the joint probability distribution of initial sizes $P\left(Z_0^1, \cdots, Z_0^n\right)$ and the generating functions in Eq. (3.18), the formula for calculating the joint probability distribution $P\left(Y_\infty^1 = y_1, \cdots, Y_\infty^n = y_n\right)$ for the total number of various types $\left(Y_\infty^1, \cdots, Y_\infty^n\right)$ can then be written as

$$
\begin{aligned}
&d_{Y_\infty}^{\mathrm{est}}\left(y_1, \cdots, y_n\right) \\
&= \sum_{\substack{z_0^1, \cdots, z_0^n = 0 \\ z_0^1 + \cdots + z_0^n \neq 0}}^{z_0^1 = y_1, \cdots, z_0^n = y_n} \left[P(Z_0^1 = z_0^1, \cdots, Z_0^n = z_0^n) \cdot c(z_0^1, \cdots, z_0^n; y_1, \cdots, y_n) \right],
\end{aligned} \tag{3.21}
$$

where the empirical joint probability distribution of the number of initial outages $\left(Z_0^1, \cdots, Z_0^n\right)$ can be obtained as

$$P\left(Z_0^1 = z_0^1, \cdots, Z_0^n = z_0^n\right) = \frac{1}{M_\mathrm{u}} \sum_{m=1}^{M_\mathrm{u}} I\left[Z_0^{m,1} = z_0^1, \cdots, Z_0^{m,n} = z_0^n \right]. \tag{3.22}$$

3.3.3   An Example for a Two-Type Branching Process

To better illustrate the method for estimating the joint probability distribution, we give an example for a two-type branching process. The empirical joint probability distribution of the number of initial outages $\left(Z_0^1, Z_0^2\right)$ can be obtained by Eq. (3.22). As in Section 3.3.1, we assume that the offspring distributions for various types of outages are all Poisson. Then the probability generating functions for a two-type branching process can be written as

$$f_1\left(s_1, s_2\right) = \sum_{u_1 = u_2 = 0}^{\infty} \frac{\lambda_{11}^{u_1} \lambda_{12}^{u_2} e^{-\lambda_{11}-\lambda_{12}}}{u_1! u_2!} s_1^{u_1} s_2^{u_2} \tag{3.23}$$

$$f_2\left(s_1,s_2\right) = \sum_{u_1=u_2=0}^{\infty} \frac{\lambda_{21}^{u_1}\lambda_{22}^{u_2}e^{-\lambda_{21}-\lambda_{22}}}{u_1!u_2!}s_1^{u_1}s_2^{u_2}, \tag{3.24}$$

where the parameters λ_{11}, λ_{12}, λ_{21}, and λ_{22} can be estimated by the method in Section 3.3.1.

In Eq. (3.20) the $n \times n$ matrix whose determinant needs to be evaluated is actually

$$\begin{bmatrix} 1 - \dfrac{s_1}{f_1}\dfrac{\partial f_1}{\partial s_1} & -\dfrac{s_1}{f_1}\dfrac{\partial f_1}{\partial s_2} \\[2ex] -\dfrac{s_2}{f_2}\dfrac{\partial f_2}{\partial s_1} & 1 - \dfrac{s_2}{f_2}\dfrac{\partial f_2}{\partial s_2} \end{bmatrix}.$$

The joint probability distribution of the two-type branching process can be obtained by evaluating Eq. (3.21) with elementary algebra. Note that the coefficients in Eqs. (3.23) and (3.24) will quickly decrease with the increase of the order of s_1 and s_2. Therefore, we can reduce the calculation burden while guaranteeing accurate enough results by only using a few terms to approximate the generating functions. In addition, the probability obtained by Eq. (3.21) will also decrease with the increase of y_1 and y_2. Thus, we do not need to calculate the negligible probability for too large blackout size. Specifically, we can only calculate the joint probability for

$$y_1 = z_0^1,\ldots,z_0^1 + \tau_1 \tag{3.25}$$

$$y_2 = z_0^2,\ldots,z_0^2 + \tau_2, \tag{3.26}$$

where τ_1 and τ_2 are integers properly chosen for a tradeoff of calculation burden and accuracy. Too large τ_1 or τ_2 will lead to unnecessary calculation for blackout sizes with negligible probability. Too small τ_1 or τ_2 will result in loss of accuracy by neglecting blackout sizes with not so small probability.

3.3.4 Validation of Estimated Joint Distribution

The empirically obtained joint distribution $P\left(Y_\infty^1 = y_1,\ldots,Y_\infty^n = y_n\right)$ can be calculated by

$$d_{Y_\infty}^{\text{emp}}\left(y_1,\ldots,y_n\right) = \frac{N\left(Y_\infty^1 = y_1,\cdots,Y_\infty^n = y_n\right)}{M}, \tag{3.27}$$

where $N\left(Y_\infty^1 = y_1,\cdots,Y_\infty^n = y_n\right)$ is the number of cascades for which there are y_1 type 1 outages, $\cdots$, y_n type n outages. It can be used to validate the estimated joint distribution $d_{Y_\infty}^{\text{est}}(y_1, \ldots, y_n)$ in Section 3.3.2. Specifically, we can perform the following comparison between $d_{Y_\infty}^{\text{est}}(y_1, \ldots, y_n)$ and $d_{Y_\infty}^{\text{emp}}(y_1, \ldots, y_n)$.

1) *Joint entropy*: We compare them by the joint entropy, which can be defined for n random variables $\left(Y_\infty^1, \ldots, Y_\infty^n\right)$ as

$$H\left(Y_\infty^1, \ldots, Y_\infty^n\right) = -\sum_{y_1} \cdots \sum_{y_n} P(y_1, \ldots, y_n) \log_2\left[P(y_1, \ldots, y_n)\right], \qquad (3.28)$$

where $P(y_1, \ldots, y_n)\log_2[P(y_1, \ldots, y_n)]$ is 0 if $P(y_1, \ldots, y_n) = 0$. The joint entropy for the estimated and the empirical joint distribution can be respectively denoted by H^{est} and H^{emp}. Then the estimated joint distribution can be validated by checking if $H^{\text{est}}/H^{\text{emp}}$ is close to 1.0.

2) *Marginal distribution*: The marginal distribution for each type of outage can be calculated after estimating the joint distribution of the total outages, which can be compared with the empirical marginal distribution directly calculated from the simulated cascades.

3) *Conditional largest possible total outage (CLO)*: We can calculate the CLO of one type of blackout size when the total outage of the other types of blackout size is known. For example, for a two-type branching process, for $i, j \in \{1, 2\}$ and $i \neq j$, given the total outage of one type of blackout size y_i we can get the total outage of another type of blackout size y_j that satisfies

$$P\left(Y_\infty^j \leq y_j \,|\, Y_\infty^i = y_i\right) = p_{\text{conf}}, \qquad (3.29)$$

where

$$P\left(Y_\infty^j \leq y_j \,|\, Y_\infty^i = y_i\right) = \sum_{k=0}^{y_j} \frac{P\left(Y_\infty^i = y_i, Y_\infty^j = k\right)}{\sum_{l=0}^{\infty} P\left(Y_\infty^i = y_i, Y_\infty^j = l\right)}, \qquad (3.30)$$

p_{conf} is the confidence level close to 1.0 and $P(A\,|\,B)$ is the conditional probability of event A given B. Given the total outage of type i as y_i, from the joint distribution, we can claim that the total outage of type j will not exceed y_j with a high probability p_{conf}. The y_j can be calculated from either the empirical joint distribution or the estimated joint distribution by branching process. We can compare them to check if the y_j from the estimated joint distribution is close to that from the empirical joint distribution.

3.3.5   Number of Cascades Needed for Multitype Branching Processes

It is useful to figure out (i) how many cascades are needed to empirically obtain a reliable joint distribution of total outages and (ii) how many cascades are enough for getting a reliable estimate of the offspring mean matrix and the joint distribution of initial outages, which can further guarantee that the estimated joint distribution of total outages is close enough to the reliable empirical joint distribution. In order to answer these questions, the following methods can be used to determine the lower bounds $M^{\min}$ and $M_u^{\min}$, respectively, for M and M_u.

Here there is an implicit assumption that all cascades are generated from the same cascading failure model or at least from similar models. If the cascades come from very different cascading failure models or are generated by very different mechanisms, it might be possible that the methods discussed next are difficult to converge or stabilize with the increase of the number of cascades.

1) Determining the Lower Bound for M

More cascades tend to contain more information about the property of cascading failure propagation. The added information brought from the additional cascades will make the joint entropy of the joint distribution empirically obtained from the cascades increase. However, the amount of information will not always grow with the increase of the number of cascades but will saturate after the number of cascades is greater than some number $M^{\min}$. This number can be determined by gradually increasing the number of cascades, recording the corresponding joint entropy of the empirical joint distribution, and finding the smallest number of cascades that can lead to the saturated joint entropy.

Assume there are a total of N_M different M's ranging from a very small number to a very large number, which are denoted by M_i, $i = 1, 2, ..., N_M$. The joint entropy of the joint distribution of the total outages obtained from M_i cascades is denoted by $H^{\mathrm{emp}}(M_i)$.

For $i = 1, ..., N_M - 2$, we let $H_i^{\mathrm{emp}} = \left[H^{\mathrm{emp}}(M_i) \cdots H^{\mathrm{emp}}(M_{N_M}) \right]$ and calculate its standard deviation $\sigma_i = \sigma\left(H_i^{\mathrm{emp}} \right)$, where $\sigma(\cdot)$ is the standard deviation of a vector. The σ_i for $i = N_M - 1$ and $i = N_M$ are not calculated because we want to calculate the standard deviation for at least three data points.

Very small and slightly fluctuating σ_i indicates that the joint entropy begins to saturate after M_i. Specifically, the M_i corresponding to $\sigma_i \le \varepsilon_\sigma$ is identified as $M^{\min}$ where ε_σ is a small real number. The $M^{\min}$ original cascades can guarantee that the accuracy on statistical values of interest is good and thus can provide a reference joint distribution of the total outages.

2) Determining the Lower Bound for M_u

When we only want a good enough estimate of the joint distribution of the total sizes, we do not need as many as $M^{\min}$ cascades but only need $M_u^{\min} \ll M^{\min}$ cascades to make sure that the information extracted from $M_u^{\min}$ cascades by the branching process can capture the general properties of the cascading failure propagation.

Since both H^{emp} and H^{est} vary with M_u, we denote them by $H^{\text{emp}}(M_u)$ and $H^{\text{est}}(M_u)$. $H^{\text{emp}}(M_u)$ can be directly obtained from the cascades by Eqs. (3.27) and (3.28) and $H^{\text{est}}(M_u)$ can be calculated by Eqs. (3.21) and (3.28).

When M_u is not large enough, it is expected that there will be a big mismatch between $H^{\text{emp}}(M_u)$ and $H^{\text{est}}(M_u)$, indicating that the estimated joint distribution from the branching process cannot capture the property of the joint distribution of the cascades very well. But with the increase of M_u more information will be obtained and thus the mismatch will gradually decrease and finally stabilize. In order to indicate the stabilization, we define

$$R\left(M_u\right) = \frac{\left|H^{\text{est}}\left(M_u\right) - H^{\text{emp}}\left(M_u\right)\right|}{H^{\text{emp}}\left(M_u\right)}. \tag{3.31}$$

We start from a small integer M_u^0 and increase it gradually by ΔM each time, and calculate the standard deviation of $R(M_u)$ for the latest three data points by

$$\tilde{\sigma}_i = \sigma\left(R_i\right), \quad i \geq 2, \tag{3.32}$$

where i denotes the latest data point and $R_i = \left[R\left(M_u^{i-2}\right) \ R\left(M_u^{i-1}\right) \ R\left(M_u^i\right) \right]$. Then $M_u^{\min}$ is determined as the smallest value that satisfies $\tilde{\sigma}_i \leq \epsilon_H$, where ε_H is used to control the tolerance for stabilization.

Decreasing ΔM may increase the accuracy of the obtained $M_u^{\min}$. But smaller ΔM will increase the number of times for calculating the joint distribution by branching processes. When more types of outages are considered, greater $M_u^{\min}$ will be needed, in which case larger ΔM can be chosen to avoid too many times of calculating the joint distribution.

The cascading outage dataset is produced by the open-loop AC OPA simulation [27, 28] on the IEEE 118-bus test system, in the same way as in Section 3.2.7. A simulation is run so as to produce $M = 50\,000$ cascades with a nonzero number of line outages at the base case load level. In each generation the number of line outages and the number of isolated buses are counted and the continuously varying amounts of the load shed are discretized as described in Section 3.2.4 to produce integer multiples of the chosen discretization unit.

For determining $M^{\min}$, we choose $N_M = 50$ and the data points are linearly scaled. The ϵ_σ is chosen as 0.002. In order to determine $M_u^{\min}$,

Table 3.5 Number of cascades needed.

No. of Types	Type	$M^{\min}$	$M_u^{\min}$
1	Line outage	18 000	1400
1	Load shed	36 000	1900
1	Isolated bus	33 000	900
2	Line outage and load shed	39 000	6500
2	Line outage and isolated bus	37 000	5500

we choose M_u^0, ϵ_H, and ΔM as 100, 0.002, and 100 for one type of outage and 1000, 0.002, and 500 for multiple types of outages. This is because the $M_u^{\min}$ for multiple outages case is expected to be greater and we need to limit the calculation burden. The determined $M^{\min}$ and $M_u^{\min}$ for different types of outages are listed in Table 3.5. It is seen that the $M_u^{\min}$ used for estimation is significantly smaller than $M^{\min}$, indicating that the branching process can help greatly improve the efficiency.

3.3.6 Estimated Parameters of Branching Processes

The ϵ in (3.16) is chosen as 0.01. The EM algorithm that estimates the offspring mean matrix of the multitype branching processes can quickly converge. The number of iterations N^{ite} is listed in Table 3.6. The estimated branching process parameters are listed in Table 3.7, where $\hat{\lambda}$ is the offspring mean for one-type branching process estimated by Eq. (3.1). It is seen that the estimated largest eigenvalue of the offspring mean matrix $\hat{\rho}$ is greater than the estimated offspring means for only considering one type of outages. This indicates that the system is closer to criticality when we simultaneously consider two types of outages. Only considering one type of outage may underestimate the extent of outage propagation. This is because different types of outages, such as line outage and the load shed, can influence each other and aggregate the cascading propagation.

Table 3.6 Number of iterations of EM algorithm.

Type	M_u	N^{ite}
Line outage and load shed	39 000	7
	6500	7
Line outage and isolated bus	37 000	4
	5500	4

Table 3.7 Estimated parameters of branching processes by using M^{min} cascades.

Type	$\hat{\lambda}$	$\hat{\rho}$	$\hat{\Lambda}$
Line outage	0.45	–	–
Load shed	0.48	–	–
Isolated bus	0.14	–	–
Line outage and load shed	–	0.55	$\begin{bmatrix} 0.45 & 0.42 \\ 0.0018 & 0.029 \end{bmatrix}$
Line outage and isolated bus	–	0.60	$\begin{bmatrix} 0.45 & 0.40 \\ 6.0\times10^{-5} & 0.0049 \end{bmatrix}$

The $\hat{\lambda}_{12}$ in $\hat{\Lambda}$ is the estimated expected discretized number of the load shed when one line is tripped, while $\hat{\lambda}_{21}$ is the estimated expected number of line outages when one discretization unit of load is shed. From the offspring mean matrix $\hat{\Lambda}$ we can see that line outages tend to have a greater influence on the load shed and the isolated buses, but the influence of the load shed or the isolated buses on line outages is relatively weak. This is reasonable because in real blackouts it is more possible for line tripping to cause load shedding or buses to be isolated. Sometimes, line outages may directly cause the load shed or isolated buses. For example, the simplest case occurs when a load is fed from a radial line.

In Table 3.7 there is some mismatch between the largest eigenvalue of the offspring mean matrix $\hat{\rho}$ estimated from Eq. (3.11) and that calculated from the estimated offspring mean matrix $\hat{\Lambda}$ by the EM algorithm. The estimator in Eq. (3.11) is the maximum likelihood estimator of the largest eigenvalue of the offspring mean matrix [31]. It does not need to make any assumption about the offspring distribution. By contrast, in order to estimate the offspring mean matrix, we have to assume a specific offspring distribution, such as the Poisson distribution used here. As mentioned in Section 3.3.1, there are general arguments suggesting that the choice of a Poisson offspring distribution is appropriate. However, the offspring distribution is only approximately Poisson but not necessarily exactly Poisson. Numerical simulation of multitype branching processes with Poisson offspring distributions shows that the estimated $\hat{\rho}$ and the largest eigenvalue of the estimated $\hat{\Lambda}$ agree with each other. Therefore, the largest eigenvalue estimated from Eq. (3.11) without any assumption of the offspring distribution is expected to be more reliable and should be used to indicate how close to criticality the system is.

The estimated parameters for branching processes by only using M_u^{min} cascades are listed in Table 3.8, which are very close to those estimated

Table 3.8 Estimated parameters of branching processes by using M_u^{min} cascades.

Type	$\hat{\lambda}$	$\hat{\rho}$	$\hat{\Lambda}$
Line outage	0.45	–	–
Load shed	0.49	–	–
Isolated bus	0.15	–	–
Line outage and load shed	–	0.56	$\begin{bmatrix} 0.45 & 0.43 \\ 0.0020 & 0.027 \end{bmatrix}$
Line outage and isolated bus	–	0.61	$\begin{bmatrix} 0.45 & 0.39 \\ 5.5 \times 10^{-5} & 0.0040 \end{bmatrix}$

from M^{min} cascades shown in Table 3.7, indicating that M_u^{min} cascades are enough to get a good estimate.

3.3.7 Estimated Joint Distribution of Total Outages

In Eqs. (3.23) and (3.24), the highest orders for both s_1 and s_2 are chosen as 4. In Eqs. (3.25) and (3.26), τ_1 and τ_2 are chosen based on the number of initial outages from the samples of cascades and the tradeoff between calculation burden and accuracy. For line outages and the load shed, τ_1 and τ_2 are chosen as 12 and 9, respectively. For line outages and isolated buses, τ_1 and τ_2 are chosen as 12 and 18, respectively.

We estimate the joint distribution of the total outages by the multitype branching process based on $M_u^{min} \ll M^{min}$ cascades and compare it with the empirical joint distribution obtained from M^{min} cascades. To have a quantitative comparison, the joint entropy is calculated and listed in Table 3.9. It is seen that the joint entropy of the estimated joint distributions is reasonably close to that of the empirical joint distributions. Also, the joint entropy of the distributions for two types of outages is significantly greater than that for one type of outages, meaning that we can get new information by jointly analyzing two types of outages.

After estimating the joint distributions, the marginal distributions for each type of outage can also be calculated. In Figure 3.9 we show the marginal distributions of the line outages and the load shed for their corresponding two-type branching process. The empirical marginal distributions of total outages (dots) and initial outages (squares) calculated from $M^{min} = 39\,000$ are shown, as well as a solid line indicating the total outages predicted by the multitype branching process from $M_u^{min} = 6500$ cascades. The branching process data is also discrete but is shown as a line for ease of comparison. It is seen that the branching process prediction with $M_u^{min} = 6500$ cascades closely matches the

Table 3.9 Joint entropy of distributions.

Type	M_u	H^{emp}	H^{est}
Line outage	18000	3.50	3.91
	1400	3.48	3.92
Load shed	36000	3.52	3.56
	1900	3.53	3.57
Isolated bus	33000	2.63	2.64
	900	2.59	2.61
Line outage and load shed	39000	6.99	7.08
	6500	6.94	7.06
Line outage and isolated bus	37000	5.33	6.45
	5500	5.30	6.44

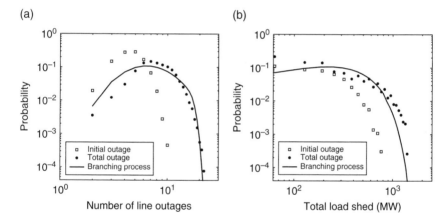

Figure 3.9 Estimated marginal probability distributions of line outages and the load shed by using $M_u^{min} = 6500$ cascades. (a) The line outages. (b) The load shed.

marginal distribution empirically obtained by using $M^{min} = 39000$ cascades. Similar results for the marginal distribution of the line outages and isolated buses are shown in Figure 3.10, for which $M^{min} = 37000$ and $M_u^{min} = 5500$.

Given the total number of line outages, the CLO defined in Section 3.3.4 can also be calculated from either the empirical joint distribution using M^{min} cascades or from the estimated joint distribution from the branching process using $M_u^{min} \ll M^{min}$ cascades. The p_{conf} in Eq. (3.29) is chosen as 0.99. The CLOs for the load shed and the isolated buses are shown in Figure 3.11. The

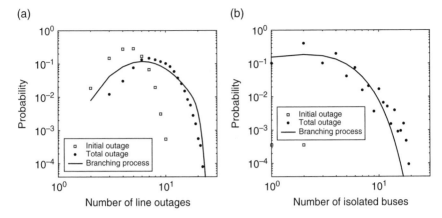

Figure 3.10 Estimated marginal probability distributions of line outages and isolated buses by using $M_u^{min} = 5500$ cascades. (a) The line outages. (b) The isolated buses.

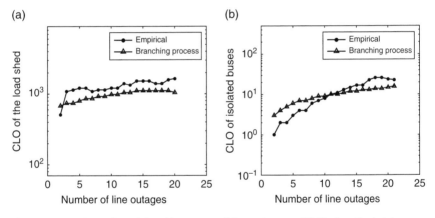

Figure 3.11 Estimated conditional largest possible total outage (CLO) when the total number of line outages is known. (a) The load shed. (b) The isolated buses.

CLO estimated by multitype branching process using a much smaller number of cascades matches the empirically obtained CLO very well.

3.3.8 Cross-Validation for Multitype Branching Processes

A thorough cross-validation is also performed. Specifically, randomly chosen M_u^{min} cascades are used to estimate the joint distribution by the multitype branching process, which is compared with the joint distribution empirically obtained from another randomly chosen M_u^{min} cascades.

The joint entropy for the empirical and estimated joint distributions is listed in Table 3.10. The marginal distributions for each type of outages are shown in Figures 3.12 and 3.13. The empirically obtained and estimated CLOs for the load shed and the isolated buses when the total number of line outages is

Table 3.10 Joint entropy of distributions in cross-validation.

Type	M_u^{min}	H^{emp}	H^{est}
Line outage and load shed	6500	6.91	7.23
Line outage and isolated bus	5500	5.33	6.38

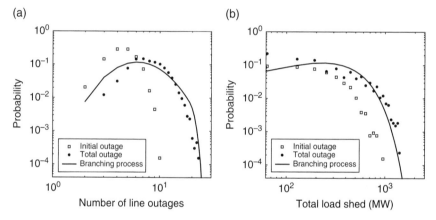

Figure 3.12 Estimated marginal probability distributions in cross-validation when line outages and the load shed are considered. (a) The line outages. (b) The load shed.

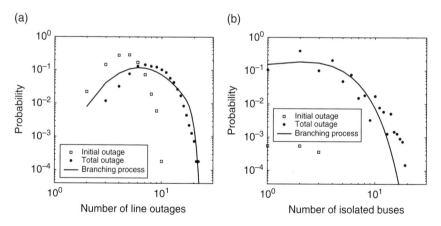

Figure 3.13 Estimated marginal probability distributions in cross-validation when line outages and isolated buses are considered. (a) The line outages. (b) The isolated buses.

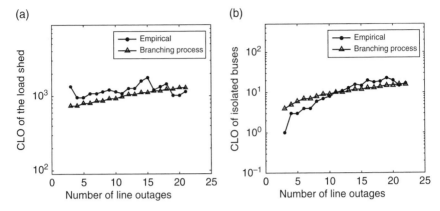

Figure 3.14 Estimated CLO in cross-validation when the total number of line outages is known. (a) The load shed. (b) The isolated buses.

known are shown in Figure 3.14. The results show that the branching process model trained by a randomly chosen subset of data is accurate in describing other subsets.

3.3.9 Predicting Joint Distribution from One Type of Outage

To further demonstrate and validate the multitype branching process model, we estimate the joint distribution of the total sizes of two types of outages by only using the predetermined offspring mean matrix and the distribution of the initial line outages.

1) The offspring mean matrix $\hat{\mathbf{\Lambda}}$ is calculated offline from M_u^{min} cascades, as shown in Table 3.8.
2) To mimic online application, M_u^{min} cascades are randomly chosen from the $M^{min} - M_u^{min}$ cascades for testing. The empirical joint distribution of line outages and the load shed (isolate buses) is calculated as a reference.
3) We estimate the joint distribution of line outages and the load shed (isolated buses) by using the $\hat{\mathbf{\Lambda}}$ in step 1 and the distribution of initial line outages, assuming there are no data about the load shed (isolated buses) whose initial outage distribution is set to be zero with probability one.
4) We compare the marginal distributions and the CLO calculated from the estimated and empirical joint distributions.

The predicted marginal probability distributions of the load shed and isolated buses are shown in Figure 3.15. The prediction is reasonably good, although we do not have the distribution of the initial load shed or isolated buses. The prediction of the load shed is very good when the blackout size is small but is not as good for a larger blackout size. By contrast, the prediction of

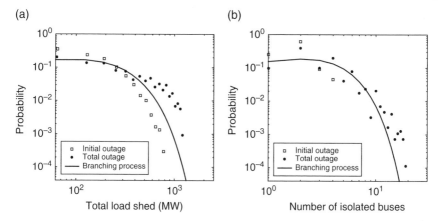

Figure 3.15 Estimated marginal probability distributions assuming there are no load shed or isolated bus data. (a) The load shed. (b) The isolated buses.

the number of isolated buses is good for a wide range of blackout sizes. This is mainly because the initial outage of the load shed can be greater than zero with a non-negligible probability. Assuming the initial outage for the load shed is zero with probability one can reduce the accuracy of the prediction. By contrast, the probability for the initial number of isolated buses to be zero or one can be very high (86.21% in our test case), since in the initial stage it is not very possible that some buses are isolated from the major part of the system. Thus, assuming the initial number of isolated buses is zero with probability one does not influence the prediction obviously.

The empirically obtained and estimated CLOs of the load shed and the isolated buses when the number of line outages is known are shown in Figure 3.16. We use $M_u^{\min}$ cascades to get the empirical and estimated CLOs. The prediction of the CLO when there are no data for the load shed or isolated buses (especially the former one) is not as good as the case with those data (the prediction of the CLO for the isolated buses is better than that for the load shed due to the same reason as that for the prediction of the marginal distribution discussed previously). However, the multitype branching processes can generate useful and sometimes very accurate predictions for those outages whose data are unavailable. This can provide important information for the operators when the system is under a cascading outage event or is in restoration. It is also seen that the estimated CLO from the branching process seems to be more statistically reliable than the empirically obtained CLO from the same number of cascades, which can oscillate as the number of line outages increases. Comparing the empirically obtained CLOs in Figure 3.16 and Figure 3.11, we can see that the oscillation in Figure 3.11 is not that obvious, mainly because it uses much more simulated cascades to obtain the empirical CLO.

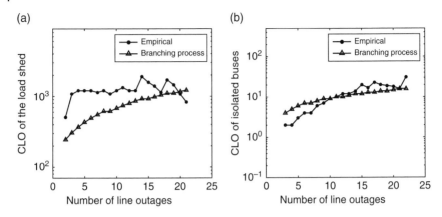

Figure 3.16 Estimated CLO when the number of line outages is known, assuming there are no load shed or isolated bus data. (a) The load shed. (b) The isolated buses.

3.3.10 Estimating Failure Propagation of Three Types of Outages

The computational complexity for estimating the joint distribution for more than two types of outages can be very high. Therefore, here we only estimate the parameters of the three-type branching process to better indicate the extent of outage propagation. When jointly considering line outages, the load shed, and the isolated buses, we determine $M^{\min} = 46\,000$ by using the method in Section 3.3.5. The EM algorithm that estimates the offspring mean matrix of the multitype branching processes converges in six steps.

The estimated largest eigenvalue of offspring mean matrix, the offspring mean matrix, and the joint entropy of the empirical joint distribution are listed in Table 3.11. It is seen that line outages tend to have a greater influence on the load shed and the isolated buses, but the influence of the load shed or the isolated buses on line outages is relatively weak. The largest eigenvalue of the offspring mean matrix is greater than that for the two-type branching processes, indicating that the system is even closer to criticality when considering the mutual influence of three types of outages. Besides, the joint entropy is also

Table 3.11 Estimated parameters for three types of outages.

Type	$\hat{\rho}$	$\hat{\Lambda}$			H^{emp}
Line outage, load shed, and isolated bus	0.64	0.44	0.39	0.39	8.54
		0.0035	0.024	0.018	
		7.4×10^{-7}	0.079	4.2×10^{-4}	

greater compared with the two-type branching process, although the increase of joint entropy from two-type to three-type is not as high as that from one-type to two-type.

3.4 Failure Interaction Analysis

The branching process model can statistically describe the cascading failure data and provide high-level statistical information about cascading failures. But it does not retain information about the network topology or power flow. It also does not attempt to specify how cascades propagate in the system in detail, such as which, where, or why transmission lines outage.

From the perspective of complex systems, the system-level failures are not caused by any specific event but by the property that the components in the system are tightly coupled and interdependent with each other [13]. Therefore, explicitly studying the interactions between components can help understand the mechanisms of cascading failures, identify key factors for their propagation, and further mitigate cascading.

In Qi et al. [12] an explicit study of the interactions between components failures obtained from detailed cascading failure simulation helps better understand the mechanisms of cascading failures. In Ju et al. [46] an interaction network is built for a Northeast Power Coordinating Council (NPCC) power system test bed, which represents the northeastern region of the Eastern Interconnection (EI) system. Then, in Ju et al. [47] a multilayer interaction graph is proposed as an extension of a single-layer interaction network. In this multilayer graph, each layer focuses on one of several aspects that are critical for the system operators' decision support, such as the number of line outages, the amount of load shedding, and the electrical distance of the outage propagation. Besides, influence graphs [48, 49] have also been proposed to extract propagation patterns. The distribution of the cascade sizes from the influence graph model is compared with that of the cascade sizes from $N-2$ contingencies in a 2896-branch test case and the two distributions are found to be similar.

By utilizing these interactions, a high-level probabilistic model called an interaction model can be used to study the influence of interactions on cascading failure risk and to support online decision-making. It will be demonstrated that it is much more time efficient to first quantify the interactions between component failures with fewer original cascades from a more detailed cascading failure model and then perform the interaction model simulation, than it is to directly simulate a large number of cascades with a more detailed model. Interaction-based mitigation measures are also suggested to mitigate cascading failure risk by weakening key links, which can be achieved in real systems by wide-area protection schemes [50–53] such as the blocking of some specific protective relays [54].

3.4.1 Estimation of Interactions between Component Failures

For power systems the transmission lines or transformers can be chosen as components. The cascades used for quantifying the interactions between component failures are called original cascades. They can come from statistical utility line outage data or simulations of detailed cascading failure models. M original cascades can be arranged as

	generation 0	generation 1	generation 2	$\cdots$
cascade 1	$F_0^{(1)}$	$F_1^{(1)}$	$F_2^{(1)}$	$\cdots$
cascade 2	$F_0^{(2)}$	$F_1^{(2)}$	$F_2^{(2)}$	$\cdots$
$\vdots$	$\vdots$	$\vdots$	$\vdots$	$\vdots$
cascade M	$F_0^{(M)}$	$F_1^{(M)}$	$F_2^{(M)}$	$\cdots$

where $F_g^{(m)}$ is the set of failed components produced in generation g of cascade m. Each cascade eventually terminates with a finite number of generations when the number of failed components in a generation becomes zero.

Assume $M_u \leq M$ original cascades are utilized to quantify the interactions. At first, we do not have enough information about which components in two consecutive generations have interactions. Thus, we assume that there are interactions between any failed component in the last generation and in this generation. By doing this, we can guarantee that no interactions will be ignored. For a system with n components, a matrix $A \in \mathbb{Z}^{n \times n}$ can be constructed whose entry a_{ij} is the number of times that component i fails in one generation before the failure of component j among all original cascades. Note that A does not depend on the order in which the cascades are processed, because it is obtained by using all M_u cascades.

The assumption based on which A is obtained actually exaggerates the interactions between component failures. It is not convincing to assert one component interacts with another one only because it fails in its last generation. For each failed component in generation one and the following generations, the failed component that most probably causes it should be determined.

Specifically, for any two consecutive generations k and $k+1$ of any cascade m, the failure of component j in generation $k+1$ is considered to be caused by a set of failed components in generation k, which can be described as

$$\left\{ i_c | i_c \in F_k^{(m)} \text{ and } a_{i_c j} = \max_{i \in F_k^{(m)}} a_{ij} \right\}. \tag{3.33}$$

Note that it is possible that two or more components in generation k are considered as the cause of the failure of component j. This is more possible

Figure 3.17 Illustration for determining the cause of component failures.

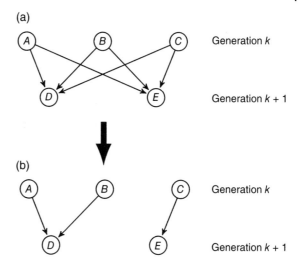

when M_u is not large enough and no component has much greater a_{ij} than the others. As an extreme case, all a_{ij} for $i \in F_k^{(m)}$ are all the same. Then it is impossible to determine which component is more likely to cause the failure of component j and all components have to be considered the cause. In this case the overestimation of the interaction by A cannot be well corrected, which will lead to the overestimation of the propagation of cascading failures. A more detailed discussion can be found in Section 3.4.5.

An illustration is shown in Figure 3.17, in which we show two consecutive generations of a cascade. If we assume that

$$a_{AD} = a_{BD} = \max_{i \in \{A,B,C\}} a_{iD} \tag{3.34}$$

$$a_{CE} = \max_{i \in \{A,B,C\}} a_{iE}, \tag{3.35}$$

we can determine the cause of component D failure as components A and B and the cause of component E failure as component C.

We do not update A when determining the most possible causes for failed components in generation one and the following generations. Therefore, the determination of the component that causes a failed component does not depend on the order in which the original cascades are processed but is completely determined by the A matrix.

After determining the cause of any component failure in generation one and the following generations for all cascades, A can be corrected to be $A' \in \mathbb{Z}^{n \times n}$, whose entry a'_{ij} is the number of times that the failure of component i causes the failure of component j. The interaction matrix $B \in \mathbb{R}^{n \times n}$ can then be calculated from A'. Its entry b_{ij} is the empirical probability that the failure of component i causes the failure of component j, which can be given by

$$b_{ij} = \frac{a'_{ij}}{N_i},\tag{3.36}$$

where N_i is the number of failures of component i.

Here the interaction matrix B is directly estimated from an approximated matrix A that is estimated based on a simple assumption that the component j failure in generation $g+1$ is caused only by the component in generation g that is in the previous generation of component j for the largest number of times. This assumption will inevitably ignore some interactions and thus can only get approximated estimations of A and B. A better approach is proposed in Qi et al. [55], in which the EM algorithm is applied to effectively deal with incomplete data about the exact causes of outages and more accurately estimate the interaction matrix.

The B matrix determines how components interact with each other. The nonzero elements of B are called links. Link $l : i \rightarrow j$ corresponds to B's nonzero element b_{ij} and starts from component i and ends with component j. By putting all links together, a directed network $\mathcal{G}(\mathcal{C},\mathcal{L})$ called an interaction network can be obtained. Its vertices $\mathcal{C}$ are components, and each directed link $l \in \mathcal{L}$ represents that a failure of the source vertex component causes the failure of the destination vertex component with a probability greater than zero.

Topological properties such as small-world [56] and scale-free [57] behaviors have been revealed in complex networks. However, it is misleading to evaluate the vulnerability of power systems based only on topological metrics [58]. It is more reasonable to discuss the property of a directed weighted interaction network generated with simulated cascades from a detailed cascading failure model that considers the physics of the system such as power flow and redispatching than directly exploring the property of the network coming from the topology of the physical system.

3.4.2 Identification of Key Links and Key Components

In order to identify key links that play important roles in cascading failure propagation, an index I_l is defined for each link $l : i \rightarrow j$ as the expected value of the number of failures that are propagated through link l. The failures propagated through link l can be directly triggered by the failure of component i or by the failure of components other than i that causes component i to fail. Therefore, in order to calculate I_l, the number of failures of its source vertex i (denoted by N_i^s) should be set to be N_i, which is the total number of failures among all of the original cascades. N_i contains not only the failures in generation zero that serve as trigger of cascading failures but also the failures caused by other component failures.

Similar to Section 3.4.1, $M_u \leq M$ original cascades are utilized to quantify the interactions and to calculate the index I_l. After obtaining the interaction network $\mathcal{G}$, we can get a directed acyclic subgraph $\mathcal{G}_j\left(\mathcal{C}_j, \mathcal{L}_j\right)$ starting with component j. The vertices represent the events of component failures, and the edges represent causal relations between events. All edges in the subgraph point in the same direction from parent to child due to the causality affecting the future, and each component is reached exactly once.

For each link there is a unique directed acyclic subgraph that can be extracted from the interaction network. It is composed of all the components influenced by this link. In Figure 3.18 we illustrate how the directed acyclic subgraph $\mathcal{G}_j$ (Figure 3.18c) can be obtained from the original subgraph (Figure 3.18a). Note that i is not in Figure 3.18a, even if there is a link from j or any other vertex to i because I_l is defined to indicate the failures link $l : i \rightarrow j$ can cause on the condition that i fails. From Figure 3.18a to Figure 3.18b, we remove the vertices for which there is no path from vertex j to them (H, I, and J denoted by dashed-line circles in Figure 3.18a). This is because we want to quantify the consequences of component j failure, and the removed vertices cannot be influenced by j. The links corresponding to the removed vertices (denoted by dotted-line arrows in Figure 3.18a) are also eliminated.

In Figure 3.18b vertex j is at level 0; the vertices that j points to that are not i are at level 1; the vertices that the level 1 vertices point to that are not i or any other vertices in the lower levels are at level 2. Because of the causal relationship between the vertices in two consecutive levels, the edges from the vertices at a higher level to those at a lower level ($D \rightarrow A$ and $F \rightarrow j$ denoted by arrows with dash-dotted lines in Figure 3.18b) are removed. Also the edges between the vertices at the same level ($C \rightarrow B$ denoted by dashed-line arrow in Figure 3.18b) are neglected since these vertices are considered independent and all fail on the condition of the failure of some component at the last level. Finally, we can get a directed acyclic subgraph $\mathcal{G}_j$ (Figure 3.18c) for which there is no loop and for each vertex (component) $c \in \mathcal{C}_j, c \neq j$ there is exactly one vertex c_s pointing to it.

The expected value of the number of failures of component j given N_i times of component i failure is

$$E_j = N_i^s b_{ij}. \tag{3.37}$$

For any other component $c \in \mathcal{C}_j, c \neq j$, the expected value of the number of failures given the times of its source vertex failure is

$$E_c = E_{c_s} b_{c_s c}. \tag{3.38}$$

(a)

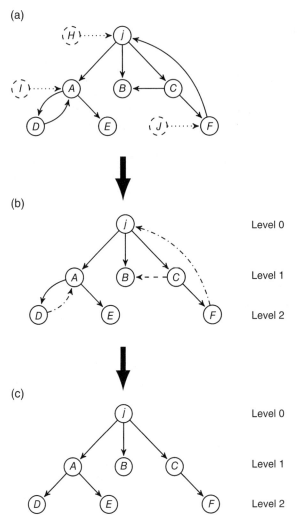

(b)

(c)

All the expected number of component failures in graph $\mathcal{G}_j$ are summated to be I_l as

$$I_l = \sum_{c \in \mathcal{C}_j} E_c. \tag{3.39}$$

I_l indicates the contribution of a link to the propagation of cascading failures. The greater the index is, the more important the link is for cascading failure propagation. Thus, the links with large I_l are defined as key links. Specifically, the set of key links $\mathcal{L}^{\text{key}}$ are those links whose weights are greater than or equal to a specified fraction of the largest link weight I_l^{max}, that is

$$\mathcal{L}^{\text{key}} = \left\{ l \middle| I_l \geq \epsilon_l I_l^{\text{max}} \right\}, \tag{3.40}$$

where ϵ_l is a value that is not too close to zero to guarantee that the weights of key links are not much smaller than the largest link weight.

With I_l as the weights of the links, the interaction network $\mathcal{G}(\mathcal{C},\mathcal{L})$ in Section 3.4.1 is now a directed weighted network. The vertex out-strength and in-strength of the interaction network are defined as follows.

$$s_i^{\text{out}} = \sum_{l \in \mathcal{L}^{\text{out}}(i)} I_l \tag{3.41}$$

$$s_i^{\text{in}} = \sum_{l \in \mathcal{L}^{\text{in}}(i)} I_l, \tag{3.42}$$

where $\mathcal{L}^{\text{out}}(i)$ and $\mathcal{L}^{\text{in}}(i)$ are, respectively, the sets of links starting from and ending with vertex i.

The out-strength and in-strength indicate how much a component contributes to failure propagation. The components with large out-strength can cause great consequences and thus are crucial for the propagation of cascading failures. Therefore, in a similar way to the key link definition, the set of key components $\mathcal{C}^{\text{key}}$ is defined as

$$\mathcal{C}^{\text{key}} = \left\{ i \middle| s_i^{\text{out}} \geq \epsilon_s s_i^{\text{out,max}} \right\}, \tag{3.43}$$

where $s_i^{\text{out,max}}$ is the largest vertex out-strength among all vertices and ϵ_s is used to guarantee that the out-strengths of the key components are not much smaller than the maximum out-strength.

3.4.3 Interaction Model

In this section we build a high-level probabilistic cascading failure model called the interaction model based on the tripping probability of each component in generation zero and the interactions between component failures. As in Section 3.4.1, we assume there are a total of M original cascades. Note that we do not necessarily need to use all the M cascades but only $M_u < M$ of them to generate the tripping probability of each component in generation zero and the interaction matrix.

It is assumed that all components are initially unfailed and each component fails with a small probability. The component failures in the same generation cause other component failures independently. The flow chart of the interaction

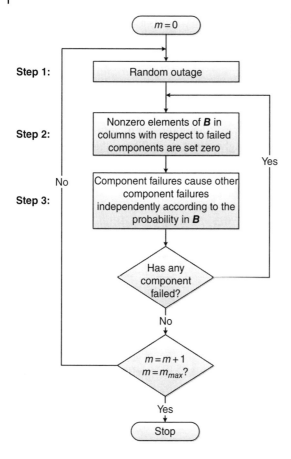

Figure 3.19 Flow chart of the interaction model.

model is shown in Figure 3.19, in which $m_{\max}$ is the number of cascades to be simulated. The model contains two loops and in each outer loop a cascade is simulated. Specifically, the model is implemented in the following three steps.

- Step 1: Accidental faults of components
  In the kth outer iteration, each component i randomly fails with probability τ_i to simulate accidental faults, and the failed components form generation zero (initial outages) of the simulated cascade. The probability that a component i fails as initial outages can be estimated by using the generation zero component failures of the M_u original cascades as

$$\tau_i = \frac{f_0^i}{M_u},\tag{3.44}$$

where f_0^i is the number of cascades for which component i fails in generation zero.

- Step 2: Modifying $\boldsymbol{B}$ after component failure
  In our model once a component fails it will remain that way until the end of the simulation. Therefore, the columns of $\boldsymbol{B}$ that correspond to the component failures are set at zero.
- Step 3: Failure propagation
  The component failures in one generation independently generate other component failures. Specifically, if component i fails in this generation it will cause the failure of any other component j with probability b_{ij}. Once it causes the failures of some components, these newly caused component failures will compose the next generation; then the model goes back to Step 2. If no component failure is caused in one generation, the inner loop stops.

By using the interaction model we can simulate as many cascades as possible. Although generated by utilizing the information of the initial outages and the interactions contained in the original cascades, the simulated cascades can reveal new rare events due to the high-level probabilistic property of the interaction model. This can help recover the missing information due to using fewer original cascades.

The interaction model can be used not only for offline study of cascading failures but can also for online decision-making support. The interaction matrix can be obtained offline from statistical utility data or simulations of detailed cascading failure models. It contains important information about the interactions between component failures. By utilizing this information, the interaction model can predict the consequences of events. If something unusual happens, the operators can apply the interaction model to quickly find out which components or areas of the system will most probably be affected. A fast response can then be performed to pull the system back to normal operating conditions and avoid or at least reduce the economic and social losses.

3.4.4 Validation of Interaction Model

To validate the interaction model, the simulated cascades from the interaction model are carefully compared with the original cascades.

1) The probability distribution of the total line outages of the original and simulated cascades are compared.
2) The probability distribution of the total line outages of the original and simulated cascades can be estimated by the branching process, and the average propagation $\hat{\lambda}$ (estimated offspring mean in Section 3.2.2) can be compared.
3) The interactions between component failures for both the original and simulated cascades are quantified and the probability distribution of the link weights and vertex out-strength and in-strength of the interaction network are compared.

4) The spreading capacities of the links quantified from the original and simulated cascades are compared by the similarity indices as defined.

Let $\mathcal{L}_1$ be the set of links shared by the original and simulated cascades. Let $\mathcal{L}_2$ and $\mathcal{L}_3$ separately denote the sets of links that are only owned by the original and simulated cascades. Denote the index of link l for the original and simulated cascades, respectively, by I_l^{ori} and I_l^{sim}. Five similarity indices are defined as

$$S_1 = \frac{\sum\limits_{l \in (\mathcal{L}_1 \cup \mathcal{L}_3)} I_l^{\text{sim}}}{\sum\limits_{l \in (\mathcal{L}_1 \cup \mathcal{L}_2)} I_l^{\text{ori}}}, \quad S_2 = \frac{\sum\limits_{l \in \mathcal{L}_1} I_l^{\text{ori}}}{\sum\limits_{l \in (\mathcal{L}_1 \cup \mathcal{L}_2)} I_l^{\text{ori}}}, \quad S_3 = \frac{\sum\limits_{l \in \mathcal{L}_1} I_l^{\text{sim}}}{\sum\limits_{l \in (\mathcal{L}_1 \cup \mathcal{L}_3)} I_l^{\text{sim}}},$$

$$S_4 = \frac{\sum\limits_{l \in \mathcal{L}_1} I_l^{\text{sim}}}{\sum\limits_{l \in \mathcal{L}_1} I_l^{\text{ori}}}, \quad S_5 = \sum\limits_{l \in \mathcal{L}_1} \left(\frac{I_l^{\text{sim}} + I_l^{\text{ori}}}{\sum\limits_{l \in \mathcal{L}_1} \left(I_l^{\text{sim}} + I_l^{\text{ori}} \right)} \frac{I_l^{\text{sim}}}{I_l^{\text{ori}}} \right).$$

S_1 is the ratio between the summation of the link weights of the simulated cascades and that of the original cascades. If S_1 is close to 1.0 the links of the original and simulated cascades have almost the same spreading capacity. S_2 and S_3 are used to indicate if the shared links play a major role among all links for the original and simulated cascades. If they are near 1.0, it means that the shared links dominate and thus the simulated cascades are similar to the original cascades. S_4 indicates the similarity between the overall spreading capacity of the shared links of the simulated cascades and that of the original cascades. $S_4 \approx 1$ suggests that the overall spreading capacity of the shared links for the simulated cascades is close to that of the original cascades.

Even when $S_4 \approx 1$, it is still possible that the weight of the same link for the original and simulated cascades can be quite different. Thus, S_5 is defined to show if the same link is close to each other. When S_5 is near 1.0, it indicates that at least the most important links of the simulated cascades have spreading capacity close to their counterparts for the original cascades.

The similarity indices defined here can be used not only to compare the similarity of the links obtained from original and simulated cascades. They can also be used to compare any two sets of links. For example, we can use them to compare the links obtained from different numbers of original cascades, which can be denoted by $I_l(M_u^1)$ and $I_l(M_u^2)$, respectively, for M_u^1 and M_u^2 original cascades. Since I_l depends on the number of cascades that are used to quantify the interactions, $I_l(M_u^1)$ and $I_l(M_u^2)$ should first be normalized before they are compared. One simple way to do so is to divide $I_l(M_u^1)$ by M_u^1/M_u^2 or to divide $I_l(M_u^2)$ by M_u^2/M_u^1.

3.4.5 Number of Cascades Needed for Failure Interaction Analysis

In Section 3.4.3, M_u out of M cascades are used to generate the tripping probability of each component in generation zero and the interaction matrix. Similar to Section 3.3.5, here we discuss how to determine the lower bounds for M and M_u.

1) Determining Lower Bound for M

   More original cascades can contain more information about the property of cascading failures of a system, or more specifically the interactions between cascading outages of the components. As the number of cascades increases, the number of identified links can also increase. However, the number of links will not always grow with the increase of the number of cascades but will saturate after the number of cascades is greater than some number M^{min}. This number can be determined by gradually increasing the number of cascades, recording the number of identified links, and finding the smallest number of cascades that leads to the saturated number of links.

   Assume there are a total of N_M different M ranging from a very small number to a very large number, which are denoted by M_i, $i = 1, 2, ..., N_M$. The number of links for M_i cascades is denoted by $card(\mathcal{L}(M_i))$, where $card(\cdot)$ denotes the cardinality of a set, which is a measure of the number of elements of a set.

   For $i = 1, 2, ..., N_M - 2$ we define $\sigma_i = \sigma(card(L)_i)$, where $\sigma(\cdot)$ is the standard deviation of a vector and $card(\mathcal{L})_i = \left[card(\mathcal{L}(M_i)) \cdots card(\mathcal{L}(M_{N_M})) \right]$. The σ_i for $i = N_M - 1$ and $i = N_M$ are not calculated because we would like to calculate the standard deviation for at least three data points. Very small and slightly fluctuating σ_i indicates that the number of links begins to saturate after M_i, and this M_i is determined as M^{min}.

   The M^{min} original cascades can guarantee that the accuracy on statistical values of interest is good and thus can provide a reference solution.

2) Determining Lower Bound for M_u

   When we only want to obtain the dominant interactions in order to generate simulated cascades that match well enough with the original cascades, we will only need $M_u^{min} \ll M^{min}$ cascades to make sure that the propagation capacity of the obtained interaction network $\mathcal{G}(\mathcal{C},\mathcal{L})$ (denoted by $PC^{\mathcal{G}}$) is consistent with that of the original cascades (denoted by PC^{ori}). Here the physical meaning of the propagation capacity is the average value of the number of caused failures in one cascade.

Since both PC^{ori} and $\text{PC}^{\mathcal{G}}$ vary with M_{u}, we denote them by $\text{PC}^{\text{ori}}(M_{\text{u}})$ and $\text{PC}^{\mathcal{G}}(M_{\text{u}})$. $\text{PC}^{\text{ori}}(M_{\text{u}})$ can be directly obtained from the original cascades by calculating the average value of the number of failures in generation one and the following generations as

$$\text{PC}^{\text{ori}}\left(M_{\text{u}}\right) = \frac{\sum_{m=1}^{M_{\text{u}}}\sum_{g=1}^{\infty}\text{card}\left(F_g^{(m)}\right)}{M_{\text{u}}}. \tag{3.45}$$

In Section 3.4.2 we have defined an index I_l for each link $l : i \rightarrow j$, which is the expected value of the number of failures that are caused through link l. To get all of the failures that are caused through the link $l : i \rightarrow j$, we set the number of failures of its source vertex i as N_i, which is the total number of failures among all of the original cascades. Note that N_i contains not only the failures in generation zero that serve as triggers of cascading failures but also the failures caused by other component failures. Here, however, we only want to calculate the expected value of the number of failures caused through link $l : i \rightarrow j$ by its source vertex i as generation zero failures. In this case, we need to set N_i^s to be $N_{i,0}$, which is the number of failures of component i in generation zero among all M_{u} original cascades. The calculated link index is denoted by I_l' to distinguish with I_l in Section 3.4.2, for which $N_i^s = N_i$. Then the propagation capacity of the interaction network can be calculated as

$$\text{PC}^{\mathcal{G}}\left(M_{\text{u}}\right) = \frac{\sum_{l \in \mathcal{L}}I_l'(M_{\text{u}})}{M_{\text{u}}}. \tag{3.46}$$

When M_{u} is not large enough, it is expected that there will be a big mismatch between $\text{PC}^{\text{ori}}(M_{\text{u}})$ and $\text{PC}^{\mathcal{G}}\left(M_{\text{u}}\right)$, because the quantified interactions between cascading outages cannot well capture the property of the cascading failure propagation. With the increase of M_{u}, more information will be obtained and the mismatch will gradually decrease. Based on this, we increase M_{u} gradually and identify $M_{\text{u}}^{\text{min}}$ as the smallest value that satisfies the following condition:

$$\left|\Delta_{\text{PC}}\left(M_{\text{u}}\right)\right| \leq \epsilon_{\text{PC}}\text{PC}^{\text{ori}}\left(M_{\text{u}}\right), \tag{3.47}$$

where $\Delta_{\text{PC}}\left(M_{\text{u}}\right) = \text{PC}^{\mathcal{G}}\left(M_{\text{u}}\right) - \text{PC}^{\text{ori}}\left(M_{\text{u}}\right)$ and ϵ_{PC} is used to determine the acceptable mismatch.

In order to get $M_{\text{u}}^{\text{min}}$ we start from very small M_{u}, such as $M_{\text{u}}^0 = 100$, and calculate the mismatch $\Delta_{\text{PC}}(M_{\text{u}})$. If the condition in (3.47) is not satisfied, M_{u} is increased by a big step ΔM_1 and $\Delta_{\text{PC}}(M_{\text{u}})$ is recalculated with the new M_{u}; otherwise M_{u} is decreased by a small step ΔM_2 until the last M_{u} for which the condition in (3.47) is still satisfied.

The number of unnecessary original cascade simulation runs M_{un} is always less than ΔM_1 and actually can be determined by $M_{un} = N_{\Delta M_2} \Delta M_2$, where $N_{\Delta M_2}$ is the number of times ΔM_2 is used for iterations. The obtained M_u^{min} is not greater than the smallest possible M_u by ΔM_2. By decreasing ΔM_1 we can decrease the upper bound of M_{un} but cannot necessarily decrease M_{un}. Differently, by decreasing ΔM_1 we can surely increase the accuracy of the obtained M_u^{min}. But smaller ΔM_1 or ΔM_2 will increase the time for getting the interaction network and quantifying I'_l for the links. The selection of ΔM_1 and ΔM_2 can be guided by ϵ_{PC}. Smaller mismatch ϵ_{PC} needs greater M_u^{min} and thus larger ΔM_1 and ΔM_2 can be chosen to avoid too many times of calculating the interaction network and quantifying I'_l.

For testing the interaction quantifying method and the interaction model, AC OPA simulation for the IEEE 118-bus test system at base case load level is run so as to produce 50 000 cascading outages with a nonzero number of line outages.

The number of identified links $\text{card}(\mathcal{L}(M))$ for different M ranging from 100 to 50 000 is shown in Figure 3.20. We can see that the number of links first grows with the increase of M and finally saturates and only fluctuates slightly when M is large enough. There are a total of $N_M = 54$ different M. In Figure 3.20 we also show σ_i for $1 \leq i \leq N_M - 2$. It is very clear that σ_i decreases with the increase of M and finally stabilizes at $M = 41\,000$. Therefore, we choose $M^{min} = 41\,000$, for which the number of identified links is 419. The largest number of links for $\text{card}(\mathcal{L}(M_i))$, $i = 1, 2, ..., N_M$ is 423. The number of identified links for $M^{min} = 41\,000$ is greater than 99% of the largest number of links.

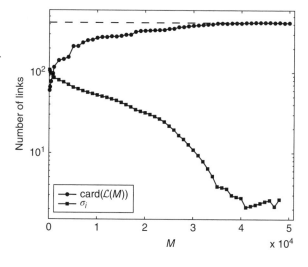

Figure 3.20 Number of links for different M. The dash horizontal line indicates $\text{card}(\mathcal{L}(M)) = 419$.

To determine M_u^{min} we choose M_u^0, ϵ_{PC}, ΔM_1, and ΔM_2 as 100, 0.01, 1000, and 100. By setting ΔM_2 at 100 we can guarantee that the obtained M_u^{min} is not greater than the smallest possible M_u by 100. After nine ΔM_1 iterations and one ΔM_2 iteration, M_u^{min} is determined as 8000, which accounts for 19.51% of M^{min}. The number of unnecessary original cascade simulation runs M_{un} in this case is only 100. If we set ϵ_{PC} as 0.05 and correspondingly decrease ΔM_1 and ΔM_2 to 200 and 10, M_u^{min} will be determined as 3680, which only accounts for 8.98% of M^{min}, after nineteen ΔM_1 iterations and two ΔM_2 iterations. The number of unnecessary original cascade simulation runs M_{un} in this case is only 20.

When we have already generated a large number of cascades we can show how the propagation capacity obtained from original cascades and the interaction network changes with the increase of M_u. This is shown in Figure 3.21, in which we can clearly see the trend of the decreasing mismatch between $PC^{ori}(M_u)$ and $PC^{\mathcal{G}}(M_u)$.

In Figure 3.21 we can also see that the quantified interaction network tends to obtain an overestimated propagation capacity for a small number of M_u. As has been tentatively discussed in Section 3.4.1, this is because for any component j no component has much greater a_{ij} than the others when only using a small number of M_u, and thus several components have to be determined as the cause. This further leads to the overestimation of the propagation of cascading failures.

To show this, we analyze how A changes with the increase of M_u in more detail. Let C^{caused} denote the set of components that are caused by other components in A, which actually corresponds to the columns of A with nonzero

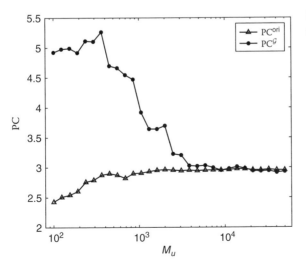

Figure 3.21 Propagation capacity for different M_u.

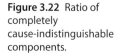

Figure 3.22 Ratio of completely cause-indistinguishable components.

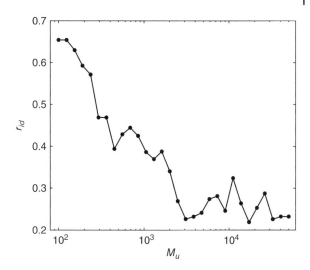

elements. For any $j \in C^{\text{caused}}$, if $\text{card}(\{i \mid a_{ij} > 0\}) > 1$ and $\max_{i \in C} a_{ij} = \overline{a}_j$, where C is all the components and $\overline{a}_j$ is the average value for all $a_{ij} > 0$, then component j is called completely cause-indistinguishable. The set of completely cause-indistinguishable components is denoted by C^{id}. Actually, when the given condition holds it is easy to prove by contradiction that all nonzero a_{ij} will be equal. Thus, in this case it is completely impossible to distinguish which cause of component j is more possible than the others. The ratio of completely cause-indistinguishable components among all caused components can be calculated as

$$r_{\text{id}} = \frac{\text{card}\left(C^{\text{id}}\right)}{\text{card}\left(C^{\text{caused}}\right)}. \tag{3.48}$$

We show r_{id} in Figure 3.22 and it is clearly seen that the completely cause-indistinguishable components account for a large proportion when M_u is small and will gradually decrease to a relatively low level with the increase of M_u. The high ratio of the completely cause-indistinguishable components for a small number of M_u leads to the overestimated interactions between component failures and further explains the overestimated propagation capacity of the interaction network.

3.4.6 Estimated Interaction Matrix and Interaction Network

There are 186 lines in the IEEE 118-bus system; thus, B is a 186×186 square matrix. Table 3.12 shows the number of components n, the number of cascades

Table 3.12 Nonzero elements in B for IEEE 118-bus system.

Model	n	M_u	card $(\mathcal{L})$	r
AC OPA	186	41 000	419	0.0121
AC OPA	186	8000	252	0.00730
AC OPA	186	3680	156	0.00450

used to quantify the interactions M_u; the number of links card$(\mathcal{L})$ which is also the number of B's nonzero elements; and the ratio of nonzero elements $r = \text{card}(\mathcal{L})/n^2$. It is seen that r is very small, indicating that the interaction matrix is very sparse and that only a small fraction of lines interact with each other.

The corresponding directed weighted interaction networks obtained from 41 000, 8000, and 3680 original cascades are shown in Figures 3.23–3.25, in which the dots denote lines in the IEEE 118-bus system and the arrows denote the links between lines. Here we do not show the weights of the links but only the topology of the interaction network. This network is different from the one-line diagram of the IEEE 118-bus system, for which the vertices are buses and the undirected links between vertices are lines.

We compare the links obtained from 8000 and 3680 original cascades with the reference 41 000 original cascades by calculating the similarity indices defined in Section 3.4.4, which are shown in Table 3.13. As is discussed in Section 3.4.4, since I_l depends on the number of cascades that are used, the index I_l of the links should first be normalized before calculating the similarity indices. Specifically, we divide the I_l for 41 000 cascades by 41 000/8000 and 41 000/3680. Results in Table 3.13 shows that the first four indices are all close to 1.0 and thus indicate the links from a smaller number of cascades are similar to those for more cascades. But the fifth index S_5, which requires more strict similarity, is not so close to 1.0, indicating that some information is missing when using fewer original cascades. However, we will show in Section 3.4.8 that the links obtained from simulating more cascades with the interaction model can be very similar to those from 41 000 original cascades.

Table 3.13 Similarity indices for 8000 and 3680 original cascades compared with 41 000 original cascades.

Model	M_u	S_1	S_2	S_3	S_4	S_5
AC OPA	8000	1.02	0.996	0.991	1.02	2.79
AC OPA	3680	1.04	0.987	0.985	1.04	6.41

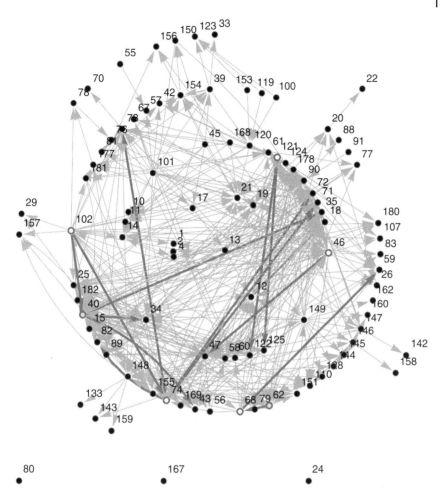

Figure 3.23 Interaction network for using 41 000 original cascades. The dots denote lines in the IEEE 118-bus system and the arrows denote the links between lines.

3.4.7 Identified Key Links and Key Components

Key links and key components that play important roles in the propagation of cascading failures are identified using the method in Section 3.4.2. Both ϵ_l and ϵ_s are chosen as 0.15 (slightly greater than 1/10) to keep the weights of all key links and all key components in the same order.

The identified key links, which are actually line pairs in the IEEE 118-bus system, and their weights I_l for the three cases separately using 41 000, 8000, and 3680 original cascades are listed in Table 3.14. They are also shown in

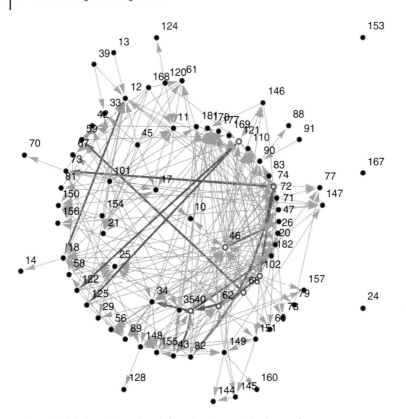

Figure 3.24 Interaction network for using 8000 original cascades.

Figures 3.23–3.25 by thick arrows. The numbers in the parentheses for $I_l(8000)$ and $I_l(3680)$ are the ranking of the key links. We can see that the identified key links for the three cases are almost the same; the ranking of the links is also quite similar. The link $(53, 54) \to (49, 51)$ is only identified for using 41 000 cascades, and the links $(65, 66) \to (45, 46)$, $(35, 36) \to (45, 46)$, and $(80, 97) \to (80, 96)$ are only identified for using 3680 cascades. Although the link $(53, 54) \to (49, 51)$ is not identified for using 8000 and 3680 cascades, the corresponding I_l for the two cases separately rank 16th and 19th and are equal to 378 and 157, which are much greater than the average values of all link indices, 109 and 82. For the three cases the number of key links are only 3.82%, 5.95%, and 11.54% of all the links, but the summation of their weights accounts for 88.08%, 85.12%, and 87.65% of the total weights of all links.

The identified key components, the corresponding lines, and their out-strengths for using 41 000, 8000, and 3680 original cascades are listed in Table 3.15, in which the numbers in the parentheses are the ranking of the key

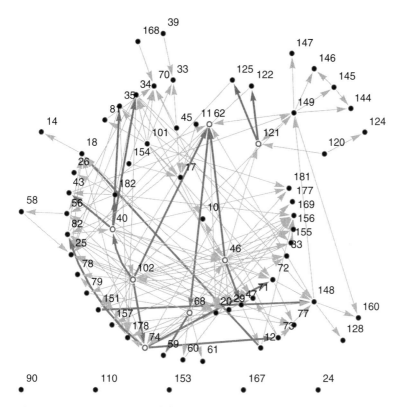

Figure 3.25 Interaction network for using 3680 original cascades.

components. The table illustrates that the identified key components and their ranking are exactly the same for the three cases. They are also highlighted in Figures 3.23–3.25 by open circles. The tripping of these lines will cause severe consequences and thus should be prevented to the greatest extent. For the IEEE 118-bus system there are a total of 186 components, and among them 97, 76, and 68 components are involved in the 41 000, 8000, and 3680 original cascades. For the three different numbers of original cascades, the number of key components are 3.76% of all components, or, 7.22%, 9.21%, and 10.29% of the involved components. The summation of the out-strengths of the key components account for 89.12%, 88.83%, and 88.65% of the total out-strengths of the involved components, respectively for the three cases.

The identified key links and key components are denoted on the one-line diagram of the IEEE 118-bus system, which is shown in Figure 3.26. It is evident from this figure that the lines corresponding to the source and destination

Table 3.14 Key links for IEEE 118-bus system.

$i \to j$	Line Pairs	I_l(41000)	I_l(8000) (Rank)	I_l(3680) (Rank)
$74 \to 72$	$(53,54) \to (51,52)$	12582	2486 (2)	1150 (1)
$74 \to 73$	$(53,54) \to (52,53)$	12469	2535 (1)	1150 (2)
$40 \to 34$	$(29,31) \to (27,28)$	11920	2305 (3)	1031 (3)
$40 \to 35$	$(29,31) \to (28,29)$	11421	2167 (4)	974 (5)
$74 \to 82$	$(53,54) \to (56,58)$	10802	2153 (5)	1000 (4)
$62 \to 68$	$(45,46) \to (45,49)$	9865	1915 (6)	940 (6)
$121 \to 122$	$(77,78) \to (78,79)$	9601	1912 (7)	862 (7)
$121 \to 125$	$(77,78) \to (79,80)$	9599	1912 (8)	862 (8)
$40 \to 182$	$(29,31) \to (114,115)$	6687	1261 (9)	592 (9)
$46 \to 47$	$(35,36) \to (35,37)$	5536	1092 (10)	502 (10)
$12 \to 18$	$(11,12) \to (13,15)$	5388	1001 (11)	450 (11)
$40 \to 43$	$(29,31) \to (27,32)$	4475	886 (12)	404 (12)
$68 \to 59$	$(45,49) \to (43,44)$	3690	706 (13)	347 (13)
$102 \to 74$	$(65,66) \to (53,54)$	3135	635 (14)	273 (14)
$74 \to 71$	$(53,54) \to (49,51)$	1977	–	–
$102 \to 40$	$(65,66) \to (29,31)$	1968	415 (15)	178 (15)
$102 \to 62$	$(65,66) \to (45,46)$	–	–	178 (16)
$46 \to 62$	$(35,36) \to (45,46)$	–	–	178 (17)
$151 \to 148$	$(80,97) \to (80,96)$	–	–	175 (18)

Table 3.15 Key components for IEEE 118-bus system.

Key Component	Line	s_i^{out}(41000)	s_i^{out}(8000)(Rank)	s_i^{out}(3680)(Rank)
74	$(53,54)$	37925	7577 (1)	3467 (1)
40	$(29,31)$	34613	6651 (2)	3015 (2)
121	$(77,78)$	19219	3833 (3)	1728 (3)
62	$(45,46)$	9892	1926 (4)	944 (4)
102	$(65,66)$	8039	1826 (5)	892 (5)
46	$(35,36)$	6645	1377 (6)	734 (6)
68	$(45,49)$	6210	1211 (7)	594 (7)

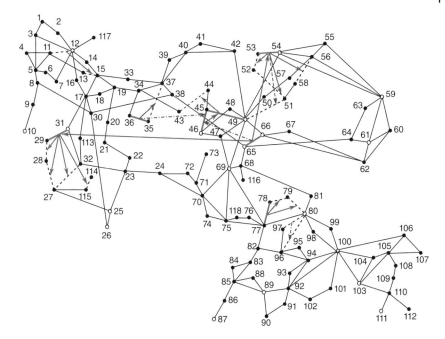

Figure 3.26 Key links and key components for IEEE 118-bus system. Key links shared by the three cases for using 41 000, 8000, and 3680 cascades are denoted by solid arrows; key links only for using 41 000 cascades are denoted by dash arrows; key links only for using 3680 cascades are denoted by dash-dotted arrows; key components are denoted by thick solid lines and the other lines involved in the key links are denoted by dash lines.

vertices of some key links can be topologically far away from each other, such as link $(65, 66) \rightarrow (53, 54)$ and $(65, 66) \rightarrow (29, 31)$, although for most key links the source and destination vertices are lines that are topologically close to each other, such as link $(53, 54) \rightarrow (51, 52)$ and $(53, 54) \rightarrow (52, 53)$. This is because the interactions and links are obtained from simulated cascades generated by the AC OPA model, which considers not only the topology of the power network but also other physics of the system, such as power flow and the operator response. These factors can also make some components tightly coupled.

3.4.8 Interaction Model Validation

The probability distributions of the total number of line outages for original cascades (triangles) and simulated cascades (dots) obtained by using 41 000, 8000, and 3680 original cascades to quantify interactions are shown in Figure 3.27. For the simulated cascades we simulate 41 000 cascades for 20 times for each case and show their average probability distribution and standard deviations (vertical lines).

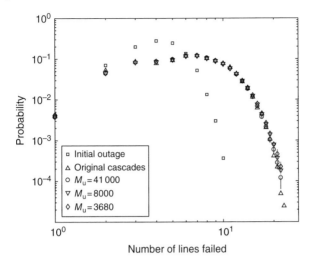

Figure 3.27 Probability distributions of the total number of line outages for original and simulated cascades. Triangles indicate the total numbers of line outages of the original cascades; ○, ▽, and ◇ separately indicate total numbers of line outages of simulated cascades for using 41 000, 8000, and 3680 original cascades to quantify interactions; vertical lines indicate standard deviations; squares indicate initial line outages.

It is seen that the distributions of total line outages of the original and simulated cascades match well, and the standard deviations of the probability distributions of the simulated cascades are small. This suggests that the interaction model can generate cascades with similar statistical properties to the original cascades. The dramatic difference between the distributions of the initial and total outages also suggests that the cascading failure is able to propagate a great amount. This is because of the interaction between component failures denoted by the sparse interaction matrix. If all elements of B are zero and the components do not interact at all, all cascades will stop immediately after initial line outages and the distribution of the total line outages will be the same as initial line outages. Thus, although being sparse, the interaction matrix does take effect.

To quantitatively compare the original cascades and the simulated cascades, the branching process is applied to estimate their average propagation. The average value m_λ and the standard deviation σ_λ of the average propagation for 20 times of simulation are listed in Table 3.16, in which M_u is the number of original cascades used to quantify the interactions between component failures and M is the number of cascades simulated by the interaction model. It is seen that the estimated average propagation of the simulated cascades is very close to that of the original cascades and the standard deviations are very small, indicating that the simulated cascades' propagation property is similar to that of the original cascades.

Table 3.16 Average propagation for original and simulated cascades.

Model	M_u	M	$m_{\hat{\lambda}}$	$\sigma_{\hat{\lambda}}$
AC OPA	–	41 000	0.402	–
Interaction	41 000	41 000	0.402	0.000969
Interaction	8000	41 000	0.410	0.000800
Interaction	3680	41 000	0.412	0.000899

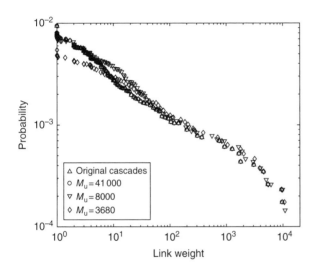

Figure 3.28 Complementary cumulative distributions (CCD) of the link weight for original and simulated cascades. Triangles indicate CCD of the link weight of original cascades; ○, ▽, and ◇ separately indicate CCD of the link weight of the simulated cascades for using 41 000, 8000, and 3680 original cascades to quantify interactions.

The complementary cumulative distributions (CCDs) of the link weights for the original and simulated cascades for M_u = 41 000, 8000, and 3680 are shown in Figure 3.28. The CCD for simulated cascades are the distribution for all the links obtained from 20 times of simulation. Note that we group the links with very big weights together and calculate the CCD for their average value to avoid the possible unreliable estimation for rarer events since the number for each of them can be very small. Also when we calculate the CCD the zero elements are considered links with zero weights. Figure 3.28 shows that the two distributions match very well and both follow obvious power laws and can range from 1 to more than 10000, suggesting that a small number of links cause much greater consequences than the others.

The CCD of the vertex out-strength and in-strength for original and simulated cascades are shown in Figures 3.29–3.30. Similar to link weights, the CCD for simulated cascades are the distribution for all the vertex out-strength and in-strength obtained from 20 times of simulation. We also group the vertices with very big out-strength or in-strength together and the components that do not appear in the interaction network are considered as vertices with zero out-strength and in-strength. The strength distributions of the original and simulated cascades match very well, indicating that the simulated cascades share similar features to the original cascades from an overall point of view. An obvious power law behavior can also be seen, which means that the failure

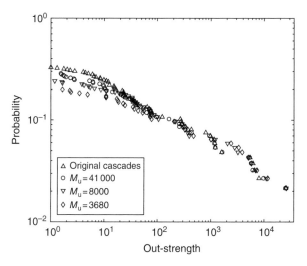

Figure 3.29 CCD of the out-strength for original and simulated cascades.

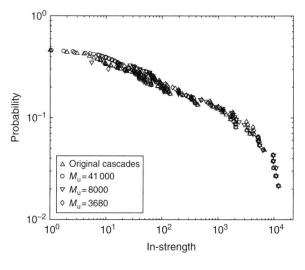

Figure 3.30 CCD of the in-strength for original and simulated cascades.

Table 3.17 Similarity indices for original and simulated cascades.

M_u	S_1 (SD)	S_2 (SD)	S_3 (SD)	S_4 (SD)	S_5 (SD)
41 000	0.994	0.994	0.973	0.974	1.01
	(0.00900)	(0.000619)	(0.0168)	(0.0194)	(0.0122)
8000	1.05	0.989	0.973	1.04	1.14
	(0.00570)	(0.00280)	(0.00930)	(0.0146)	(0.0379)
3680	1.07	0.987	0.944	1.03	1.17
	(0.00540)	(0.00240)	(0.00180)	(0.00690)	(0.0207)

of most vertices (components) have small consequences while a small number of them have much greater impact.

The five similarity indices defined in Section 3.4.4 for the quantified links from the simulated and the original 41 000 cascades are listed in Table 3.17. It is seen that all five indices are close to 1.0 and thus the links obtained from the original and simulated cascades are actually quite similar. The standard deviations (SDs) of the similarity indices for 20 times of simulation are listed in the parentheses and are all very small. The similarity indices between the quantified links from 8000 and 3680 original cascades and those from 41 000 cascades are listed in Table 3.13. By comparing the results in Table 3.13 and Table 3.17 we can see that the first four indices are only slightly different, but the fifth index significantly improves for simulating 41 000 cascades with the interaction model by using fewer original cascades to quantify interactions, indicating that the missing information for fewer original cascades can be recovered to a great extent by simulating more cascades with the interaction model, which has high-level probabilistic property and can reveal more rare events by performing a large number of simulations. It is also not surprising that using fewer original cascades cannot get as good results as using the whole 41 000 original cascades, since fewer original cascades will inevitably miss some information.

3.4.9 Cascading Failure Mitigation

The system-level failures of a complex system are actually due to the strong interaction or coupling of the components. Thus, cascading risks can be reduced by weakening some key links between component failures, which will possibly stop the propagation of cascading failures at an early stage. Here, weakening key links means reducing the corresponding element in the interaction matrix B.

After the interaction model in Section 3.4.3 is properly validated, it can be applied to study how the interactions between component failures influence the cascading failure risk and efficiently validate the effectiveness of the mitigation measures based on the weakening of key links.

In real systems the weakening of key links can be implemented by blocking some specific protective relays. A zone 3 relay blocking method called an adaptive distance relay scheme has been discussed in Lim et al. [54]. The relays are blocked under the condition of the tripping of the lines corresponding to the source vertices of the key links. Since the key links can cause a tremendous expected number of failures and thus play crucial roles in the propagation of cascading failures, it should be beneficial to the overall security of the system to stop the propagation from the source vertices of key links to the destination vertices by blocking the operation of the relay of the destination vertices, thus securing time for the operators to take remedial actions, such as redispatching the generation or even shedding some loads, and finally helping mitigate catastrophic failures.

This relay blocking strategy under the condition of some specific line tripping can be considered as a wide-area protection scheme, which can be simulated in the AC OPA model by adding a relay blocking module. Note that overloaded lines can be tripped not only by zone 3 relay but can also by other causes, such as tree flashover. The tripping of some lines in two significant outages in the western North America in 1996 [59], the U.S.-Canadian blackout on August 14, 2003 [60], and the outage in Italy on September 28, 2003 [61] can all be attributed to tree flashover to some extent, which has been discussed in Qi et al. [16]and Henneaux et al. [62].

To simulate the implementation of the mitigation strategy by weakening some links in real systems and also to compare with the results from the interaction model, we add a relay blocking module in the AC OPA model, which decreases the tripping probability of some overloaded lines and thus simulates the weakening of key links. When the line corresponding to the source vertex of a key link is tripped and further causes the overloading of the line corresponding to its destination vertex, the destination vertex line will be tripped with a reduced probability to simulate the part of the role played by blocking of its relay in preventing the line tripping. Due to the reduced tripping probability, AC OPA will probably go to its next inner iteration without tripping this destination vertex line; thus, AC optimal power flow (OPF) will be calculated and generation redispatching and load shedding will be performed to eliminate the overloading of the destination vertex line.

We assume that 90% of the tripping of overloaded lines is due to the operation of zone 3 relays. For the interaction model, the weakening of the key links is simulated by reducing the corresponding elements in the interaction matrix by 90%. For each of the three cases in which $M_u = 41\,000$, $M_u = 8000$, and $M_u = 3680$, the key links identified in Table 3.14 are weakened by reducing the corresponding elements in $\boldsymbol{B}(M_u)$ by 90%. By doing so, we get $\boldsymbol{B}_{int}(M_u)$. For comparison, the same number of randomly chosen links are also weakened in the same way and $B_{\mathrm{rand}}(M_{\mathrm{u}})$ is obtained. Cascading failures are separately simulated with the interaction model by using $B_{\mathrm{int}}(M_{\mathrm{u}})$ and $B_{\mathrm{rand}}(M_{\mathrm{u}})$. The two mitigation strategies are respectively called *intentional mitigation* and *random mitigation*. Each case under each mitigation strategy is simulated for 20 times.

Figure 3.31 shows the probability distributions of the total number of line outages under the two mitigation strategies for $M_{\mathrm{u}} = 41\,000$, $M_{\mathrm{u}} = 8000$, and $M_{\mathrm{u}} = 3680$. It is seen that the risk of large-scale cascading failures can be significantly mitigated by weakening only a small number of key links. By contrast, the mitigation effect is minor if the same number of randomly chosen links is weakened.

This can be explained by the power law distribution of the link weights. Most links have small weights and only a small number of links have much greater weights. When randomly weakening links, it is more possible to choose small-weight links, and thus the total weights of the weakened links for random mitigation can be significantly weaker than that for the intentional mitigation. The probabilities for exactly choosing the key links and thus getting the largest possible total link weights for $M_{\mathrm{u}} = 41\,000$, $M_{\mathrm{u}} = 8000$, and $M_{\mathrm{u}} = 3680$ are respectively as low as 3.10×10^{-29}, 1.91×10^{-24}, and 5.93×10^{-24}. The summation of the weights of weakened links for intentional and random mitigation are listed in Table 3.18, in which $m_{I_{c^{\mathrm{key}}}}$ and $\sigma_{I_{c^{\mathrm{key}}}}$ separately denote the average value and standard deviation of the total weights of the weakened links. The total

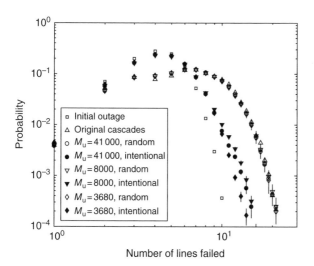

Figure 3.31 Probability distributions of the total number of line outages under two mitigation strategies. Triangles indicate total numbers of line outages of the original cascades under no mitigation; ○, ▽, and ◇ (●, ▼, and ◆) indicate total numbers of line outages of the simulated cascades under random (intentional) mitigation for using 41 000, 8000, and 3680 original cascades to quantify interactions; vertical lines indicate the standard deviations; squares indicate initial line outages.

weights of the weakened links for intentional mitigation is more than an order greater than those for random mitigation. The standard deviations of the total weights of the weakened links for random mitigation are big because the distribution of the link weights follows power law and the link weights can vary significantly. The big standard deviations cannot be decreased by increasing the number of simulations, which indicates that the random mitigation strategy is not stable and the mitigation effects between different random mitigation can be quite different.

To quantitatively compare the effects of different mitigation measures, the branching process is applied to estimate the average propagation of the original and simulated cascades under two mitigation strategies. The average value and standard deviation of the estimated average propagation are listed in Table 3.19. It is seen that the average propagation decreases dramatically under intentional mitigation but only a little under random mitigation. Also the relative standard deviations of the average propagation for random mitigation are much higher than those for intentional mitigation, which can be explained by the large standard deviation of the total weights of the weakened links for random mitigation.

Table 3.18 Link weights for different mitigation strategies.

Model	Mitigation strategy	M_u	$m_{l_{C^{key}}}$	$\sigma_{l_{C^{key}}}$
Interaction	Intentional	41 000	121 114	0
Interaction	Random	41 000	6126	6624
Interaction	Intentional	8000	23 380	0
Interaction	Random	8000	1737	1570
Interaction	Intentional	3680	11 247	0
Interaction	Random	3680	1043	696

Table 3.19 Average propagation for different mitigation strategies.

Model	Mitigation Strategy	M_u	M	$m_{\hat{\lambda}}$	$\sigma_{\hat{\lambda}}$
Interaction	Intentional	41 000	41 000	0.0966	0.000657
Interaction	Random	41 000	41 000	0.391	0.0128
Interaction	Intentional	8000	41 000	0.113	0.00110
Interaction	Random	8000	41 000	0.394	0.0154
Interaction	Intentional	3680	41 000	0.0965	0.000711
Interaction	Random	3680	41 000	0.391	0.0143

To simulate the implementation of the mitigation strategy by weakening some links in real systems, we add a relay blocking module in the AC OPA model. For intentional or random mitigation, when the source vertices of the predetermined links fail and the destination vertices of corresponding links become overloaded and will be tripped by protective relays, the probability of the operation of the relays is reduced by 90%. AC OPA will then have a much greater chance to go to the next inner iteration without tripping this overloaded destination vertex line, in which AC OPF is performed to simulate the redispatching of generation and some loads are shed if necessary to eliminate the violation of the line limits. In this way the AC OPA simulations can get cascades under mitigation strategies.

In Figure 3.32 we compare the probability distributions of the total numbers of the line outages for the AC OPA model and interaction model under intentional mitigation for M_u = 41 000, M_u = 8000, and M_u = 3680. Note that the identified key links for M_u = 41 000, M_u = 8000, and M_u = 3680 are separately denoted by $\mathcal{L}^{key}$(41 000), $\mathcal{L}^{key}$(8000), and $\mathcal{L}^{key}$(3680), and the AC OPA model simulation is performed three times, each of which respectively weakens $\mathcal{L}^{key}$(41 000), $\mathcal{L}^{key}$(8000), and $\mathcal{L}^{key}$(3680). It is seen that for all three cases using different M_u the distributions for both models match well under the intentional strategies. Results for random mitigation are similar and thus are not given.

The branching process is also applied to estimate the average propagations of the AC OPA and interaction model under intentional mitigation strategy; these

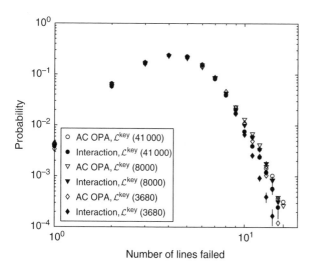

Figure 3.32 Probability distributions of the total number of line outages under intentional mitigation for AC OPA and interaction model simulations. ○, ▽, and ◇ indicate total numbers of line outages for AC OPA model simulation; ●, ▼, and ◆ indicate total number of line outages for simulated cascades using 41 000, 8000, and 3680 original cascades to quantify interactions; vertical lines indicate standard deviations.

Table 3.20 Average propagation for AC OPA and interaction model.

Model	Mitigation Strategy	M_u	M	$m_{\hat{\lambda}}$	$\sigma_{\hat{\lambda}}$
AC OPA	Intentional ($\mathcal{L}^{key}(41\,000)$)	–	41 000	0.0944	–
Interaction	Intentional ($\mathcal{L}^{key}(41\,000)$)	41 000	41 000	0.0966	0.000657
AC OPA	Intentional ($\mathcal{L}^{key}(8000)$)	–	41 000	0.109	–
Interaction	Intentional ($\mathcal{L}^{key}(8000)$)	8000	41 000	0.113	0.00110
AC OPA	Intentional ($\mathcal{L}^{key}(3680)$)	–	41 000	0.0998	–
Interaction	Intentional ($\mathcal{L}^{key}(3680)$)	3680	41 000	0.0965	0.000711

are listed in Table 3.20. The average propagations for two models match each other closely and the standard deviations $\sigma_{\hat{\lambda}}$ of the average propagation are very small, indicating that first quantifying the interaction by using fewer original cascades (such as 8000 or 3680 cascades) and then performing the interaction model simulation to get 41 000 cascades can get consistent and almost the same results as directly simulating 41 000 cascades with the AC OPA model.

3.4.10 Efficiency Improvement by Interaction Model

Both the AC OPA and the interaction model are implemented with Matlab and all tests are carried out on a 3.4 GHz Intel(R) Core(TM) i7-3770 based desktop. The timing for simulating 41 000 cascades for AC OPA simulation and the interaction model simulation based on different number of original cascades M_u is shown in Table 3.21, in which T_1, T, and T_2 are, respectively, the time for AC OPA simulation, for calculating the probability that each component fails in generation zero and the interaction matrix, and for interaction model simulation. We can see that it is more time efficient to first quantify the interactions between the component failures with $M_u \ll M$ original cascades and then perform the interaction model simulation than it is to directly simulate M cascades with the AC OPA model.

When many rounds of simulation need to be performed, the advantage of first quantifying the interactions and then doing the interaction model simulation will become more obvious. Assume it takes t_1 and t_2 to generate one

Table 3.21 Efficiency improvement of the interaction model.

Model	M_u	M	T_1 (hour)	T (second)	T_2 (second)
AC OPA	–	41 000	57	0	0
Interaction	41 000	41 000	57	95	29
Interaction	8000	41 000	11	15	29
Interaction	3680	41 000	5	7	29

cascade from the AC OPA model and the interaction model and we have $t_1 \gg t_2$. The time for calculating the probability that each component fails in generation zero and the interaction matrix by using M_u cascades is denoted by $T(M_u)$. To get N sets of M cascades, the ratio between the simulation time for the AC OPA model and that for first quantifying the interactions and then performing interaction model simulation is

$$R = \frac{NMt_1}{M_u t_1 + T(M_u) + NMt_2} \tag{3.49}$$

and by letting $NM \to \infty$ we can get

$$\lim_{NM \to \infty} R = \lim_{NM \to \infty} \frac{t_1}{\dfrac{M_u t_1}{NM} + \dfrac{T(M_u)}{NM} + t_2} = \frac{t_1}{t_2}, \tag{3.50}$$

which indicates a significant efficiency improvement for the interaction model simulations compared with the AC OPA simulations.

In our case $t_1 \approx 5.04$ and $t_2 \approx 0.0007$ seconds. Thus, $\lim_{NM} \to \infty R \approx 7200$. By letting $M = M^{\min} = 41\,000$, we show how R changes with N for $M_u = 41\,000$, $M_u = 8000$, and $M_u = 3680$ in Figure 3.33, in which we can see that R first quickly increases and then finally saturates at about 7200. Also as expected, when N is not large enough the efficiency improvement for $M_u = 3680$ is better than that for $M_u = 8000$ because smaller M_u will lead to shorter time for obtaining original cascades from AC OPA simulation and also shorter time for quantifying the interactions.

The efficiency improvement can be reflected in studying the effects of mitigation measures. The interaction model can generate cascades and study the influence of component interactions on cascading failure risk far more time efficiently while reserving most of the general properties of the cascades. In Section 3.4.9, we have shown that the simulated cascades from the interaction model are consistent with the AC OPA simulation under the same mitigation strategy. However, in order to obtain the cascades under a specific mitigation strategy, AC OPA simulation will need about 57 hours to obtain 41 000 cascades, while the interaction model simulation only requires about 29 seconds

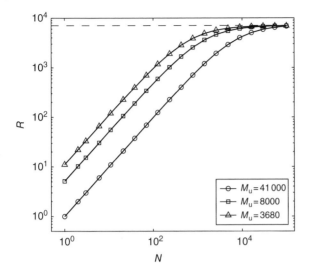

Figure 3.33 R for different N. Blue, green, and red dots indicate the efficiency improvement R for using 41 000, 8000, and 3680 original cascades. The dashed horizontal line indicates $R = 7200$.

to obtain the same number of cascades by simply changing some elements in the already obtained interaction matrix. This efficiency improvement is important if we would like to quickly find out the impact of a mitigation measure.

Further, the interaction model can be used for online decision-making support by rapidly predicting the consequences of the events happening in the system. For online operation there is not enough time to perform detailed cascading failure simulations. But we can obtain the interaction matrix offline from simulations of detailed cascading failure models or from statistical utility data and then apply the interaction model to quickly find out which components or areas of the system will most probably be affected; thus, a fast response can be performed to pull the system back to normal conditions and to avoid or at least reduce the economic and social losses. The efficiency improvement of the interaction model reflected in this aspect is a great advantage compared with many other cascading failure models.

Although in this book most results are based on the analysis of simulated data from detailed cascading failure models, the discussed approaches can be easily applied to real utility data. For example, in [30] the branching process has been applied to the transmission line outage data which include 10512 outages in the transmission availability data system (TADS) recorded by a North American utility over a period of 12.4 years [63]. The statistical utility data can be grouped into different cascades and then into different generations within each cascade based on the outages' timing. Similarly, the multitype branching process discussed in Section 3.3 and the failure interaction analysis discussed in Section 3.4 can also be readily applied to real utility data or even real data capturing cascading failures in systems other than power grids.

References

1 Bak, P., Tang, C., and Wiesenfeld, K. (1987). Self-organized criticality: an explanation of the 1/f noise. *Physical Review Letters* 59 (4): 381.
2 Bak, P., Tang, C., and Wiesenfeld, K. (1988). Self-organized criticality. *Physical Review A* 38 (1): 364.
3 Bak, P. and Tang, C. (1989). Earthquakes as a self-organized critical phenomenon. *Journal of Geophysical Research* 94 (15): 635–615.
4 Drossel, B. and Schwabl, F. (1992). Self-organized critical forest-fire model. *Physical Review Letters* 69 (11): 1629.
5 Goh, K.I., Lee, D.S., Kahng, B., and Kim, D. (2003). Sandpile on scale-free networks. *Physical Review Letters* 91 (14): 148701.
6 Izmailian, N.S., Papoyan, V.V., Priezzhev, V., and Hu, C.K. (2007). Self-organizing behavior in a lattice model for co-evolution of virus and immune systems. *Physical Review E* 75 (4): 041104.
7 Athreya, K.B. and Ney, P.E. (2012). *Branching Processes*, Vol. 196. Springer Science & Business Media.
8 Harris, T.E. (2002). *The Theory of Branching Processes*. Courier Corporation.
9 Kim, J., Wierzbicki, K.R., Dobson, I., and Hardiman, R.C. (2012). Estimating propagation and distribution of load shed in simulations of cascading blackouts. *IEEE Systems Journal* 6 (3): 548–557.
10 Qi, J., Dobson, I., and Mei, S. (2013). Towards estimating the statistics of simulated cascades of outages with branching processes. *IEEE Transactions on Power Systems* 28 (3): 3410–3419.
11 Qi, J., Ju, W., and Sun, K. (2017). Estimating the propagation of interdependent cascading outages with multi-type branching processes. *IEEE Transactions on Power Systems* 32 (2): 1212–1223.
12 Qi, J., Sun, K., and Mei, S. (2015). An interaction model for simulation and mitigation of cascading failures. *IEEE Transactions on Power Systems* 30 (2): 804–819.
13 Perrow, C. (2011). *Normal Accidents: Living with High Risk Technologies*. Princeton University Press.
14 Carreras, B.A., Lynch, V.E., Dobson, I., and Newman, D.E. (2002). Critical points and transitions in an electric power transmission model for cascading failure blackouts. *Chaos: An Interdisciplinary Journal of Nonlinear Science* 12 (4): 985–994.
15 Dobson, I., Carreras, B.A., Lynch, V.E., and Newman, D.E. (2007). Complex systems analysis of series of blackouts: cascading failure, critical points, and

self-organization. *Chaos: An Interdisciplinary Journal of Nonlinear Science* 17 (2): 026103.

16 Qi, J., Mei, S., and Liu, F. (2013). Blackout model considering slow process. *IEEE Transactions on Power Systems* 28 (3): 3274–3282.

17 Pastor-Satorras, R., Vázquez, A., and Vespignani, A. (2001). Dynamical and correlation properties of the internet. *Physical Review Letters* 87 (25): 258701.

18 Dorogovtsev, S. and Mendes, J. (2003). *Evolution of Networks: From Biological Nets to the Internet and WWW*. Oxford University Press.

19 Stanley, H. and Mantegna, R. (2000). *An Introduction to Econophysics*. Cambridge: Cambridge University Press.

20 Mei, S., Zhang, X., and Cao, M. (2011). *Power Grid Complexity*. Springer Science & Business Media.

21 Olami, Z., Feder, H.J.S., and Christensen, K. (1992). Self-organized criticality in a continuous, nonconservative cellular automaton modeling earthquakes. *Physical Review Letters* 68 (8): 1244.

22 Bröker, H.M. and Grassberger, P. (1997). Random neighbor theory of the Olami–Feder–Christensen earthquake model. *Physical Review E* 56 (4): 3944.

23 d'Humieres, D., Lallemand, P., and Frisch, U. (1986). Lattice gas models for 3D hydrodynamics. *EPL (Europhysics Letters)* 2 (4): 291.

24 Carreras, B.A., Newman, D.E., Dobson, I., and Poole, A.B. (2004). Evidence for self-organized criticality in a time series of electric power system blackouts. *IEEE Transaction on Circuits Systems -I, Regular Papers* 51 (9): 1733–1740.

25 Weng, X., Hong, Y., Xue, A., and Mei, S. (2006). Failure analysis on China power grid based on power law. *Journal of Control Theory and Applications* 4 (3): 235–238.

26 Dobson, I., Carreras, B.A., Lynch, V.E., and Newman, D.E. (2001). An initial model for complex dynamics in electric power system blackouts. In: *Proceedings of the Annual Hawaii International Conference on System Sciences*, Citeseer, p. 51.

27 Mei, S.W., Weng, X.F., Xue, A.C. et al. (2006). Blackout model based on OPF and its self-organized criticality. In: *2006 Chinese Control Conference*, IEEE, 1673–1678.

28 Mei, S., Ni, Y., Wang, G., and Wu, S. (2008). A study of self-organized criticality of power system under cascading failures based on AC-OPF with voltage stability margin. *IEEE Transactions on Power Systems* 23 (4): 1719–1726.

29 Dobson, I., Kim, J., and Wierzbicki, K.R. (2010). Testing branching process estimators of cascading failure with data from a simulation of transmission line outages. *Risk Analysis* 30 (4): 650–662.

30 Dobson, I. (2012). Estimating the propagation and extent of cascading line outages from utility data with a branching process. *IEEE Transactions on Power Systems* 27 (4): 2146–2155.

31 Guttorp, P. (1991). *Statistical Inference for Branching Processes*, Vol. 122. Wiley-Interscience.

32 Dion, J.P. (1975). Estimation of the variance of a branching process. *The Annals of Statistics* 1183–1187.

33 Janev, N.M. (1976). On the statistics of branching processes. *Theory of Probability & Its Applications* 20 (3): 612–622.

34 Ren, H. and Dobson, I. (2008). Using transmission line outage data to estimate cascading failure propagation in an electric power system. *IEEE Transactions on Circuits and Systems* 55 (9): 927–931.

35 Mei, S., He, F., Zhang, X. et al. (2009). An improved OPA model and blackout risk assessment. *IEEE Transactions on Power Systems* 24 (2): 814–823.

36 Buldyrev, S.V., Parshani, R., Paul, G. et al. (2010). Catastrophic cascade of failures in interdependent networks. *Nature* 464 (7291): 1025–1028.

37 Parandehgheibi, M., Modiano, E., and Hay, D. (2014) Mitigating cascading failures in interdependent power grids and communication networks. In *2014 IEEE International Conference on Smart Grid Communications (SmartGridComm)*, IEEE, pp. 242–247.

38 Shahidehpour, M., Fu, Y., and Wiedman, T. (2005). Impact of natural gas infrastructure on electric power systems. *Proceedings of the IEEE* 93 (5): 1042–1056.

39 Li, T., Eremia, M., and Shahidehpour, M. (2008). Interdependency of natural gas network and power system security. *IEEE Transactions on Power Systems* 23 (4): 1817–1824.

40 Adachi, T. and Ellingwood, B.R. (2008). Serviceability of earthquake-damaged water systems: effects of electrical power availability and power backup systems on system vulnerability. *Reliability Engineering & System Safety* 93 (1): 78–88.

41 Amin, M. (2000). Toward self-healing infrastructure systems. *Computer* 33 (8): 44–53.

42 Rinaldi, S.M., Peerenboom, J.P., and Kelly, T.K. (2001). Identifying, understanding, and analyzing critical infrastructure interdependencies. *IEEE Control Systems* 21 (6): 11–25.

43 Maaouia, F. and Touati, A. (2005). Identification of multitype branching processes. *The Annals of Statistics* 33 (6): 2655–2694.

44 Dempster, A.P., Laird, N.M., and Rubin, D.B. (1977). Maximum likelihood from incomplete data via the EM algorithm. *Journal of the Royal Statistical Society. Series B (Methodological)* 39 (1): 1–38.

45 Good, I.J. (1960). Generalizations to several variables of Lagrange's expansion, with applications to stochastic processes. In: *Mathematical Proceedings of the Cambridge Philosophical Society*, Vol. 56, 367–380, Cambridge University Press.

46 Ju, W., Qi, J., and Sun, K. (2015). Simulation and analysis of cascading failures on an NPCC power system test bed. In *2015 IEEE Power Energy Society General Meeting*, 1–5.

47 Ju, W., Sun, K., and Qi, J. (2017). Multi-layer interaction graph for analysis and mitigation of cascading outages. *IEEE Journal on Emerging and Selected Topics in Circuits and Systems* 7 (2): 239–249.

48 Hines, P.D.H., Dobson, I., Cotilla-Sanchez, E., and Eppstein, M. (2013). "Dual graph" and "random chemistry" methods for cascading failure analysis. In *2013 46th Hawaii International Conference on System Sciences*, 2141–2150.

49 Hines, P., Dobson, I., and Rezaei, P. (2017). Cascading power outages propagate locally in an influence graph that is not the actual grid topology. *IEEE Transactions on Power Systems* 32(2): 958–967.

50 Begovic, M., Novosel, D., Karlsson, D. et al. (2005). Wide-area protection and emergency control. *Proceedings of the IEEE* 93 (5): 876–891.

51 Terzija, V., Valverde, G., Cai, D. et al. (2011). Wide-area monitoring, protection, and control of future electric power networks. *Proceedings of the IEEE* 99 (1): 80–93.

52 Sun, K., Zheng, D.Z., and Lu, Q. (2003). Splitting strategies for islanding operation of large-scale power systems using OBDD-based methods. *IEEE Transactions on Power Systems* 18 (2): 912–923.

53 Sun, K., Hur, K., and Zhang, P. (2011). A new unified scheme for controlled power system separation using synchronized phasor measurements. *IEEE Transactions on Power Systems* 26 (3): 1544–1554.

54 Lim, S.I., Liu, C.C., Lee, S.J. et al. (2008). Blocking of zone 3 relays to prevent cascaded events. *IEEE Transactions on Power Systems* 23 (2): 747–754.

55 Qi, J., Wang, J., and Sun, K. (2018). Efficient estimation of component interactions for cascading failure analysis by EM algorithm. *IEEE Transactions on Power Systems* 33(3): 3153–3161.

56 Watts, D.J. and Strogatz, S.H. (1998). Collective dynamics of 'small-world' networks. *Nature* 393 (6684): 440–442.

57 Barabási, A.L. and Albert, R. (1999). Emergence of scaling in random networks. *Science* 286 (5439): 509–512.

58 Hines, P., Cotilla-Sanchez, E., and Blumsack, S. (2010). Do topological models provide good information about electricity infrastructure vulnerability? *Chaos: An Interdisciplinary Journal of Nonlinear Science* 20 (3): 033122.

59 North America Electric Reliability Council (2002). 1996 System Disturbances.

60 U.S.-Canada Power System Outage Task Force (2004). Final report on the August 14th blackout in the United States and Canada.

61 Union for the Co-ordination of Transmission of Electricity (UCTE) (2003). Interim Report of the Investigation Committee on the 28 September 2003 Blackout in Italy.

62 Henneaux, P., Labeau, P.E., and Maun, J.C. (2013). Blackout probabilistic risk assessment and thermal effects: impacts of changes in generation. *IEEE Transactions on Power Systems* 28 (4): 4722–4731.

63 Bonneville Power Administration Transmission Services Operations & Reliability. [Online]. Available: http://transmission.bpa.gov/Business/ Operations/Outages.

4

Strategies for Controlled System Separation

4.1 Questions to Answer

When cascading failures propagate in a power system to cause overloads of transmission lines and out-of-step conditions of generators, some protective relays may react to trip lines and generators. These local protective actions, while saving their protected equipment, may gradually worsen the situation due to lack of coordination at the system level and eventually result in collapse of the transmission network into electrical islands.

When loss of integrity of the transmission network becomes unavoidable, controlled system separation (CSS) as introduced in Chapter 1 is considered the final resort to save the system against widespread power outages. When CSS is performed, the control center proactively separates the transmission network into sustainable islands in a controlled manner. Each island is strategically formed with stable, matched generation that continues to support its load through a sub-system within the island. This is a more advisable operation than waiting for the network to collapse by itself in an unpredictable way. As a result of CSS, generated electricity of the system continues to be delivered to most of customers of the system in parallel islands although the network separates. In other words, a power blackout is effectively prevented.

Once all failures are corrected, the whole system can be restored by resynchronization of those electrical islands. Such an action of system resynchronization is usually easier than a black-start process because most of loads have been saved by CSS. In comparison, a black-start process has to restart the portion of the system in the power outage area following a time-consuming procedure. As described later in Chapter 6, the procedure's steps include starting black-start generating units to crank other non-black-start units, energizing transmission lines, and picking up critical and noncritical loads, which usually takes several hours.

Power System Control Under Cascading Failures: Understanding, Mitigation, and System Restoration, First Edition. Kai Sun, Yunhe Hou, Wei Sun and Junjian Qi.
© 2019 John Wiley & Sons Ltd. Published 2019 by John Wiley & Sons Ltd.
Companion website: www.wiley.com/go/sun/cascade

Studies on CSS need to determine when and where to separate the power grid in order to mitigate out-of-step conditions. Even with the timing and locations of separation determined, CSS still needs to be performed by a sophisticated scheme to successfully stop cascading failures with the minimum loss of loads. For example, generation rejection in a generation-rich island or load shedding in a load-rich island may be included in the scheme of CSS to maintain a power balance and stabilize generators. Therefore, a CSS scheme needs to answer three important questions:

- Where to separate?
- When to separate?
- How to separate?

The first question is about the locations of separation, that is, which transmission lines to open in order to form sustainable electrical islands. The set of lines to be opened is actually a cut set to partition the network and should satisfy the constraints critical to survival of islands, such as generation-load balance and transmission capacity limits in each island.

The second question concerns the timing of separation on a set of determined locations of separation. We do not want to open the lines at those locations too late to miss the best timing of separating the generators that are going to lose synchronism. Under extreme events such as cascading failures, the system condition may change any time in an unpredictable manner. For instance, new unstable generators or overloaded lines may occur in an electrical island. Generally speaking, the sooner we disconnect electrical islands at the determined locations, the more likely we will mitigate cascading failures and save the system.

The last question considers implementation issues associated with CSS, including what hardware devices are supposed to execute separation, how to coordinate them at multiple locations, and what additional control actions to take after CSS. Those control actions include necessary generation rejection or load shedding in some of the electrical islands formed.

Note that the three questions "where," "when," and "how" are coupled with each other. However, they are rarely studied together under one framework. Most existing studies focus on solving either "where" or "when" independently and give less consideration to question "how."

4.2 Literature Review

Among the three questions, "where" is the most studied in literature. Sun et al. and Zhao et al. [1–3] apply graph theory to model and simplify a power network and formulate a problem on this question: finding separation points addressing several criteria about CSS. Here, the criteria include generator coherency, generation-load balance, and transmission capacity constraints in each island. Then, ordered binary decision diagram (OBDD)-based methods are introduced to obtain feasible separation points. Further simulation studies and implementation strategies on OBDD-based methods are presented in [4, 5]. The graph-theory and OBDD-based methods will be introduced in detail in this chapter. Questions "when" and "how" related to the OBDD-based methods are studied by the two following papers: Jin et al. [6] proposes an online CSS scheme for formation of stable islands that integrates an OBDD-based method for determination of separation points with a single-machine-equivalent-based technique – such as the Extended Equal Area Criterion (EEAC) [7] and SIngle Machine Equivalent (SIME) method [8] – for transient stability assessment; Sun et al. [9] proposes a real-time event-based separation scheme that identifies a separation strategy matching best the current situation from a table of CSS strategies maintained online based on simulations of critical contingencies.

Four papers [10–13] present slow-coherency-based CSS techniques, which first identify the dynamically weakest connections between groups of coherent generators using a slow-coherency-based grouping algorithm and then isolate each group of generators into one island with matched loads. Intuitively, without a rigorous proof, generators in each island formed may easily maintain synchronism and stability. Accordingly, separation points between islands are optimized by either an exhaustive search or minimum net flow method. Xu and others [14, 15] apply a spanning tree algorithm to find separation points for slow-coherency-based CSS and conduct tests on North American Western Interconnection power system models. Graph theory-based spectral clustering and multilevel partitioning techniques are utilized in several papers [16–23] to answer the question "where."

A variety of heuristic searching algorithms are used by researchers [24–29], such as the Particle Swarm Optimization, Genetic Algorithm, Ant Search, and Tabu Search, to find separation points. Mixed Integer Linear Programming is utilized in Trodden et al. and Ding et al. [30–33] to optimize separation points. There are also data driven methods answering question "where" such as in Wang et al. and Raak et al. [34, 35].

Of these studies answering the question "where," only a few also address the questions "when" or "how" [6, 9, 11, 15, 23]. Out-of-step relays placed in advance on tie lines between control areas are traditionally applied to separate two out-of-step control areas once detecting an out-of-step condition from power swings [36]. Some papers suggest more sophisticated methods addressing these two questions. For instance, Li et al. [23] calculates the maximal Lyapunov exponent of power swing measured by synchrophasors to predict an out-of-step condition and accordingly decide the timing of separation. Diao et al. and Senroy et al. [37, 38] propose feeding real-time synchrophasor measurements to data mining tools such as Decision Trees, which are trained beforehand using offline data to decide when formation of any predesigned island becomes needed. To stabilize each island formed after system separation, which is related to the question "how," load-shedding schemes for postseparation conditions are studied [10, 39], where the rate of change of frequency is used as an important index to decide the amount of load to shed. Finally, out-of-step protection systems will be the key hardware devices to perform the final action of CSS on determined separation points at the right timing.

The rest of this chapter will focus on formulating and solving the problem regarding the question "where". In Chapter 5, online techniques addressing three questions will be introduced, and then a unified CSS scheme proposed in Sun et al. [40] will be presented, which for the first time suggests a paradigm addressing the questions "where", "when", and "how" in one framework.

4.3   Constraints on Separation Points

To answer the question "where," we must determine a separation strategy that specifies a set of separation points, that is, transmission lines, to disconnect the power transmission network. Such a set of separation points can be either an optimal solution or an acceptable solution. For the former, the islands formed by disconnection at those points should enable optimization (local or global) of a defined objective function; for the latter, the islands formed need to satisfy all given constraints. Correspondingly, answering of this question can be formulated as either an optimization problem or a satisfiability checking problem. The resulting strategies respectively correspond to an optimal separation strategy and an acceptable separation strategy. In the following, we first discuss what constraints or criteria the separation points for CSS should meet. Then, two ways of formulating this problem in mathematics will be introduced.

In a power system, all synchronous generators must be in synchronous operations; that is, the electrical speed of each generator is basically the same as the average electrical speed or angular frequency of the entire system. Also, the total generation of generators in operations should meet the total load of the system so that the system frequency is maintained at the rated frequency within an acceptable tolerance. Another consideration is that the transmission lines should not be loaded so heavily as to violate their transmission capacity limits. Since each electrical island needs to be operated independently as an isolated power system, these dynamic and steady-state criteria have to be satisfied. Therefore, a properly selected set of separation points should avoid, or at least minimize, violations of those criteria in every island. If violation is inevitable, it should be mitigated by the least amount of control.

Traditionally, out-of-step relays are installed by power companies on predetermined lines based on offline power system stability and protection studies.

Those fixed separation points always separate the system at the same set of locations whenever separation is required. But in fact, fixed separation points do not adapt themselves well to variations in either the system operating condition or the disturbance that causes an out-of-step condition.

In practice, a set of fixed separation points may be designed to form generation-load balanced islands for a typical operating condition as the reference condition. However, if a new operating condition's power flow profile is significantly different from that of the reference condition, the set of separation points may not be applicable anymore. For example, consider a case of CSS forming two electrical islands, No. 1 and No. 2. The region of Island 1 may have a load much higher than its generation if economic dispatch makes it import a large amount of generation from the region of Island 2. If the two islands are formed as planned, a large amount of load would have to be shed to mitigate underfrequency issues in the load-rich Island 1, while Island 2 would have to reject about the same amount of generation to avoid overfrequency issues. That can definitely be avoided if separation points are adjusted according to the power exchange and load shift between two islands. One approach could be to let separation points penetrate the load-rich Island 1 to relieve its imbalance while reducing the surplus of generation in Island 2.

Moreover, the disturbances triggering cascading failures may differ case by case. A fixed set of separation points is not adaptive to the ways in which cascading failures happen and propagate. First, the location where cascading failures initiate is not easy to predict in the offline planning stage. If the network is disconnected at fixed separation points, the generation-load balance of an electrical island formed is hard to maintain. Second, the pattern in which generators become out-of-step may change due to either the operating condition or the disturbance. If the power system is separated always at fixed separation points, some generators being out-of-step may be assigned into one island and then become unstable. To stabilize such an island, unstable generators may have to be tripped with consequent load shedding to maintain the frequency and voltage levels of the island. Comparatively, the separation points that are adjustable to match the actual out-of-step mode can reduce the chances of unstable generators and consequently reduce loss of generation.

To overcome the disadvantages from fixed separation points, a more advisable method is to use adaptive separation points. For instance, an approach is to install redundant separation devices at potential separation points identified by planning studies, and then the control center or the central controller of CSS may select an optimal set of separation devices to disconnect the network while blocking the others. In comparison with fixed separation points, adaptive separation points have better adaptability to changes in the system condition and disturbance so as to limit the generation-load imbalance in each formed island and the size of power outage. However, a drawback is that redundant separation devices are required, which means additional installation and maintenance costs. Also, it requires a real-time algorithm to optimize the set of separation points and send communication signals including tripping signals and blocking signals to all separation devices at the right timing.

In order to determine separation points, either fixed or adaptive, the following constraints need to be satisfied towards creation of stable and sustainable islands. Namely, the islands formed need minimum additional actions to stabilize generators and secure frequency and voltage levels.

- *Generation coherency constraint* (GCC): All generators after being separated into each island are stable or, in other words, can keep their synchronism. This constraint addresses the dynamic performance of each formed island while the rest of the constraints regard steady-state conditions. For most cases when CSS becomes needed, generators are oscillating and some generators may potentially become out of step from the rest of generators. The patterns in which generators group together need to be considered during the determination of island boundaries; that is, separation points. If all generators in one island belong to one coherent group of generation before the island is formed, they have a better chance of being stable spontaneously or are easily stabilized.
- *Power balance constraint* (PBC): Generation-load imbalance in each island is minimized or limited under a predetermined threshold. This constraint ensures that the supply and demand in every island formed are approximately equal within a tolerance not causing its frequency and bus voltages to drift out of acceptable ranges. In practice, the frequency or the balance in real power is more critical for each island to maintain while the bus voltages or reactive powers can be controlled using local compensation devices after the formation of the island.
- *Transmission capacity constraint* (TCC): Transmission capacity limits in each island cannot be violated. This constraint is to avoid further overloads and trips of transmission lines after islands are formed.

The first constraint, that is, the generation coherency constraint, is the most critical one because it maximizes the amount of generation of each island by avoiding unstable or out-of-step generators after CSS. According to the size of generation, a matched amount of load can be allocated. Any CSS that fails to meet the generation coherency constraint may result in unstable generators in some islands. Trips of these generators can cause a reduction of generation and a surplus of load in the island to break their balance, that is, violation of the power balance constraint. Other consequences of loss of generation are changes in the power flow profile to cause overloading of some lines, that is, violation of the transmission capacity constraint.

The satisfaction to the power balance constraint is mainly to define and enforce a threshold of real power imbalance when allocating balanced generation and load. Also, sufficient reactive power reserves need to be planned in each potential island. Here, "generation" is composed of actual generation and spinning reserve that can quickly turn into true generation to balance the surplus of load and support the frequency; "load" may contain dispatchable loads that can flexibly be shed to match generation. Reactive power reserves include shunt capacitors, reactors, and dynamic VAR sources such as SVCs (static var compensators) and STATCOMs (static synchronous compensators), which can control voltages of the island formed. The frequency and voltage levels of each island will be indicators of the imbalances in real and reactive powers, so frequency and voltage criteria should be designed for each potential island to trigger the aforementioned frequency and voltage control using spinning reserves, dispatchable loads, and reactive power reserves.

If the transmission capacity constraint is violated in an island formed, or in other words, some transmission line is overloaded, such a line will further be tripped, and that may cause cascading failures or unintentional system separation to continue in the island. Because the network in an island often becomes more vulnerable than it has been in the original system, any additional line outage may be fatal to the survival of the island.

In addition to those three constraints, other constraints may be added, such as

- *Small disruption constraint*: When the system is separated, it is better to avoid cutting off branches that are carrying large power flows. Otherwise, the power disruptions on these branches will result in significant power flow redistribution in the network. Hence, that may cause large impacts on the stability of the resulting sub-system isolated to an island as well as the rest of the system.
- *Minimum-cut constraint*: The set of lines to disconnect should be the minimum cut set of the network toward the desired island formation. Namely, we need to avoid cutting any additional line that does not contribute to separation of the targeted islands from each other.

With these constraints defined, the problem of where to separate the system can be formulated in two ways as follows:

- Finding one or multiple *acceptable solutions* that satisfy specific constraints from those mentioned. Usually the generation coherency, power balance, and transmission capacity constraints are the constraints that have to be satisfied. More specifically for the power balance constraint, the threshold of the imbalance in each island needs to be predefined based on the acceptable level of loss of load after system separation.
- Finding the *optimal solution* that minimizes the generation-load imbalance (i.e., satisfaction to the power balance constraint) in each island and satisfies the other constraints.

In fact, the first problem is a relaxation of the second problem because if solutions to the first problem exist, the optimal solution of the second problem is necessarily a solution of the first problem. In case the predefined threshold on the generation-load imbalance is too small for an acceptable solution to exist, the optimal solution of the second problem still provides the best candidate strategy in terms of minimization of the generation-load imbalance. From the computational complexity point of view, both problems fall into a class of most complex problems, that is, NP-hard problems, which will be explained later.

The constraints, except for the generation coherency constraint, are in fact steady-state constraints. In practice, given a set of separation points, system planners can more easily check these steady-state constraints. Comparatively, the verification of the generation coherency constraint involves the observation on the dynamics of generators and prediction of stability and synchronism of generators under the postseparation condition, so checking this constraint is related to answering a more difficult question on power system dynamics.

Besides these constraints, other steady-state or dynamics constraints can be defined. For example, another constraint can require that all islands should easily be resynchronized in the system restoration stage. Here, easily can be interpreted based on a specific power restoration plan.

4.4 Graph Models of a Power Network

In this section, graph models of a power network are introduced as mathematical tools for analyzing the answers of question "where." In fact, that question is to determine a set of separation points satisfying given constraints. The difficulty of such a problem depends on the scale and topological complexity of the power network, or in other words, it depends on the dimensions of the space where to search for possible separation points. Even if given a certain set of separation points, answering the question may still be challenging because the verification of all given constraints within a desired short time could be complicated. Thus, a key factor in deciding the difficulty of the problem is how many sets of possible separation points exist for selection or optimization.

A real-life interconnected power system is conventionally planned and operated as multiple predefined control areas connected by tie lines. Therefore, for most cases, the selection of separation points may give high priorities to tie lines to largely reduce the dimension of the search space, and the resulting electrical islands may be some of control areas being isolated. However, consider another case, which is a control area rich in load with insufficient generation to be self-balanced after being isolated. Power flows on its tie lines will be vitally important for its load to survive. If such an area is islanded, it has to connect with generation from one or multiple neighboring control areas to support its excessive load. For such a case, separation points need to penetrate its neighboring areas that can provide the required generation. Thus, the problem may significantly increase its difficulty, especially when its neighboring areas have topologically meshed networks. Also, islanding a generation-rich control area has similar concerns.

This analysis uncovers a fact that the difficulty of addressing the question "where" largely depends on the number of lines being considered as candidate separation points rather than the total number of lines in the system. Therefore, we need to use mathematical tools to model a power network and reduce the components that are irrelevant to the selection of separation points.

Using graph theory in mathematics, the question "where" can be translated into a graph partitioning problem. A graph is made up of nodes (also called vertices) connected by edges (also called arcs) and may represent a network system such as a power network, where buses are modeled by nodes, and branches (including lines and transformers) are modeled by edges in the graph. Depending on what characteristics of the physical network system are modeled, a graph may be either undirected, meaning that there is no distinction between the two nodes associated with each edge, or directed if each edge is directed from one node to the other.

In the following, two types of graph models are introduced to represent a power network: an undirected node-weighted graph and a directed edge-weighted graph.

4.4.1 Undirected Node-Weighted Graph

A power network can be represented by an undirected node-weighted graph G, which can be written as a 3-tuple $G(V, E, W)$. Here, $V = \{v_1, ..., v_n\}$ is the set of n nodes and each node v_i represents a bus in the power network, and $W = \{w_1, ..., w_n\}$ is the set of the weights of all nodes and will be explained next. E is the set of edges, and each element connects two nodes, representing a branch, that is, a transmission line or transformer, between two buses of the power network. We use e_{ij} to denote the edge between nodes v_i and v_j. Since each edge is undirected, e_{ji} is actually the identical edge. In the following, if $i < j$, we only use e_{ij}, that is, pair (v_i, v_j), as the notation of an edge between the two nodes, for example, e_{12} instead of e_{21}. If buses i and j are connected by multiple branches, for the convenience of analysis for CSS, we use only one edge e_{ij} to represent an aggregated equivalent branch of all these multiple branches. Thus, opening e_{ij} means disconnection of all corresponding branches between nodes v_i and v_j. For the buses that directly connect to at least one generator, they are generator buses and their corresponding nodes in the graph are called generator nodes. The other nodes are called load nodes.

Each node v_i is assigned with a weight w_i, which is equal to the net power injected to the network from all devices (generators or loads) connected to bus i.

If only real power is concerned in the power balance constraint, its weight w_i can be defined by

$$w_i = P_{G,i} - P_{L,i}, \tag{4.1}$$

where $P_{G,i}$ and $P_{L,i}$ are, respectively, the total real power generation and real power load at bus i. Note that for a generator node, $w_i > 0$ only when $P_{G,i} > P_{L,i}$, that is, the bus having surplus generation contributed to the network. Therefore, in the graph model, a generator node unnecessarily has a positive weight.

If the total real power generation and load of the power network represented by the graph are balanced, ideally w_i should satisfy Eq. (4.2) for a lossless power network.

$$\sum_i w_i = 0 \tag{4.2}$$

If network power losses are considered, we may alternatively define node weights as follows to still make Eq. (4.2) satisfied.

First, estimate the total power loss P_{loss} by

$$P_{loss} = \sum_i P_i = \sum_i \left(P_{G,i} - P_{L,i} \right). \tag{4.3}$$

Identify the buses with positive injections, that is, $P_i > 0$, to the network and subtract a portion of the total power loss from each of those buses that is proportional to its injection. Thus, the weight of node v_i is redefined as

$$w_i = \begin{cases} P_i \cdot \left(1 - \dfrac{P_{loss}}{P_+} \right), & \text{if } P_i > 0 \\ P_i, & \text{if } P_i \leq 0 \end{cases} \quad \text{where } P_+ = \sum_{k, P_k > 0} P_k. \tag{4.4}$$

It can easily be proved that the node weights defined by Eq. (4.4) satisfy Eq. (4.2).

Alternatively, we may identify the buses having positive absorption, that is, $P_i < 0$, to the network and add a portion of the total power loss to each of those buses that is proportional to its absorption. For this case, the weight of node v_i is redefined as

$$w_i = \begin{cases} P_i, & \text{if } P_i \geq 0 \\ P_i \cdot \left(1 + \dfrac{P_{loss}}{P_-} \right) & \text{if } P_i < 0 \end{cases} \quad \text{where } P_- = \sum_{k, P_k < 0} |P_k|. \tag{4.5}$$

It also satisfies Eq. (4.2). These two definitions are actually equivalent.

Let V_G be the set of generator nodes and V_L be the set of load nodes. We have

$$V = V_G \cup V_L. \tag{4.6}$$

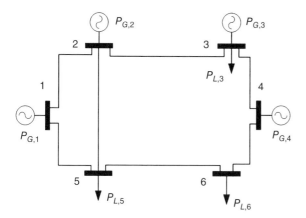

Figure 4.1 Six-bus power system.

Assume that when some faults occur, generators separate into K coherent groups. Let $V_{G,1}, ..., V_{G,K}$ denote the node sets that correspond to these groups of generator nodes. Then, we have

$$V_G = V_{G,1} \cup V_{G,2} \cup \cdots \cup V_{G,K}. \tag{4.7}$$

If we separate these groups of generators with matched loads from the rest of the system in a proper way, the generation coherency constraint and power balance constraint may be satisfied. This provides a graph model for studying CSS of a power network.

Consider a six-bus power system as shown in Figure 4.1. Its node-weighted graph model is given in Figure 4.2. Its graph has six nodes and seven edges. The white nodes denote the generator nodes and black nodes denote load nodes. There are $V_G = \{v_1, v_2, v_3, v_4\}$ and $V_L = \{v_5, v_6\}$. The weights of six nodes are calculated by Eq. (4.1) if losses are ignored.

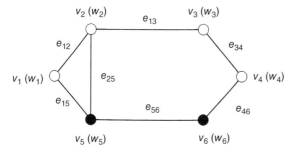

Figure 4.2 Node-weighted graph model of the six-bus power system.

4.4.2 Directed Edge-Weighted Graph

Alternatively, a power network can also be formulated using a directed, edge-weighted graph $G(V, E, W)$. V and E are still the node set and edge set, respectively, representing the buses and branches, but W here is the set of the weights of all edges. Unlike the undirected graph, each edge e_{ij} in E corresponds to an ordered pair of nodes (i, j), and the ordering of i and j indicates the direction of the real power flow of the branch connecting buses i and j. Thus, weight w_{ij} of edge e_{ij} is calculated by

$$w_{ij} = \frac{|P_{ij}| + |P_{ji}|}{2} = \frac{|P_{ij} - P_{ji}|}{2}. \tag{4.8}$$

Here P_{ij} and P_{ji} are the real power flows in the branch measured respectively from its two ends, bus i and bus j. If the power loss is ignored, $P_{ji} = -P_{ij}$ and hence $w_{ij} = |P_{ij}|$.

We may also include $w_{ji} = -w_{ij}$ into the set W of edge weights; that is, each edge is associated with a pair of opposite weights. The positive weight has its subscripts i and j ordered in the same direction as that of the actual real power flow. To model a power network, we may use either one to represent the edge weight as long as the "from" bus and "to" bus are specified.

For the six-bus power system in Figure 4.1, Figure 4.3 depicts its edge-weight graph model assuming certain directions of line power flows as indicated by arrows. In each "( )," only the positive edge weight is given, indicating the actual direction of the real power flow in the line.

For this six-bus power system, if we ignore power losses on transmission lines and consider the following operating condition in per unit with the power base equal to 100 MVA:

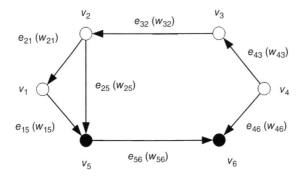

Figure 4.3 Edge-weighted graph model of the six-bus power system.

$$P_{G,1} = 0.20, \qquad P_{G,2} = 0.30, \qquad P_{G,3} = 0.20, \qquad P_{L,3} = 0.10,$$

$$P_{G,4} = 0.30, \qquad P_{L,5} = 0.50, \qquad P_{L,6} = 0.40,$$

$$P_{21} = 0.10, \qquad P_{15} = 0.30, \qquad P_{25} = 0.45, \qquad P_{32} = 0.25,$$

$$P_{43} = 0.15, \qquad P_{46} = 0.15, \qquad P_{56} = 0.25.$$

The node-weighted graph model has the node set $V = \{v_1, v_2, v_3, v_4, v_5, v_6\}$, edge set $E = \{e_{12}, e_{15}, e_{23}, e_{25}, e_{34}, e_{46}, e_{56}\}$ and $W = \{w_1, w_2, w_3, w_4, w_5, w_6\}$. From Eq. (4.5), node weights are:

$$w_1 = 0.20, \qquad w_2 = 0.30, \qquad w_3 = 0.20 - 0.10 = 0.10, \qquad w_4 = 0.30,$$

$$w_5 = -0.50, \qquad w_6 = -0.40$$

The edge-weighted graph model has the node set $V = \{v_1, v_2, v_3, v_4, v_5, v_6\}$, edge set $E = \{e_{21}, e_{15}, e_{32}, e_{25}, e_{43}, e_{46}, e_{56}\}$ and $W = \{w_{12}, w_{21}, w_{15}, w_{51}, w_{23}, w_{32}, w_{25}, w_{52}, w_{43}, w_{46}, w_{64}, w_{56}, w_{65}\}$. From Eq. (4.8), edge weights are:

$$w_{21} = -w_{12} = 0.10, \; w_{15} = -w_{51} = 0.30, \; w_{25} = -w_{52} = 0.45, \; w_{32} = -w_{23} = 0.25,$$

$$w_{43} = w_{34} = 0.15, \; w_{46} = -w_{64} = 0.15, \; w_{56} = -w_{65} = 0.25$$

4.5 Generator Grouping

When generators start to oscillate under disturbances, a general observation is that generators are clustered as two or more groups. The generators having strong interconnections, geographical or electrical, tend to have dynamical coherency and naturally swing together.

The ways in which generators tend to group under disturbances are critical to the success of CSS. As desired by the generation coherency constraint, all generators allocated into one island should keep their synchronism if no other measures are available to guarantee their stabilities. Thus, dynamical coherency of these generators before CSS can be an important indicator on how easily they can be synchronized after CSS.

In the planning stage for CSS strategies, the planning engineers may study which generators tend to form a coherent group under a credible critical contingency. Generators of a coherent group usually belong to the same or adjacent power plants. To simplify the problem, it is unnecessary for CSS to separate the generators of the same group if they keep coherency at a high probability. Thus, as a result, a strongly coherent group of generators may be merged to an equivalent generator, that is, modeled as a single generator bus in the power network. After this simplification, separation points can be found only on the branches between identified strongly coherent generator groups. In real-time power system operations, the monitoring or predicting of out-of-step conditions may be focused only on the scenarios among those identified strongly coherent groups using wide-area measurements from synchrophasors. If any generator within an originally coherent group goes out-of-step, it will be tripped by its protective relay. Such an approach of merging generators by coherent groups may not prevent every generator from instability but significantly simplifies the problems addressing the questions "where," "when," and "how."

When generators of an interconnected power system start oscillating as two or more coherent groups, it is often observed that the oscillations between coherent groups are associated with slow inter-area electro-mechanical oscillation modes at low frequencies typically lower than 0.7 Hz, whereas the generators within each coherent group, if not perfectly behaving the same, may oscillate in fast local modes at higher frequencies. The slow inter-area oscillation modes and fast local modes, respectively, indicate the dynamically weak and strong connections between generators. Therefore, generator clustering may be studied by modal analysis on the system model.

In the following, a widely used modal analysis technique, the slow coherency analysis method, is introduced for offline identification of coherent generator groups. For large-scale power systems, the software program DYNRED (Dynamic Model Reduction) developed by EPRI can be applied to solve coherent groups of generators. The slow coherency analysis method and related engineering techniques are applied by DYNRED.

4.5.1 Slow Coherency Analysis

The slow coherency method focuses on solving the problem of identifying theoretically the weakest connection in a complex power network by finding the slowest mode [41]. The slow coherency method has originally been applied to perform dynamic equivalents for large-scale power systems for the purpose of transient stability studies. Because dynamic coherency of generators in groups suggests the way of system separation that creates islands with higher probabilities of being stable, this method has been applied to study where to separate a power system.

Usually, the following assumptions are made in system separation studies for identification of coherent groups [42]:

- "The coherent groups of generators are independent of the size of the disturbance, so that the linearized model can be used to determine the coherency".
- "The coherent groups are independent of the level of detail used in modeling the generating units, so that a classical generator model can be considered".

These assumptions come from the observations on electro-mechanical oscillations. The coherency behaviors of a power system's generators are not significantly changed as the duration time of a specific fault is increased. Although using a simplified generator model in simulations can affect the shapes of each simulated rotor swing curve, the manner in which generators tend to group may not radically change. The assumptions are reliable for a power system that in topology has obvious regions of generation interconnected via not many tie lines or transmission paths. Oscillations among those regions more easily develop to systemwide stability issues while local oscillations within each region have limited impacts on the overall system. Even for a power system with meshed topology, its generators also tend to group in a limited number of manners that can be identified offline by slow coherency analysis together with accurate time-domain simulations.

Consider an n-machine, N-bus power system. Represent each machine by the classical model with a zero damping coefficient. Represent each load as a constant impedance. Thus, the whole system can be modeled by n second-order differential equations about the generator rotor angles as shown in Eq. (4.9), where H_i is the inertia (in seconds) of generator i and $\omega_0 = 2\pi \times 60$ (rad/s). All the other quantities are per unit: P_{mi} is the input mechanical power, P_{ei} is the output electrical power, E_i is the electromotive force (EMF) behind a transient reactance x'_{di}, and G_{ij} and B_{ij} are, respectively, the real and imaginary parts of the (i, j) entry of the reduced admittance matrix about the EMFs of generators.

$$\frac{2H_i}{\omega_0}\ddot{\delta}_i = P_{mi} - P_{ei}$$

$$= P_{mi} - \sum_{j=1}^{n}\left[E_i'E_j'B_{ij}\sin\left(\delta_i - \delta_j\right) + E_i'E_j'G_{ij}\cos\left(\delta_i - \delta_j\right)\right] \ i = 1,\ldots,n \tag{4.9}$$

Linearize Eq. (4.9) about a steady-state condition having rotor angles $\boldsymbol{\delta}_0$:

$$\frac{\mathbf{M}}{\omega_0}\Delta\ddot{\boldsymbol{\delta}} = \mathbf{K}\Delta\boldsymbol{\delta} \quad \Leftrightarrow \quad \Delta\ddot{\boldsymbol{\delta}} = \omega_0\mathbf{M}^{-1}\mathbf{K}\Delta\boldsymbol{\delta} \triangleq \mathbf{A}\Delta\boldsymbol{\delta}, \tag{4.10}$$

where $\mathbf{M}$ is the diagonal matrix whose diagonal elements are $2H_1$ to $2H_m$ and matrix $\mathbf{K}$ has entries

$$K_{ij} = E_i'E_j'B_{ij}\cos\left(\delta_i - \delta_j\right) - E_i'E_j'G_{ij}\sin\left(\delta_i - \delta_j\right)\Big|_{\delta = \delta_0} \quad j \neq i$$

$$K_{ii} = -\sum_{j=1,j\neq i}^{n} K_{ij} \tag{4.11}$$

Entries K_{ij} are known as the synchronizing torque coefficients and are the key to tell the coherency of generators or, in other words, the dynamical linkages between generators.

Let us divide n generators into a given number (say r) of coherent groups. As discussed before, oscillations between groups are also associated with slow oscillatory modes. The idea of the slow-coherency analysis method is to first identify r slowest oscillatory modes and then cluster all generators regarding these modes to obtain r coherent groups. The algorithm given in reference [41] suggests the following steps:

1) Find r slowest modes, i.e. r eigenvalues $\lambda_1, ..., \lambda_r$ of matrix $\mathbf{A}$ in Eq. (4.10) that have the smallest absolute values, including the zero eigenvalue.
2) Compute the eigen-basis matrix $\mathbf{V}_{n \times r}$ satisfying $\mathbf{AV} = \mathbf{V\Lambda}$, where $\mathbf{\Lambda} = \text{diag}(\lambda_1, ..., \lambda_r)$.
3) Apply Gaussian elimination with complete pivoting to $\mathbf{V}$ and obtain the set of reference generators, respectively, for r groups. Each generator group will include one and only one reference generator. Suppose that $\mathbf{V_r}$ is an $r \times r$ matrix composed of $\mathbf{V}$'s r rows that correspond to reference generators.
4) Compute $\mathbf{L} = \mathbf{VV_r}^{-1}$. Find the rows corresponding to the reference generators to form an $r \times r$ identity matrix. Determine generator grouping by comparing each of the other $(n-r)$ nonreference rows with those r reference rows.

For example, consider the six-bus power system in Figure 4.1 and divide its four generators into two (i.e., $r = 2$) coherent groups. Assume

$$
\mathbf{K} = \begin{bmatrix} -12.1205 & 11.6041 & 0.0990 & 0.4174 \\ 11.6041 & -14.1060 & 2.0835 & 0.4184 \\ 0.0990 & 2.0835 & -12.2812 & 10.0987 \\ 0.4174 & 0.4184 & 10.0987 & -10.9344 \end{bmatrix}, \quad \mathbf{M} = \begin{bmatrix} 150 & 0 & 0 & 0 \\ 0 & 50 & 0 & 0 \\ 0 & 0 & 50 & 0 \\ 0 & 0 & 0 & 150 \end{bmatrix}.
$$

The matrix $\mathbf{A} = \omega_0 \mathbf{M}^{-1}\mathbf{K}$ has four eigenvalues sorted by magnitude in ascending order:

$$\lambda_1 = 0, \quad \lambda_2 = -9.836, \quad \lambda_3 = -108.973, \quad \lambda_4 = -138.089.$$

The first one is a nonoscillatory mode related to the common dynamic of all generators and the other three are oscillatory modes indicating three tendencies that one group oscillates against the rest at the following frequencies:

$$\omega_2 = \sqrt{\lambda_2} = \pm j3.136 \text{ rad/s}, \quad \omega_3 = \sqrt{\lambda_3} = \pm j10.439 \text{ rad/s}, \omega_4 = \sqrt{\lambda_4} = \pm j11.751 \text{ rad/s}, \text{ or}$$
$$f_2 = 0.499 \text{ Hz}, f_3 = 1.661 \text{ Hz}, f_4 = 1.870 \text{ Hz}.$$

A set of normalized eigenvectors corresponding to the four eigenvalues are

$$
v_1 = \begin{bmatrix} -0.5 \\ -0.5 \\ -0.5 \\ -0.5 \end{bmatrix}, \quad
v_2 = \begin{bmatrix} -0.5647 \\ -0.4233 \\ 0.4315 \\ 0.5619 \end{bmatrix}, \quad
v_3 = \begin{bmatrix} 0.1559 \\ -0.4220 \\ -0.8517 \\ 0.2687 \end{bmatrix}, \quad
v_4 = \begin{bmatrix} 0.2312 \\ -0.8536 \\ 0.4563 \\ -0.0988 \end{bmatrix}.
$$

Eigenvalues λ_1 and λ_2 correspond to the two slowest modes, so their eigenvectors make up matrix $\mathbf{V}$, whose four rows correspond to generators 1–4 on buses 1–4, respectively.

$$
\mathbf{V} = \begin{bmatrix} -0.5 & -0.5647 \\ -0.5 & -0.4233 \\ -0.5 & 0.4315 \\ -0.5 & 0.5619 \end{bmatrix}
\begin{array}{l} \text{Generator 1} \\ \text{Generator 2} \\ \text{Generator 3} \\ \text{Generator 4} \end{array}
$$

The entry having the largest absolute value is -0.5647 in the first row, so the first reference is Generator 1. Move that entry to the first column by a column exchange as shown by the second matrix $\mathbf{V}^{(1)}$ below, and then eliminate the rest of its entries in the first column to obtain $\mathbf{V}^{(2)}$.

$$
\mathbf{V} = \begin{bmatrix} -0.5 & -0.5647 \\ -0.5 & -0.4233 \\ -0.5 & 0.4315 \\ -0.5 & 0.5619 \end{bmatrix}
\rightarrow \mathbf{V}^{(1)} = \begin{bmatrix} -0.5647 & -0.5 \\ -0.4233 & -0.5 \\ 0.4315 & -0.5 \\ 0.5619 & -0.5 \end{bmatrix}
\rightarrow \mathbf{V}^{(2)} = \begin{bmatrix} -0.5647 & -0.5 \\ 0 & -0.1252 \\ 0 & -0.8821 \\ 0 & -0.9976 \end{bmatrix}
\begin{array}{l} \text{Generator 1} \\ \text{Generator 2} \\ \text{Generator 3} \\ \text{Generator 4} \end{array}
$$

In the second to fourth rows, the entry -0.9976 has the largest magnitude, so Generator 4 is the second reference. So $\mathbf{V_r}$ is composed of the first and fourth rows of $\mathbf{V}$. Then, matrix $\mathbf{L}$ is calculated as follows.

$$
\mathbf{V_r} = \begin{bmatrix} -0.5 & -0.5647 \\ -0.5 & 0.5619 \end{bmatrix}, \quad
\mathbf{L} = \mathbf{V}\mathbf{V_r}^{-1} = \begin{bmatrix} 1 & 0 \\ 0.8745 & 0.1255 \\ 0.1157 & 0.8843 \\ 0 & 1 \end{bmatrix}
\begin{array}{l} \text{Generator 1} \\ \text{Generator 2} \\ \text{Generator 3} \\ \text{Generator 4} \end{array}
$$

The first and fourth rows correspond to two reference generators. By comparing the Euclidean distances from them to the second and third rows, it is easy to conclude that Generators 1 and 2 belong to one coherent group and Generators 3 and 4 belong to the other group.

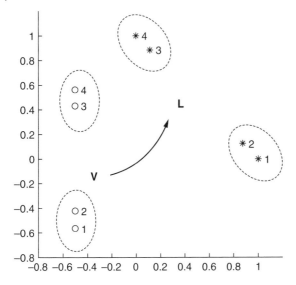

Figure 4.4 Generator grouping with the six-bus power system.

Using the entries in the first and second columns of **V** and **L** as the coordinates, Figure 4.4 gives the locations of Generators 1–4: the four little circles are the locations of generators according to **V** and the stars are their locations according to **L** after transformed by $\mathbf{Vr}^{-1}$. There are two apparent clusters, which are two coherent groups {1, 2} and {3, 4}.

4.5.2 Elementary Coherent Groups

Strictly speaking, generator coherency may vary, although insignificantly, with respect to changes in the system's operating condition including load and topological changes. Therefore, coherent groups identified under one specific operating condition may not be preserved under another condition. For example, one previously coherent group may collapse into smaller groups, or on the contrary, one group may join another to become a bigger coherent group.

A practical approach is to select a number of typical operating conditions to discover potential scenarios of generator grouping by, for example, slow coherency analysis. These scenarios are then considered in the offline studies on separation points. Alternatively, generator grouping can also be analyzed online based on the operating condition estimated from the real-time state estimator.

The concept of elementary coherent group is introduced next as a practical compromise of simplifying the identification of generation grouping while considering uncertainties in the real-time operating condition. In order to give enough considerations to probable changes in generator grouping under a wide range of credible operating conditions, all generators are separated into a number of small groups called elementary coherent groups (ECGs). It is expected that all combinations or separations among these ECGs can cover a majority of probable scenarios of generator grouping or those grouping scenarios of interests. ECGs can be determined by slow coherency analysis. In addition, the ECGs may be verified and modified by simulation results on a list of credible "N-k" contingencies, which likely trigger extreme conditions with increased risks of system separation. If simulation tells that under a scenario one ECG will lose synchronism, this ECG may further be split into two or more smaller ECGs addressing that scenario, or the ECG should be equipped with a remedial action designed to enforce its synchronization in case that scenario really happens.

In fact, a reason that causes the ambiguity in the boundaries of actual coherent groups is the existence of "fuzzy" generators that are easier to be disturbed to move from one group to another when the operating condition changes. Generally speaking, such a "fuzzy" generator does not have much stronger coherency with one group than the others. This, in fact, increases the probability of stabilizing it with a preselected adjacent coherent group if they together form one island. Thus, a "fuzzy" generator can be handled as follows:

- If the generator has a big capacity and its tripping after CSS is unacceptable, it should be treated as an ECG by itself.
- If the generator is a small unit, it can be treated as an "unimportant generator" and assigned to an adjacent ECG that is close or needs additional generation to balance its nearby loads. In case that small generator is not synchronous with the ECG after CSS, it may be tripped.

In practice, system planners have ideas about potential boundaries of system separation and the number of islands after considering all related engineering or regulatory factors. Accordingly, a number of ECGs can be defined. Then the answering of "where" and "when" may focus on addressing out-of-step conditions among those defined ECGs, and synchronism or instability within an ECG may not be concerned until CSS is performed. In case unstable generators really occur within any ECG, postseparation control actions can, for example, trip unstable generators, as part of the solution addressing the question "how".

Real-time oscillations among ECGs may be monitored using synchrophasors. At least, one synchrophasor needs to be available at a bus inside or close to each ECG. For more reliable monitoring, it is recommended to place multiple synchrophasors near the terminal buses of main generators of each ECG. Figure 4.5 shows an example of a power system with three ECGs, that is ECG-1, ECG-2, and ECG-3, as circled by broken lines. Assume that they have, respectively, three, two, and four generators that are each monitored by one

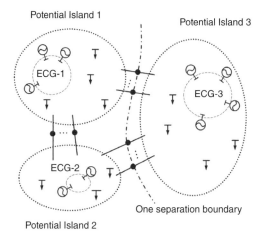

Potential Island 1

Potential Island 3

One separation boundary

Potential Island 2

Figure 4.5 A power system with three ECGs.

synchrophasor. The offline studies on CSS need to consider the possibility that each ECG individually forms a potential island together with matched load. In the real-time operations, the separation boundary depends on the actual out-of-step condition of the ECGs. As illustrated by the figure, if ECG-3 goes out of step from ECG-1 and ECG-2, the system should be separated into two islands.

4.6 Finding Separation Points

4.6.1 Formulations of the Problem

The problem of where to separate can be formulated as a graph partition problem. In the following, five ways to formulate such a problem are introduced mainly addressing the generation coherency constraint and power balance constraint for CSS.

We first define the mathematical model of a separation strategy, which is a K-cut of a node-weighted graph model or an edge-weighted graph model for the power network.

Definition 1 K-cut of a node-weighted graph
For an undirected node-weighted graph $G(V, E, W)$, an edge set $E_0 \subset E$ is called a K-cut of G if it partitions G into K sub-graphs $G_1(V_1, E_1, W_1), ..., G_K(V_K, E_K, W_K)$, where, for each $k = 1, ..., K$, V_k, E_k, and W_k are, respectively, the node set, edge set, and node weight set of subgraph G_k as defined before, and the following are held:

$$V = V_1 \cup V_2 \cup \cdots \cup V_K$$
$$W = W_1 \cup W_2 \cup \cdots \cup W_K$$
$$E = E_0 \cup E_1 \cup E_2 \cup \cdots \cup E_K$$

$$\forall i \in \{1,\ldots,K\}, E_i \cap E_0 = \varnothing$$
$$\forall j \in \{1,\ldots,K\}, \text{if } j \neq i, \text{ then } V_i \cap V_j = W_i \cap W_j = E_i \cap E_j = \varnothing.$$

Here, each edge of the K-cut E_0 has two nodes respectively belonging to two different node sets of V_1 to V_K.

Definition 2 K-cut of an edge-weighted graph

For a directed edge-weighted graph $G(V, E, W)$, an edge set $E_0 \subset E$ is called a K-cut of G if it partitions G into K subgraphs $G_1(V_1, E_1, W_1)$, ..., $G_K(V_K, E_K, W_K)$, where, for each $k = 1, \ldots, K$, V_k, E_k, and W_k are, respectively, the node set, edge set, and node weight set of subgraph G_k as defined before, and the following are held:

$$V = V_1 \cup V_2 \cup \cdots \cup V_K$$
$$W = W_0 \cup W_1 \cup W_2 \cup \cdots \cup W_K$$
$$E = E_0 \cup E_1 \cup E_2 \cup \cdots \cup E_K$$

$$\forall i \in \{1,\ldots,K\}, E_i \cap E_0 = W_i \cap W_0 = \varnothing$$
$$\forall j \in \{1,\ldots,K\}, \text{if } j \neq i, \text{ then } V_i \cap V_j = W_i \cap W_j = E_i \cap E_j = \varnothing.$$

Here, W_0 is the set of edge weights of all edges in E_0.

The following Problem 1a, Problem 1b, Problem 2a, Problem 2b, and Problem 3 give five ways to formulate where to separate. The first two problems are formulated on a node-weighted graph and the last three problems are formulated on an edge-weighted graph.

Problem 1a Balanced K-way graph partition

For an undirected node-weighted graph $G(V, E, W)$, given subsets $V_{G,1}, \ldots, V_{G,K}$, and a small positive number $\varepsilon_p > 0$, find a K-cut E_0 that partitions G into K subgraphs $G_1(V_1, E_1, W_1)$, ..., $G_K(V_K, E_K, W_K)$ and satisfies

$$V_{G,1} \subset V_1, \ldots, V_{G,K} \subset V_K \text{ (generation coherency constraint)}$$

$$\left| \sum_{w_i \in W_k} w_i \right| \leq \varepsilon_p, k = 1,\ldots,K \text{ (power balance constraint)}. \tag{4.12}$$

Here, ε_p reflects the tolerance of real power imbalance in each island formed after CSS. It needs to be small enough to guarantee that the frequency offset in each island after CSS still meets reliability criteria on frequency. For example, in North America power grids, a sustained, absolute frequency offset >0.7 Hz may trigger overfrequency generation trip or automatic underfrequency load shedding [43], so we may let ε_p be 0.5 Hz or less for a safety margin.

Problem 1a belongs to a type of *satisfiability* problems that find acceptable solutions satisfying a set of given constraints [44]. The resolution of Problem 1a can be achieved by tackling the following related *optimization* Problem 1b.

Problem 1b Minimally unbalanced K-way graph partition
For an undirected node-weighted graph $G(V, E, W)$ given subsets $V_{G,1}, ..., V_{G,K}$, find a K-cut E_0 that partitions G into K subgraphs $G_1(V_1, E_1, W_1), ..., G_K(V_K, E_K, W_K)$ and satisfies

$$V_{G,1} \subset V_1, ..., V_{G,K} \subset V_K \text{ (generation coherency constraint)}$$

and minimizing

$$\sum_{k=1}^{K} \left| \sum_{w_i \in W_k} w_i \right| \text{ (objective of power balance)}. \tag{4.13}$$

If a solution of Problem 1a exists, then the solution of Problem 1b is necessarily a solution of Problem 1a; if the solution of Problem 1b cannot meet the inequality constraint $\left| \sum_{w_i \in W_k} w_i \right| \leq \varepsilon_p$ for every $k = 1, ..., K$, Problem 1a has no solution.

For a lossless power network, the following Problem 2a and Problem 2b formulated on an edge-weighted graph are respectively equivalent to Problem 1a (satisfiability) and Problem 1b (optimization).

Problem 2a K-cut with an acceptable total weight
For a directed edge-weighted graph $G(V, E, W)$, given subsets $V_{G,1}, ..., V_{G,K}$, and a small positive number ε_p, find a K-cut E_0 that partitions G into K subgraphs $G_1(V_1, E_1, W_1), ..., G_K(V_K, E_K, W_K)$, and satisfies

$$V_{G,1} \subset V_1, ..., V_{G,K} \subset V_K \text{ (generation coherency constraint)}$$

$$\left| \sum_{\substack{\forall w_{ij} \in W_0 \\ v_i \in V_k, v_j \notin V_k}} w_{ij} \right| \leq \varepsilon_p, k = 1, ..., K \text{ (power balance constraint)} \tag{4.14}$$

Problem 2b   K-cut of an edge-weighted graph
For a directed edge-weighted graph $G(V,E)$, given subsets $V_{G,1}, ..., V_{G,K}$, find a K-cut E_0 that partitions G into K subgraphs $G_1(V_1, E_1, W_1), ..., G_K(V_K, E_K, W_K)$, satisfies

$$V_{G,1} \subset V_1, ..., V_{G,K} \subset V_K \text{ (generation coherency constraint)}$$

and minimizes

$$\sum_{k=1}^{K} \left| \sum_{\substack{\forall w_{ij} \in W_0 \\ v_i \in V_k, v_j \notin V_k}} w_{ij} \right| \text{ (objective of power balance).} \tag{4.15}$$

In some literature, the corresponding optimal graph partition problem of system separation is formulated as follows, instead.

Problem 3   K-cut with the minimum total absolution weight
For an edge-weighted graph $G(V,E)$, subsets $V_{G,1}, ..., V_{G,K}$, find a K-cut E_0 that partitions G into K subgraphs $G_1(V_1, E_1, W_1), ..., G_K(V_K, E_K, W_K)$, satisfies

$$V_{G,1} \subset V_1, ..., V_{G,K} \subset V_K \text{ (generation coherency constraint),}$$

and minimizes

$$\sum_{k=1}^{K} \left(\sum_{\substack{\forall w_{ij} \in W_0 \\ v_i \in V_k, v_j \notin V_k}} |w_{ij}| \right). \tag{4.16}$$

Note that the graph here does not need to be a directed graph since only the absolute values of edge weights are used in the formulation. Compared to the formulation of Problem 2b, although the difference is just to move the absolute operator from the sum of edge weights to each individual edge weight, the minimization of objective function Eq. (4.16) *may not guarantee satisfaction to the power balance constraint.* That will be explained using the six-bus power system as follows.

For convenience of analysis, Figure 4.6 shows all node weights w_1 to w_6 and the positive edge weights $w_{ij} > 0$; that is, the actual real power flow is from bus i to bus j as indicated by the arrow. Because the system is lossless, $w_1 + w_2 + w_3 + w_4 + w_5 + w_6 = 0$. Assume that Generators 3 and 4 together will become one group going out of step from Generators 1 and 2 under disturbances. There are $V_{G1} = \{v_1, v_2\}$ and $V_{G2} = \{v_3, v_4\}$. Set the power imbalance tolerance $\varepsilon_p = 0.05$. Problem 1 has only one solution, that is, the 2-cut $E_0 = \{e_{23}, e_{56}\}$ because it partitions $V_1 = \{v_1, v_2, v_5\}$ and $V_2 = \{v_3, v_4, v_6\}$ (the generation coherency constraint is

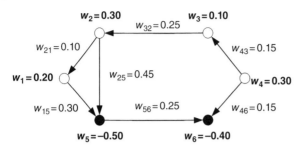

Figure 4.6 Graph model of the six-bus power system.

satisfied), and $w_1 + w_2 + w_5 = w_3 + w_4 + w_6 = 0 < \varepsilon_p$ indicating both resulting subgraphs to be balanced (the power balance constraint is satisfied, too). Thus, by removing those two edges e_{23} and e_{56}, two perfectly balanced subgraphs (islands) are formed. Obviously, $\{e_{23}, e_{56}\}$ is also the optimal solution of the Problem 1b.

Then, a next question is whether or not the 2 cut $E_0 - \{e_{23}, e_{56}\}$ (or $\{e_{32}, e_{56}\}$ in the edge-weighted graph model) is also a solution of Problem 2a (with the same ε_p) or Problem 2b. The answer is yes. In fact, E_0 is the only satisfactory solution of Problem 2a and is also the optimal solution of Problem 2b because $|w_{23} + w_{56}| = 0 < \varepsilon_p$ (note that v_2 and $v_5 \in V_1$ and v_5 and $v_6 \in V_2$). This means that the total real power exchanges between the two subgraphs is zero.

However, if we solve Problem 3, the solution is not $\{e_{23}, e_{56}\}$. In fact, the optimal graph partition strategy for Problem 3 is $\{e_{23}, e_{46}\}$ instead. The former has an objective function equal to $|w_{23}| + |w_{56}| = 0.50$ while the latter has an objective function value equal to $|w_{23}| + |w_{46}| = 0.40$, 20% less.

From this example, the objective function Eq. (4.16) with Problem 3 is actually to minimize the power-flow interruption; that is, cutting the lines that carry smallest absolute power flows to reduce the impact of system separation. In the next section, we will prove that Problem 1a, Problem 1b, Problem 2a, and Problem 2b are NP (nondeterministic polynomial-time) complete while Problem 3 may not have the same level of complexity depending on the topology of the graph and the number of subgraphs to generate.

For a power network, its edge-weighted graph model and node-weighted graph model are actually equivalent to each other. In other words, there is a duality between problems about "where to separate" formulated respectively on its edge-weighted graph model and node-weighted graph model. Therefore, in the rest of the chapter, node-weighted graphs are used to model power networks.

4.6.2 Computational Complexity

In the following, we are going to first introduce a well-known NP-complete problem, the 0–1 *KNAPSACK* problem, and then prove that Problem 1 is an NP-compete problem. The complexity of Problem 3 will also be discussed.

Problem 4   0–1 *KNAPSACK*

Given non-negative integers c_j ($j = 1, ..., n$) and $C > 0$, is there a subset S of the set of indices $\{1, ..., n\}$ such that $\sum_{j \in S} c_j = C$?

Remark: The 0–1 *KNAPSACK* problem given here is slightly different from the one given in [45]. We require $c_j \geq 0$. However, this does not change the NP-completeness nature of the problem.

An NP problem is a problem such that each of its candidate solutions can be tested to be true or false within polynomial time. NP-complete problems form the set of the most difficult NP problems. NP-hard problems are more general problems and unnecessarily NP problems. They are called NP-hard because they are at least as difficult as the NP-complete problems. NP-completeness and NP-hardness of a problem are concepts used to quantify and compare the computational complexities of problems. The purpose here is to justify theoretically the fact that the large number of possible ways to select CSS strategies contributes essentially to the difficulty in answering the question "where." In other words, facing the fundamental complexity of this problem in computation, a successful method will have to efficiently reduce the huge search space that grows exponentially with the size of the problem. All candidate CSS strategies have to be effectively narrowed down by the method. Formal definitions of the terminologies and concepts related to computational complexity can be found in [45].

Theorem 1   Problem 1a is NP-complete.

Proof: Obviously, this problem is an NP problem; that is, if a solution of the problem is given, the solution can be verified by direct calculation of $\sum_{j \in S} c_j = C$, which is a polynomial time algorithm.

Next, we transform the 0–1 *KNAPSACK* problem to Problem 1a for a special case with $K = 2$. Given any instance of $c_1, ..., c_n$, and C, we construct the following instance of Problem 1a. The undirected, node-weighted complete graph $G(V, E, W)$ with $n + 2$ nodes is fully connected with node weights $w_i = -c_i$ ($i = 1, ..., n$), $w_{n+1} = \sum_{j=1}^{n} c_j - C$, and $w_{n+2} = C$. Let $V_{G1} = \{v_{n+1}\}$, $V_{G2} = \{v_{n+2}\}$, and $\varepsilon_p = 1$. An instance of $n = 4$ is shown in Figure 4.7. We claim that there is a subset S of $\{1, 2, ..., n\}$ such that $\sum_{j \in S} c_j = C$ if and only if there exists a 2-cut E_0 that partitions G into two subgraphs, $G_1(V_1, E_1, W_1)$ and $G_2(V_2, E_2, W_2)$, satisfying

$$v_{n+1} \in V_1, v_{n+2} \in V_2$$

$$\left| \sum_{w_i \in W_1} w_i \right| < 1 \text{ and } \left| \sum_{w_j \in W_2} w_j \right| < 1.$$

If: Because w_i, $i = 1, ..., n + 2$ are non-negative integers, we have the equivalence

$$\left| \sum_{w_j \in W_2} w_j \right| = 0 \Leftrightarrow \sum_{v_j \in V_2 \setminus \{v_{n+2}\}} c_j = C.$$

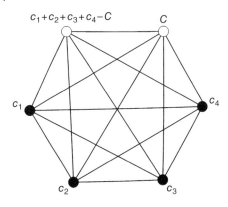

Figure 4.7 Balanced K-way graph partition problem converted to a 0–1 KNAPSACK problem.

It follows directly that $S = V_2 \backslash \{v_{n+2}\}$.

Only if: Suppose that $\sum_{j \in S} c_j = C$ for some $S \subseteq \{1, 2, \dots, n\}$. Let $V_1 = \{v_i \mid i \in S\} \cup \{v_{n+2}\}$ and $V_2 = V \backslash V_1$. Obviously, $v_{n+2} \in V_2$. Immediately, there is $\sum_{w_i \in W_1} w_i = \sum_{w_j \in W_2} w_j = 0$.

So E_0 is just the 2-cut that partitions G into G_1 and G_2. This concludes the proof.

Remark: In the formulation of Problem 1a, all edges are candidates to be included in the K-cut. In practice, additional information on possible separation points is helpful to restrict the search space for a proper K-cut.

Obviously, solutions of the problem "where to separate" are necessarily the solutions of Problem 1a because Problem 1a or Problem 1b does not check the transmission capacity constraint. Thus, the problem of finding separation strategies addressing "where to separate" is at least as complex as Problem 1a.

Corollary 1 The problem of finding separation strategies satisfying three constraints is NP-hard. Those three constraints include the generation coherency constraint, power balance constraint, and transmission capacity constraint. Equivalently, the problem of minimizing the real power imbalance in each island while satisfying the generation coherency constraint and transmission capacity constraint is NP-hard.

In Problem 3, the number K of sets of nodes $V_{G,1}, \dots, V_{G,K}$ to be partitioned is specified. Namely, K is fixed. In other words, at least K nodes (called terminals) respectively from those K sets of nodes should be specified. Thus, this problem is actually equivalent to a type of minimum K-cut problem regarding specified K nodes. For $K = 2$, this problem is the well-known minimum *s-t* cut problem (transformable to the *max-flow min-cut* problem), which is solvable by the Stoer–Wagner algorithm in polynomial time $O(|V||E| + |V|^2 \log |V|)$ [46]. For $K \geq 3$, this problem is NP-hard in general, but if the graph is planar, it can be solved in polynomial time $O(|V|^3 \log|V|)$ [47]. To summarize, Problem 3 may not be as difficult as Problem 1a, Problem 1b, Problem 2a, and Problem 2b.

An NP-hard problem cannot be solved in polynomial time. The size of the search space for solutions exponentially explodes with the growth of the number of variables, that is, the size of the problem. For the problem of finding separation strategies, the number of variables depends on the number of branches in the power network. Thus, for a large and topologically complex power grid, the best separation points are not as obvious as those in the six-bus power system because the search space for possible separation points grows exponentially with the number of branches of the grid. For example, even for the IEEE 30-bus test system with only 41 lines, there are in total $2^{41} \approx 2.2 \times 10^{12}$ possible strategies of separation points. For the IEEE 118-bus test system with 186 lines, the number of possible strategies of separation points becomes $2^{186} \approx 9.8 \times 10^{55}$!

4.6.3 Network Reduction

For a realistic large power grid, it is impossible to find the optimal strategy of separation points or even a strategy satisfying given constraints by an exhaustive search unless effective network reduction is performed before strategy search. An effective way to reduce a power network is to merge groups of nodes to equivalent nodes, respectively. Besides, the graph model of the power network can be simplified by graph theory. The following four measures can be used to reduce a power network:

- *Measure I: Merging an area containing a coherent group of generators to an equivalent generator bus*
  Coherency of generation and potential out-of-step modes can be studied offline under a variety of disturbances. From the results, a group of generators may have the tendency to be coherent or their coherency is preferred due to some operational considerations. Thus, an area containing such a group of generators can be manually reduced to one equivalent generator bus. Its generation equals the total generation of the group minus the total load of that area.
- *Measure II: Merging a geographical area to a generator or load bus whose integrity is desired by system planners*
  System planners may expect to maintain the connection of a particular geographical area, for example, a load center representing a metropolitan area. Then, the whole area can be regarded as one equivalent bus. The type of the bus depends on whether or not major generating units exist in the area. If yes, stabilities of these generating units should be maintained after CSS, so the bus will be marked as a generator bus in order to address its coherency with other generators; otherwise, the bus can be regarded as a load bus.

- *Measure III: Merging a cluster of load buses to an equivalent load bus*
  A power network often has many buses connected to small loads. Usually, it is not worth optimizing the allocation of each specific small load into an island. Thus, many load buses may be clustered into load areas and then the buses in each load area can be merged to an equivalent load bus. The size of the load area will influence the power balance of the island it belongs to and hence will need to be limited to avoid causing a large frequency offset in the island. Let P_{Area} denote the upper limit of the total real power load of each load area; P_{Area} can be chosen as follows. Use Δf to represent the maximum acceptable frequency offset in each island, which usually cannot be much higher than 0.5 Hz. The composite frequency response characteristic (also called frequency bias factor) of a potential island is expected to be not less than a threshold denoted by β to maintain a frequency regulation capability against additional disturbances. Then, there are the following inequalities about P_{Area}, ε_p, β, and Δf:

$$P_{Area} \leq \varepsilon_p < \beta \Delta f. \tag{4.17}$$

Here, the first inequality means that the total load of an area to be merged should not exceed the tolerance of power imbalance ε_p so as to still keep enough room to decide which island it belongs to considering the power balance constraint. The second inequality is to ensure an acceptable frequency offset in each island satisfying the power balance constraint.

- *Measure IV: Simplification using the graph model*
  A realistic power network can be reduced first using these measures. Then, a graph model can be created based on the reduced network. The following three simplification measures can be applied to further simplify the graph model [1].

Measure IV.a: Merging a zero- or small-weight node into a neighboring node as illustrated by Figure 4.8 from (a) to (b), where the broken line denotes an arbitrary number (zero, one, or more) of edges. If a separation point is decided to be on the edge between Node 1 and Node 2 of the reduce graph (b), the separation point could be on either Edge 1–3 or Edge 2–3, which can generate equivalent results.

Measure IV.b: Merging a one-degree node into its neighbor as illustrated by Figure 4.9. A one-degree node has only one edge connecting to it. If its

(a) Original graph   (b) Reduced graph

Figure 4.8 Merge a zero-weight node into its neighbor.

(a) Original graph   (b) Reduced graph

Figure 4.9 Merge a one-degree node into its neighbor.

(a) Original graph   (b) Reduced graph

Figure 4.10 Remove a redundant edge.

isolation is not expected, a one-degree node can be merged into its neighbor with its weight. It should be noted that Measure III should avoid generating any one-degree node like v_4 since it will be merged into its neighbor.

Measure IV.c: Remove an edge whose connection and disconnection completely depends on other edges of the graph, as illustrated by the solid line in Figure 4.10. Nodes v_1 and v_2 have weights w_1 and w_2, respectively. They are connected by edge e_{12} and they both connect to nodes of a subgraph G_S. Either v_1 or v_2 is connected to the rest of the graph as indicated by the thin broken lines. The simplest case is that v_1 and v_2 are directly connected by two edges. The thick broken line connecting v_1 or v_2 with G_S denotes an arbitrary nonzero number of edges. In such a case, edge e_{12} is redundant in judging the connectivity of v_1 and v_2 and can be reduced. The reason is that if it is not expected to isolate G_S as one island, whether to disconnect e_{12} completely depends on whether v_1 and v_2 should belong to the same island, and that can be judged by whether to open the edges of G_S and the thick broken lines. Thus, e_{12} can be reduced. If, finally, buses corresponding to v_1 and v_2 need to be in the same island, all branch(es) associated with e_{12} should not be disconnected.

Figure 4.11 gives an example on the use of Measure IV.c. If $\{v_4, v_{12}\}$ is not expected to be isolated to form an island, e_{56} becomes a redundant edge and is reduced. Similarly, if $\{v_9, v_{10}, v_{11}\}$ is not expected to be isolated, e_{78} can be reduced. If isolation of $\{v_{10}, v_{11}\}$ is forbidden, e_{79} can be reduced. Also, e_{13} or e_{23}, may be reduced if $\{v_2\}$, or $\{v_1\}$, is not expected to be isolated.

Figure 4.12 gives another example of the IEEE 30-bus power system. Figure 4.12a shows the graph model, where generator buses, load buses, and the buses without generator or load are differentiated by white, black, and gray

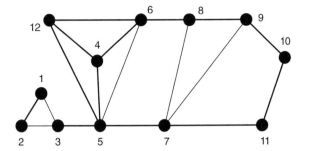

Figure 4.11 Example of which edges can be reduced.

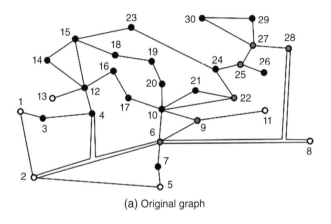

(a) Original graph

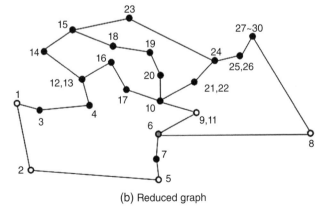

(b) Reduced graph

Figure 4.12 IEEE 30-bus network.

nodes, respectively. The graph model can be simplified to reduce the number of nodes from 30 to 23 and reduce the number of edges from 41 to 26, as shown by Figure 4.12b.

The graph model of the IEEE 118-bus power system is shown in Figure 4.13, where the 19 white nodes denote 19 generator buses and black nodes denote load buses.

Consider this out-of-step mode: generators 10, 12, 25, 26, and 31 become the first group, generators 87, 89, 100, 103, and 111 become the second group, and the other generators become the third group. As shown in Figure 4.14a, the areas containing these three groups of generators can, respectively, be merged to be three equivalent generator buses. For load buses, assume that the buses in each small gray area need to be clustered. Then, each of such load areas can be merged to be an equivalent load bus.

Thus, a simplified graph is generated as shown in Figure 4.14b. That graph can be further simplified based on graph theory. For example, all the broken lines can be reduced according to Measure III.c shown in Figure 4.10. Then, the final reduced graph is given in Figure 4.14c. Solving its separation points for an operating condition becomes much easier. A possible strategy of separation points is indicated by three broken lines.

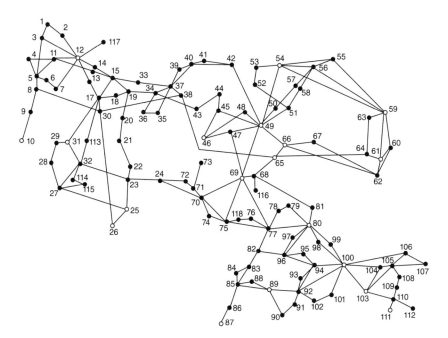

Figure 4.13 IEEE 118-bus power system and its graph model.

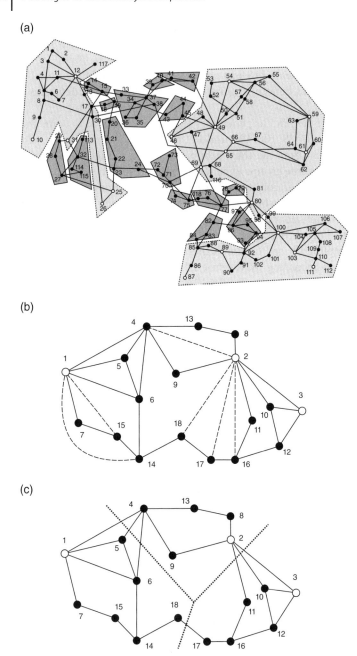

Figure 4.14 Simplification of the IEEE 118-bus power system. (a) Original network with generation areas each containing a coherent generator group and load areas having clusters of load buses. (b) Reduction generator and load areas to equivalent generator and load buses. (c) Final reduced graph and potential separation strategies.

4.6.4 Network Decomposition for Parallel Processing

High-performance parallel computers can significantly decrease the time cost of answering "where to separate." A large-scale power network may be decomposed into a number of subnetworks so that analyses, computations, or optimizations can be performed simultaneously on these subnetworks in parallel. Then, the results on subnetworks can be combined to give the solution of the whole power network. In general, a power network can naturally be decomposed according to its geographical zones or operational control areas. In the following, a graph-model-based network decomposition approach is introduced for studies on CSS and more specifically answering the question "where to separate."

The graph model $G(V, E, W)$, of a power network is decomposed into a number of subgraphs that denote as many subnetworks. Some adjacent subgraphs are linked through one small subgraph called an "interface graph" (denoted by G^I) directly connected with all of them. To parallelize the analysis on separation strategies for all subnetworks, we require that an interface graph should have as little load as possible. Figure 4.15 illustrates two cases of graph-model decomposition, respectively, through one and two interface graphs.

The graph-model-based network decomposition approach has four steps:

1) In $G(V, E, W)$, select one interface graph denoted by $G^I(V^I, E^I, W^I)$, by which G is decomposed into subgraphs including the interface graph itself and a number (say m) of subgraphs $G_1(V_1, E_1, W_1), ..., G_m(V_m, E_m, W_m)$ as illustrated by Figure 4.15a. The part of the power network represented by G^I should have as little load as possible. Regarding those subgraphs, the following conditions are satisfied:
   a) There is no edge in E that directly connects two nodes respectively from any two subgraphs from $G_1(V_1, E_1, W_1), ..., G_m(V_m, E_m, W_m)$, and each of

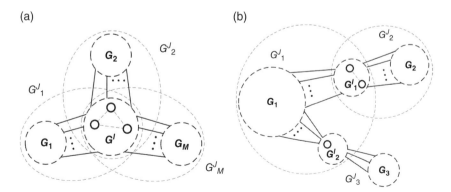

Figure 4.15 Graph-based network decomposition. (a) Decomposition by one interface graph. (b) Decomposition by two interface graphs.

these m subgraphs has at least one node directly connected to a node in V^I by an edge. That requires satisfactions to both of these propositions:

$$\forall v_i \in V_k \quad \text{if } \exists e_{ij} \in E \quad \text{then } v_j \in V_k \cup V^I$$

$$\forall V_k \quad \exists v_i \in V_k \text{ and } e_{ij} \in E \quad \text{such that } v_j \in V^I.$$

b) If two nodes of any edge are in the node set of the same subgraph, that is, one from $G_1(V_1, E_1, W_1)$, ..., $G_m(V_m, E_m, W_m)$ and $G^I(V^I, E^I, W^I)$, that edge should also be in the edge set of that subgraph. Namely, both of the following two propositions hold:

$$\forall e_{ij} \in E \text{ if } v_i \in V_k \text{ and } v_j \in V_k \text{ then } e_{ij} \in E_k$$

$$\forall e_{ij} \in E \text{ if } v_i \in V^I \text{ and } v_j \in V^I \text{ then } e_{ij} \in E^I.$$

Note: any edge connecting two nodes respectively from two different subgraphs from $G_1(V_1, E_1, W_1)$, ..., $G_m(V_m, E_m, W_m)$ and $G^I(V^I, E^I, W^I)$ are not in either subgraph.

2) Construct G's subgraphs $G^J_k(V^J_k, E^J_k, W^J_k)$ ($k = 1, ..., m$) satisfying

$$V^J_k = V_k \cup V^I, \quad E^J_k = E_k \cup E^I \cup \{e_{ij} | \, v_i \in V_k \text{ and } v_j \in V^I, \text{ or } v_j \in V_k \text{ and } v_i \in V^I\}.$$

Namely, each G^J_k is composed of nodes and edges from both G_k and G^I and the edges directly connecting nodes from the two subgraphs. Thus, G^J_1 to G^J_m all contain the nodes and edges of interface graph G^I as shown in Figure 4.15a.

3) Partition the node weights of G^I into G^J_1 to G^J_m according to this rule: if a node v_i of G^I has weight w_i, then its corresponding weights in G^J_1 to G^J_m (denoted by $w_{i,1}, ..., w_{i,m}$) satisfy Eqs. (4.18) and (4.19). Eq. (4.19) means each subgraph is still balanced.

$$\sum_{k=1,...,m} w_{i,k} = w_i, \quad \forall v_i \in V^I \tag{4.18}$$

$$\sum_{v_i \in V_k} w_l + \sum_{v_i \in V^I} w_{i,k} = 0, \quad k = 1, ..., m \tag{4.19}$$

4) Steps 1–3 may be recursively applied to any of G_1 to G_m that is still too complex to solve separation strategies. Finally, assume that there are M subgraphs G^J_k ($k = 1, ..., M$) and L interface graphs $G^I_1, ..., G^I_L$.

As an example, the graph model of the IEEE 118-bus system can be decomposed into three subgraphs G^J_1 to G^J_3 by two interface graphs G^I_1 and G^I_2, as illustrated by Figure 4.15b, where G^J_1 is made up by G_1, G^I_1 and G^I_2, G^J_2 is made

Table 4.1 Decomposition of the graph model of the IEEE 118-bus system.

Subgraphs	G_1	G_2	G_3	G'_1	G'_2
Serial numbers of nodes in the original G	1–48, 68–99, 101, 102, 113–118	50–64, 66, 67	103–112	49, 65	100

up by G_2 and G'_1, and G'_3 is made up by G_3 and G'_2. Table 4.1 gives the series numbers of nodes in each subgraph.

An advantage of this way of decomposition is that every interface graph only has generator nodes. Node weights $w_{100,3}$, $w_{49,2}$, and $w_{65,2}$ are first determined to make G'_2 and G'_3 have a zero sum of weights. Then $w_{100,1}$, $w_{49,1}$, and $w_{65,1}$ are calculated by

$$w_{100,1} = w_{100} - w_{100,3}$$

$$w_{49,1} = w_{49} - w_{49,2}$$

$$w_{65,1} = w_{65} - w_{65,2}.$$

After this network decomposition, the largest subgraph G_1 has 91 nodes, which may either continue to be decomposed to smaller subgraphs or be further reduced by measures in Section 4.6.3. Figures 4.16a and b respectively gives details of G'_1 and G'_2 after they are simplified using the measures in Section 4.6.3 and their nodes are renumbered. Gray nodes represent those having flexibility to be put into any islands. As shown in Figure 4.16c, G'_3 keeps all details of the original graph decomposed by node 100 since it is small.

4.6.5 Application of the Ordered Binary Decision Diagram

In the following, we will first describe the generation coherency constraint and power balance constraint by Boolean algebra. Then we formulate Problem 1a as a Boolean satisfiability problem (the so-called SAT problem) to solve a Boolean equation about the separation strategies satisfying both generation coherency constraint and power balance constraint. Consequently, an effective tool called the ordered binary decision diagram, or OBDD, is introduced to solve Problem 1a.

4.6.5.1 Boolean Algebraic Model

We use $\mathbf{A}$ to denote the adjacency matrix of the N-node graph model $G(V, E)$ of a power network, which could be the original network or a reduced network using the measures in Section 4.6.3. If there is an edge $e_{ij} \in E$ (assuming $i < j$), the elements $(\mathbf{A})_{ij}$ and $(\mathbf{A})_{ji}$ of $\mathbf{A}$ have an identical Boolean variable denoted by b_{ij};

(a)

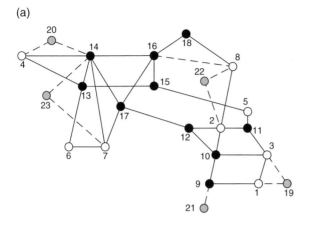

(b)

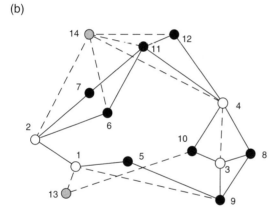

(c)

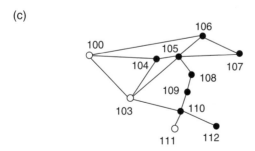

Figure 4.16 Graph-based network decomposition of the IEEE 118-bus system. (a) Subgraph G^J_1. (b) Subgraph G^J_2. (c) Subgraph G^J_3.

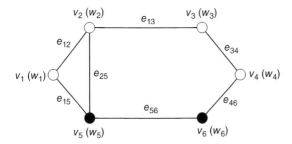

Figure 4.17 Node-weighted graph model of the six-bus power system.

otherwise $(\mathbf{A})_{ij} = (\mathbf{A})_{ji} = 0$. For example, the six-bus power system mentioned previously is redrawn in Figure 4.17 and has the following $\mathbf{A}$

$$
\mathbf{A} = \begin{bmatrix}
0 & b_{12} & 0 & 0 & b_{15} & 0 \\
b_{12} & 0 & b_{23} & 0 & b_{25} & 0 \\
0 & b_{23} & 0 & b_{34} & 0 & 0 \\
0 & 0 & b_{34} & 0 & 0 & b_{46} \\
b_{15} & b_{25} & 0 & 0 & 0 & b_{56} \\
0 & 0 & 0 & b_{46} & b_{56} & 0
\end{bmatrix}.
\tag{4.20}
$$

In fact, to decide whether $b_{ij} = 0$ or 1, that is, "open" or "closed" status of e_{ij}, is simply to decide whether $e_{ij} \in E_0$. Consequently, resolution of Problem 1a or 1b is to decide all Boolean variables b_{ij}.

Here we treat logic "AND" as a Boolean multiplication operation and "OR" as a Boolean addition operation. In the following, we respectively denote "AND" and "OR" by "$\otimes$" (often omitted) and "$\oplus$," to be differentiated from conventional multiplication and addition operations. Note: in many text books, symbol "$\oplus$" is conventionally used for logic "EXCLUSIVE OR" operation, which outputs 1 only when one or the other is true but not both are true. In this section, we assign "$\oplus$" to Boolean addition, instead, and use another symbol, "$\overline{\oplus}$," for "EXCLUSIVE OR."

Then it is easy to draw the following conclusions from Boolean matrix theory [48].

- Element $(\mathbf{A}^k)_{ij}$ is a polynomial regarding Boolean multiplication and addition if there is a path of length $\leq k$ from v_i to v_j in G; otherwise $(\mathbf{A}^k)_{ij} = 0$. Dually, all paths of length $\leq k$ from v_i to v_j in G are determined uniquely by $(\mathbf{A}^k)_{ij}$.
- If we define

$$
\mathbf{A}^* \stackrel{\text{def}}{=} I \oplus \mathbf{A} \oplus \mathbf{A}^2 \oplus \cdots \oplus \mathbf{A}^L,
\tag{4.21}
$$

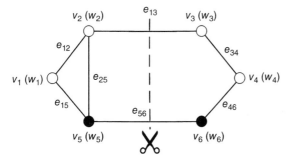

Figure 4.18 Graph model of the six-bus system having paths from nodes 1–4 cut off.

where L is the longest path in G and I is identity matrix, then $\mathbf{A}^{\ast}$ can determine the connection of arbitrary two nodes after arbitrary graph partition.

For instance, consider the $\mathbf{A}^{\ast}$ of the six-bus power system. All four paths from v_1 to v_4 that do not repeat any edge in Figure 4.18 can be known from

$$\left(\mathbf{A}^{\ast}\right)_{14} = b_{12}b_{23}b_{34} \oplus b_{15}b_{56}b_{46} \oplus b_{12}b_{25}b_{56}b_{46} \oplus b_{15}b_{25}b_{23}b_{34}, \qquad (4.22)$$

whose four terms, respectively, correspond to paths

$$b_{12}b_{23}b_{34}: v_1 \rightarrow v_2 \rightarrow v_3 \rightarrow v_4$$

$$b_{15}b_{56}b_{46}: v_1 \rightarrow v_5 \rightarrow v_6 \rightarrow v_4$$

$$b_{12}b_{25}b_{56}b_{46}: v_1 \rightarrow v_2 \rightarrow v_5 \rightarrow v_6 \rightarrow v_4$$

$$b_{15}b_{25}b_{23}b_{34}: v_1 \rightarrow v_5 \rightarrow v_2 \rightarrow v_3 \rightarrow v_4.$$

If the system separates into two islands by cutting off e_{23} and e_{56}, then all the paths will become invalid. It follows that $(\mathbf{A}^{\ast})_{14} = 0$, which can also be obtained by setting both $b_{23} = 0$ and $b_{56} = 0$ in Eq. (4.22), since each of the four terms contains either b_{23} or b_{56}. This is illustrated by Figure 4.18.

4.6.5.2 Boolean Expressions on the Generation Coherency and Power Balance Constraints

Let $V_{G,1}, \ldots, V_{G,K}$ denote the node sets of K groups of coherent generators to be separated into K islands. Define $I_{G,1}$ to $I_{G,K}$ as the sets of their node indices. The set of all node indices is

$$I_G = I_{G,1} \cup I_{G,2} \cup \cdots \cup I_{G,K}. \qquad (4.23)$$

Let N_G denote the total number of generator nodes. Arbitrarily select K node indices denoted by k_1 to k_K, respectively, from index sets $I_{G,1}$ to $I_{G,K}$. Then the generation coherency constraint can be expressed as a Boolean function of all the Boolean variables b_{ij}s that correspond to all edges:

$$f_{GCC} = \prod_{j=1,\ldots,K}^{\otimes} \left[\prod_{\forall i \in I_{G,j}}^{\otimes} \left(A^* \right)_{i,k_j} \right] \otimes \prod_{\forall l \notin I_G}^{\otimes} \left[\sum_{j=1,\ldots,K}^{\overline{\oplus}} \left(A^* \right)_{l,k_j} \right], \qquad (4.24)$$

where "$\overline{\oplus}$" denotes "EXCLUSIVE OR."

The right-hand side of Eq. (4.24) is equal to the Boolean multiplication of all terms such as $\prod_{\forall i \in I_{G,j}}^{\otimes} (A^*)_{i,k_j}$ and $\sum_{j=1,\cdots,K}^{\overline{\oplus}} (A^*)_{l,k_j}$, so the equation actually describes that $f_{GCC} = 1$ requires all terms in such two forms to be equal to 1. The terms in these two forms are explained in the following:

- **Term** $\sum_{\forall i \in I_{G,j}}^{\otimes} (A^*)_{i,k_j}$ **equal to 1 is equivalent to that all nodes of $V_{G,j}$ are connected.** In fact, that term equals the Boolean multiplication of the elements of A^* that are on the k_j-th column and the rows of indices in $I_{G,j}$. Since each element of A^* represents the connection between two nodes whose indices are, respectively, the indices of the column and row of that element, that term is equal to 1 if and only if all nodes in $V_{G,j}$ are connected with an arbitrarily selected node k_j, or in other words, they are all connected.

- **Term** $\sum_{j=1,\ldots,K}^{\otimes} (A^*)_{l,k_j}$ **equal to 1 is equivalent to that every node (with an arbitrary index l) of the graph is connected with one and only one of nodes k_1 to k_K.** That can easily be understood since as defined before, operation $\overline{\oplus}$ of all $(A^*)_{l,k_j}$'s outputs 1 when one and only one of $(A^*)_{l,k_j}$'s is true.

Then, the Boolean function on power balance constraint can be expressed as

$$f_{PBC} = \prod_{i \in N_G}^{\otimes} \langle \, | (A^*)_{i*} \cdot W | \leq \varepsilon_p \rangle \qquad (4.25)$$

where operator "$\langle \ \rangle$" about a proposition takes the value of 1 if and only if the proposition is true. $(A^*)_{i*}$ is the i-th row of A^* and $W = [w_1, \ldots, w_N]^T$ is a column vector of all node weights, so $(A^*)_{i*} \cdot W$ for a node having index i is equal to the total weight of all nodes connected with that node. Thus, Eq. (4.25) describes that $f_{PBC} = 1$ requires all nodes connected with each generator node to have a small total weight less than the tolerance.

If we assume $f_{GCC} = 1$ (since this condition has to be satisfied anyway), f_{PBC} can be rewritten as

$$f_{PBC} = \prod_{j=1,\ldots,K}^{\otimes} \left\langle \left| \sum_{i \notin I_G} \left(A^* \right)_{i,k_j} \cdot w_k + \sum_{l \in I_{G,j}} w_l \right| \leq \varepsilon_p \right\rangle. \qquad (4.26)$$

Equation (4.26) describes that the total weight of the generator nodes in each coherent group, that is, $\sum_{l \in I_{G,j}} w_l$, plus the total weight of all load nodes connected to one generator node from that group, that is, $\sum_{i \notin I_G} (\mathbf{A}^*)_{i,k_j} \cdot w_k$, is less than the tolerance.

A benefit of using Eq. (4.26) instead of Eq. (4.25) is that we do not need to compute all $N \times N$ elements of $\mathbf{A}^*$ anymore. Note that each element is a Boolean expression about b_{ij}s. The use of Eq. (4.26) only requires computing the k_1-th, k_2-th, ..., k_K-th rows (or columns) of $\mathbf{A}^*$, which are reduced to $K \times N$ elements.

The Boolean function representation of the power balance constraint can be obtained by two approaches.

- The first approach (called Approach I) is to directly solve Eq. (4.26) and write all its solutions in one Boolean expression by "$\otimes$" and "$\oplus$."
- The second approach (Approach II) is to apply binary encoding, which is briefly introduced as follows.

Approach II

Step 1: Represent w_1, ..., w_N and ε_p by binary codes. Thus, they must be replaced by integers $\lceil \lambda \cdot w_1 \rceil$, ..., $\lceil \lambda \cdot w_N \rceil$ and $\lceil \lambda \cdot \varepsilon_p \rceil$, where λ is a big enough integer to decrease rounding errors and "$\lceil \ \rceil$" is the ceil operator obtaining the nearest integer larger than a real number. In order to reduce the binary bits used in encoding, denoted by n_b, we apply modular arithmetic to represent all integers (including negative integers) by 0, 1, ..., $2^{n_b} - 1$. Thus, any integer x is replaced by $x \pmod{2^{n_b}}$. Since all $I_{G,1}$ to $I_{G,K}$ must be separated finally, n_b can be selected by

$$n_b \geq \log_2 \left\lceil \lambda \cdot \max_{k=1,\ldots,K} \left(\sum_{i \in I_{G,k}} w_i \right) \right\rceil. \tag{4.27}$$

Step 2: The integer operations in Eq. (4.25) or (4.26) are translated into equivalent logic operations.

For example, consider a common example of addition operations: $X + Y = Z$, where X, Y, and Z are all integers less than 8 (i.e., having no more than 8 discretized units). Assuming that the 3-bit binary encoding strings of X, Y, and Z are $[x_3 x_2 x_1]$, $[y_3 y_2 y_1]$ and $[z_3 z_2 z_1]$, we have

$$[x_3 x_2 x_1] + [y_3 y_2 y_1] = [z_3 z_2 z_1]. \tag{4.28}$$

There is the following relationship:

$$z_i = (x_i \overline{\oplus} y_i) \overline{\oplus} c_{i-1}, \text{ where } c_{i-1} = (x_{i-1} \otimes y_{i-1}) \oplus [c_{i-2} \otimes (x_{i-1} \oplus y_{i-1})] \text{ and } c_0 = 0. \tag{4.29}$$

From Eq. (4.25) or (4.26), the Boolean function representation f_{PBC} on the power balance constraint can be obtained easily. In fact, many OBDD software packages support OBDD vector representation. They can encode any integer expressions and build their OBDDs by similar approaches. Given n_b and a variable ordering, most of them can directly build an OBDD from Eq. (4.25) or (4.26).

As an illustration, consider the condition of the above six-bus power system, whose node-weighted graph model is shown in Figure 4.17. Assume the following node weights. We find separation strategies to separate Generators 1 and 2 from Generators 3 and 4.

$$w_1 = 0.20, \quad w_2 = 0.30, \quad w_3 = 0.10, \quad w_4 = 0.30, \quad w_5 = -0.50, \quad w_6 = -0.40$$

Selecting $k_1 = 1$ and $k_2 = 4$, we have Eq. (4.30) from Eq. (4.24),

$$f_{GCC} = \left(\mathbf{A}^*\right)_{12} \otimes \left(\mathbf{A}^*\right)_{34} \otimes \left[\left(\mathbf{A}^*\right)_{15} \overline{\oplus} \left(\mathbf{A}^*\right)_{45}\right] \otimes \left[\left(\mathbf{A}^*\right)_{16} \overline{\oplus} \left(\mathbf{A}^*\right)_{46}\right] \quad (4.30)$$

where

$$\left(\mathbf{A}^*\right)_{12} = b_{12} \oplus b_{15}b_{25} \oplus b_{15}b_{56}b_{46}b_{34}b_{23}, \quad \left(\mathbf{A}^*\right)_{34} = b_{34} \oplus b_{23}b_{25}b_{56}b_{46} \oplus b_{23}b_{12}b_{25}b_{56}b_{46},$$

$$\left(\mathbf{A}^*\right)_{15} = b_{15} \oplus b_{12}b_{25} \oplus b_{12}b_{23}b_{34}b_{46}b_{56}, \quad \left(\mathbf{A}^*\right)_{45} = b_{46}b_{56} \oplus b_{34}b_{23}b_{25} \oplus b_{34}b_{23}b_{12}b_{15},$$

$$\left(\mathbf{A}^*\right)_{16} = b_{15}b_{56} \oplus b_{12}b_{25}b_{56} \oplus b_{12}b_{23}b_{34}b_{46} \oplus b_{15}b_{25}b_{23}b_{34}b_{46},$$

$$\left(\mathbf{A}^*\right)_{46} = b_{46} \oplus b_{34}b_{23}b_{25}b_{56} \oplus b_{34}b_{23}b_{12}b_{15}b_{56}$$

From Eq. (4.26), we have

$$f_{PBC} = \left\langle \left| w_1 + w_2 + \left(\mathbf{A}^*\right)_{15} \cdot w_5 + \left(\mathbf{A}^*\right)_{16} \cdot w_6 \right| \leq \varepsilon_p \right\rangle \otimes \left\langle \left| w_3 + w_4 + \left(\mathbf{A}^*\right)_{45} \cdot w_5 + \left(\mathbf{A}^*\right)_{46} \cdot w_6 \right| \leq \varepsilon_p \right\rangle$$

$$(4.31)$$

If we apply Approach I, these equations need to be solved:

$$\begin{cases} \left| 0.2 + 0.3 - 0.5 \cdot \left(\mathbf{A}^*\right)_{15} - 0.4 \cdot \left(\mathbf{A}^*\right)_{16} \right| \leq 0.1 \\ \left| 0.1 + 0.3 - 0.5 \cdot \left(\mathbf{A}^*\right)_{45} - 0.4 \cdot \left(\mathbf{A}^*\right)_{46} \right| \leq 0.1. \\ \left(\mathbf{A}^*\right)_{15}, \left(\mathbf{A}^*\right)_{16}, \left(\mathbf{A}^*\right)_{45} \text{ and} \left(\mathbf{A}^*\right)_{46} \in \{0,1\} \end{cases} \quad (4.32)$$

The only solution turns out to have $(\mathbf{A}^k)_{15} = 1$, $(\mathbf{A}^k)_{16} = 0$, $(\mathbf{A}^k)_{45} = 0$, and $(\mathbf{A}^k)_{46} = 1$. So we have

$$f_{PBC} = \left(\mathbf{A}^*\right)_{15} \otimes \left(\mathbf{A}^*\right)_{46} \otimes \overline{\left(\mathbf{A}^*\right)_{16}} \otimes \overline{\left(\mathbf{A}^*\right)_{45}}. \quad (4.33)$$

Alternatively, we apply Approach II. Let $\lambda = 10$. $n_b \geq 3$ from Eq. (4.27), so let $n_b = 3$. Applying modular arithmetic, we have

$$\lceil \lambda \cdot w_1 \rceil (\mathrm{mod}\ 8) = 2, \quad \lceil \lambda \cdot w_2 \rceil (\mathrm{mod}\ 8) = 3, \quad \lceil \lambda \cdot w_3 \rceil (\mathrm{mod}\ 8) = 1,$$

$$\lceil \lambda \cdot w_4 \rceil (\mathrm{mod}\ 8) = 3, \quad \lceil \lambda \cdot w_5 \rceil (\mathrm{mod}\ 8) = 3, \quad \lceil \lambda \cdot w_6 \rceil (\mathrm{mod}\ 8) = 4$$

$$\lceil \lambda \cdot p \rceil (\mathrm{mod}\ 8) = 1.$$

Thus, we can rewrite the power balance constraint as

$$
\begin{aligned}
f_{\mathrm{PBC}} &= \left\langle 2 + 3 + 3 \cdot (\mathbf{A}^*)_{15} + 4 \cdot (\mathbf{A}^*)_{16} = 0 \right\rangle \otimes \left\langle 1 + 3 + 3 \cdot (\mathbf{A}^*)_{45} + 4 \cdot (\mathbf{A}^*)_{46} = 0 \right\rangle \\
&= \left\langle 3 \cdot (\mathbf{A}^*)_{15} + 4 \cdot (\mathbf{A}^*)_{16} = 3 \right\rangle \otimes \left\langle 3 \cdot (\mathbf{A}^*)_{45} + 4 \cdot (\mathbf{A}^*)_{46} = 4 \right\rangle \\
&= \left\langle \left[0, (\mathbf{A}^*)_{15}, (\mathbf{A}^*)_{15} \right] + \left[(\mathbf{A}^*)_{16}, 0, 0 \right] = [011] \right\rangle \otimes \left\langle \left[0, (\mathbf{A}^*)_{45}, (\mathbf{A}^*)_{45} \right] + \left[(\mathbf{A}^*)_{46}, 0, 0 \right] = [100] \right\rangle
\end{aligned}
$$

$$(4.34)$$

Then, according to Eq. (4.29), we can obtain same result as Eq. (4.33). For more detailed information about binary encoding, see Sun et al. [1].

4.6.5.3 OBDDs on the Generation Coherency and Power Balance Constraints

To yield a compact OBDD representing f_{GCC} and f_{PBC}, all b_{ij}s need to be appropriately ordered. In fact, the problem of finding the optimal ordering of all b_{ij}s for the most compact OBDD itself is an NP-hard problem. However, there is some guidance to determine their ordering for an OBDD with an acceptable size. For CSS, we may use this guidance: the edges whose "open" or "closed" status is more critical to meet the generation coherency and power balance constraints should have higher priorities in OBDDs; these critical edges are such as those either directly connected with or close to generator nodes.

The following algorithm can be adopted to determine the ordering of all edges:

Step 1: Assign new indices 1 to N_G to all generator nodes from 1 to N_G in ascending order of their weights; let $k = 1$ and $O = N_G + 1$.

Step 2: If any node connected with the kth node has not been assigned a new index yet, assign index $= O$ to it and let $O = O + 1$; otherwise, let $k = k + 1$. If k exceeds the total number of nodes N, then go to Step 3; otherwise, perform Step 2 again.

Step 3: Order all b_{ij}s in ascending order of $i' \times N + j'$, where i' and j' satisfying $i' < j'$ are the reordered indices of two nodes of each b_{ij}.

Variable orderings selected by this approach consider the structural information of G, so they are more efficient than random ordering in general. Then

OBDDs of f_{GCC} and f_{PBC} can be easily built by OBDD software. Two OBDDs can be combined by the operation "AND" to obtain the final OBDD on $f_{\text{GCC}} \otimes f_{\text{PBC}}$. One of its solutions can be obtained by the "*satisfy one*" procedure. We even can obtain all solutions or the number of solutions by the "*satisfy all*" and "*satisfy count*" procedures. They can be implemented by OBDD software and their original algorithms can be found in Bryant [49]. In an OBDD with M Boolean variables, any path that passes m different variable nodes from root node to "1" node corresponds to 2^{M-m} solutions of the problem.

For example, by this approach, the ordering of the Boolean variables on seven edges of the six-bus power system is

$$b_{12}, \quad b_{23}, \quad b_{25}, \quad b_{34}, \quad b_{46}, \quad b_{15}, \quad b_{56}.$$

We can construct the OBDD of $f_{\text{GCC}} \otimes f_{\text{PBC}}$, as shown in Figure 4.19.

This OBDD has three paths from the root node to the terminal "1" node, which correspond to all four solutions of the problem as follows:

$$b_{12}\overline{b}_{23}b_{25}b_{34}b_{46}\overline{b}_{56}: \quad \{e_{23}, e_{56}\} \text{ and } \{e_{23}, e_{56}, e_{15}\};$$

$$b_{12}\overline{b}_{23}\overline{b}_{25}b_{34}b_{46}b_{15}\overline{b}_{56}: \quad \{e_{23}, e_{25}, e_{56}\};$$

$$\overline{b}_{12}\overline{b}_{23}b_{25}b_{34}b_{46}b_{15}\overline{b}_{56}: \quad \{e_{12}, e_{23}, e_{56}\}.$$

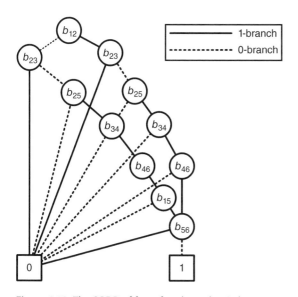

Figure 4.19 The OBDD of $f_{\text{GCC}} \otimes f_{\text{PBC}}$ about the six-bus power system.

Obviously, the $\{e_{23}, e_{56}\}$ is the minimum cut and is contained by the other three solutions, which cut one additional edge, that is e_{15}, e_{25}, or e_{12}, without breaking the connection between nodes 1, 2, and 5 for the first island. The minimum cut has prominence compared to the others, for example, minimizing line outages in creating the same islands. However, under some conditions, for example when a line becomes overloaded in a formed island, disconnecting a line not contributing to separate islands may be needed, such as e_{15}, e_{25}, or e_{12} in this example.

4.6.5.4 Minimum-Cut Strategies

If the minimum cut is expected, the minimum-cut constraint (MCC) may be considered, which can be formulated by

$$f_{MCC} = \prod_{\forall i,j,i<j}^{\otimes} \left\langle \left(\mathbf{A}^*\right)_{ij} \rightarrow b_{ij} \right\rangle = \prod_{\forall i,j,i<j}^{\otimes} \left[\overline{\left(\mathbf{A}^*\right)_{ij}} \oplus b_{ij} \right], \qquad (4.35)$$

where "$\rightarrow$" denotes logic operation "implication." This constraint says that the connectivity of e_{ij} is the same as the connectivity between v_i and v_j through the graph, or, in other words, once v_i and v_j are assigned to the same island, e_{ij} should never open.

For the six-bus power system, the OBDD on $f_{GCC} \otimes f_{PBC} \otimes f_{MCC}$ is given in Figure 4.20 and it only has one solution $\{e_{23}, e_{56}\}$, which is the same as illustrated in Figure 4.18.

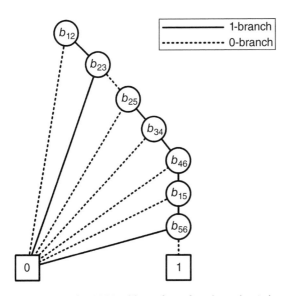

Figure 4.20 The OBDD of $f_{GCC} \otimes f_{PBC} \otimes f_{MCC}$ about the six-bus power system.

4.6.6 Checking the Transmission Capacity and Small Disruption Constraints

Separation strategies satisfying the generation coherency and power balance constraints (or the minimum-cut constraint) can be solved as *k* cuts of the graph model. Each *k*-cut solution can easily be translated back into separation points on the original power network. Then power flows of all branches can be calculated for the separated network with each solution. Accordingly, the transmission capacity constraint and small disruption constraint can be checked to filter out solutions violating these two constraints. Finally, the solutions satisfy all constraints are given as so-called feasible separation strategies.

Note that if the problem of finding separation points is formulated as an optimization problem, one of constraints (typically the power balance constraint) will be formulated as the objective function to be minimized while the others remain as constraints. The optimal separation strategies are given by solving this optimization problem.

Because of complexity and urgency of the situation when CSS is needed in real time, there is no sufficient time to simulate whether a separation strategy satisfying all the aforementioned constraints will necessarily ensure transient stabilities of all generators after islands are formed. In practice, we may relax the definition of a successful system separation to be a feasible separation strategy. After a feasible separation strategy is performed against a detected out-of-step condition, the generators of each island can keep synchronism except for few small, unimportant generators, which can be tripped if losing stability. Thus, it is worth investigating what separation strategies obtained offline have higher chances to lead to a successful system separation, or in other words, to generate fewer unstable generators. In the following, we first introduce one way to formulate the small disruption constraint proposed in Sun et al. [3, 4] in order to increase the chances of satisfactory strategies leading to stable generators. Then, the procedure for checking all constraints is summarized.

4.6.6.1 Small Disruption Constraint

In fact, system separation, either controlled or not, is a disturbance to the power network. If a separation strategy cuts many lines that carry heavy line flows, a large disturbance will be brought by system separation since redistribution of power flows in the power network will significantly impact each formed island. Two indices are defined next to reflect the degree of the power-flow disruption brought by a separation strategy *S*.

$$\gamma_{\text{system}}\left(S\right) \overset{\text{def}}{=} \sum_{e_{ij} \in S} \left|P_{ij}\right| \Big/ \sum_{P_l > 0, v_l \in V} P_l \leq \Gamma_{\text{system}} \tag{4.36}$$

$$\gamma_{\text{island}}(S) \overset{\text{def}}{=} \max_{k=1,\dots,K} \left(\sum_{\substack{e_{ij} \in S, \\ v_i, v_j \in V^k}} |P_{ij}| \Bigg/ \sum_{\substack{P_l > 0, v_l \in V^k}} P_l \right) \leq \Gamma_{\text{island}}, \tag{4.37}$$

where P_l and P_{ij} are, respectively, the real power from the l-th bus and the real power flow of line i-j before separation, K is the number of islands that S produces, and $V^1, \dots, V^K$ are, respectively, the node sets on the K islands.

$\gamma_{\text{system}}(S)$ and $\gamma_{\text{island}}(S)$, respectively, evaluate the system-wise and island-wise percentage power disruptions due to system separation, and Γ_{system} and Γ_{island} are their thresholds determined by offline studies. Sun et al. [4] studies how Γ_{system} and Γ_{island} influence the stabilities of islands formed by separation strategies satisfying the small disruption constraint as well as the generation coherency, power balance, and transmission capacity constraints. The main findings are introduced next.

Randomly select N separation strategies satisfying the generation coherency, power balance, and transmission capacity constraints given by the OBDD-based method. Assume that N_T strategies among all N strategies also satisfy the small disruption constraint for given thresholds Γ_{system} and Γ_{island}, and a ratio R_T of N_T strategies cannot lead to successful separation. N_T depends on values of Γ_{system} and Γ_{island}. Therefore, both N_T and R_T are functions of Γ_{system} and Γ_{island}. For instance, in Sun et al. [4] 500 separation strategies for the IEEE 118-bus power system are studied for the selection of Γ_{system} and Γ_{island}. The coherent groups of generators are assumed to be {10, 12, 25, 26} and {31, 46, 49, 54, 59, 61, 65, 66, 69, 80, 87, 89, 100, 103, 111}. Let Γ_{system} and Γ_{island} both change in 0.0–0.5; the characteristics of $N_T(\Gamma_{\text{system}}, \Gamma_{\text{island}})$ and $R_T(\Gamma_{\text{system}}, \Gamma_{\text{island}})$ are shown in Figure 4.21.

The following are observations from the figure:

1) Larger Γ_{system} and Γ_{island} lead to larger N_T, and usually larger R_T as well.
2) If $\Gamma_{\text{system}} < 0.205$ or $\Gamma_{\text{island}} < 0.025$, then $N_T = 0$, meaning that no strategy satisfies the small disruption constraint.
3) There is a region in the "Γ_{system}-Γ_{island}" plane where the corresponding small disruption constraint makes $N_T > 0$ and $R_T = 0$, meaning that at least one separation strategy satisfying the small disruption constraint leads to successful system separation.

Thus, projection of plots of $N_T(\Gamma_{\text{system}}, \Gamma_{\text{island}})$ and $R_T(\Gamma_{\text{system}}, \Gamma_{\text{island}})$ onto the "Γ_{system}-Γ_{island}" plane determines some regions with different success rates of separation strategies that satisfy the small disruption constraint (as well as the generation coherency constraint, power balance constraint, and transmission capacity constraint), as illustrated in Figure 4.22.

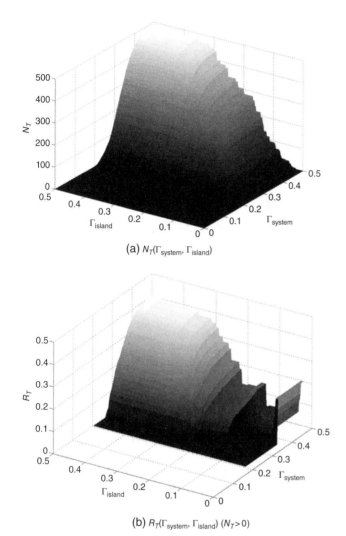

(a) $N_T(\Gamma_{\text{system}}, \Gamma_{\text{island}})$

(b) $R_T(\Gamma_{\text{system}}, \Gamma_{\text{island}})$ $(N_T > 0)$

Figure 4.21 Characteristics of N_T and R_T for 500 separation strategies.

- Region A_{-1}: no separation strategy satisfies the small disruption constraint.
- Region A_1: all separation strategies that satisfy the small disruption constraint are successful separation strategies.
- Region $A_{0.8}$: 80% of separation strategies that satisfy the small disruption constraint are successful separation strategies.
- Region A_0: 0% of separation strategies that satisfy the small disruption constraint are successful separation strategies.

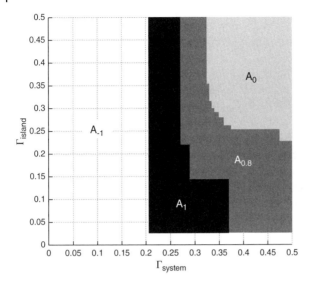

Figure 4.22 Regions in the "Γ_{system}–Γ_{island}" plane indicating the success rate of separation strategies that satisfy the small disruption constraint.

More regions like $A_{0.8}$ may be defined between A_1 and A_0 for different success rates. It is likely that under some conditions the size of A_1 is zero or very small. It means that no matter what threshold values are taken for the small disruption constraint, there are always strategies at some probabilities to cause unsuccessful system separation. Thus, it is suggested that the values of Γ_{system}, Γ_{island} be selected in region A_1 if it exists or at least a region with a high success rate. The relationships between the strategies satisfying the generation coherency constraint, power balance constraint, and transmission capacity constraint, the strategies also satisfying the small disruption constraint, and the strategies leading to successful system separation are illustrated by Figure 4.23. By properly selecting the values of Γ_{system} and Γ_{island}, the separation strategies satisfying the generation coherency, power balance, transmission capacity, and small disruption constraints may have a high probability to lead to successful system separation.

If the minimum-cut constraint is not required, it is possible that some separation strategies satisfying the generation coherency, power balance, and transmission capacity constraints are not the minimum cut of the graph model, or in other words, they open some lines not contributing to separation of islands. Then, there are some measures to reduce computation associated with verification of the transmission capacity constraint and small disruption constraint.

For any given separation strategy S_1, let $\mathbf{A}^{*}(S_1)$ be a constant matrix whose entry on the i-th row and j-th column takes 0 or 1 depending on whether nodes v_i and v_j are still connected after the graph model is partitioned by S_1. Consider

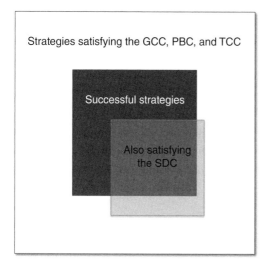

Figure 4.23 Relationships between the strategies satisfying the generation coherency constraint (GCC), power balance constraint (PBC), and transmission capacity constraint (TCC); the strategies also satisfying the small disruption constraint (SDC); and the strategies leading to successful system separation.

another separation strategy S_2 that creates exactly the same islands. We call S_2 a level-k offspring strategy of S_1, and hence S_1 is called a parent strategy of S_2. If S_2 is the i-th ($i \geq 1$) offspring strategy at the same level, we denote it as $S_2 = (S_1)^k_i$. Obviously, there are

$$\mathbf{A}^* (S_2) = \mathbf{A}^* (S_1) \tag{4.38}$$

$$\gamma_{\text{island}} (S_2) \geq \gamma_{\text{island}} (S_1), \ \gamma_{\text{system}} (S_2) > \gamma_{\text{system}} (S_1). \tag{4.39}$$

Two immediate conclusions are

- A minimum-cut separation strategy (satisfying the minimum-cut constraint), say $S_{C,i}$, and its all offspring strategies have the identical "$\mathbf{A}^*(\ )$" matrix, that is, equal to $\mathbf{A}^*(S_{C,i})$.
- For any values of Γ_{system} and Γ_{island}, if a separation strategy does not satisfy the small disruption constraint, then none of its offspring strategies satisfy this constraint, either. Thus, there is no need to check them with the small disruption constraint at all.

Thus, an idea is to first check the small disruption constraint for each minimum-cut separation strategy. Moreover, checking this constraint is much simpler than checking the transmission capacity constraint since the latter needs power-flow calculation. Therefore, if the small disruption constraint is

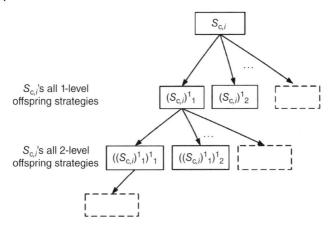

Figure 4.24 Tree structure for checking the small disruption constraint and transmission capacity constraint starting from a minimum-cut strategy.

concerned, the transmission capacity constraint is checked only when the small disruption constraint has been satisfied. The following steps are recommended to check these two constraints starting from N_C separation strategies satisfying the generation coherency constraint, power balance constraint and minimum-cut constraint (i.e., minimum-cut strategies), denoted by $S_{C,1}$ to $S_{C,Nc}$:

Step 1: For each $S_{C,i}(i = 1, ..., N_C)$, calculate $\mathbf{A}^*(S_{C,i})$ and check the small disruption constraint. Only if it is satisfied does the transmission capacity constraint need to be checked. If both constraints are satisfied, $S_{C,i}$ is suggested as a separation strategy, which does not need to trip any additional line to make the transmission capacity constraint satisfied, and thus end the procedure.

Step 2: If none of $S_{C,1}$ to $S_{C,Nc}$ satisfies the small disruption constraint, continue the same checking in Step 1 for their offspring strategies from the lowest level to highest level, but ignore any strategy whose parent strategy violates the small disruption constraint. End the procedure until a strategy satisfying both the small disruption constraint and transmission capacity constraint is found.

This checking progress can be performed on a tree structure as illustrated in Figure 4.24. N_C parallel processors can be used to simultaneously perform such searching procedures respectively from N_C minimum-cut strategies.

4.6.7 Checking All Constraints in Three Steps

Finally, Figure 4.25 suggests a three-stage procedure for online checking of the generation coherency constraint, power balance constraint, minimum-cut constraint, small disruption constraint, and transmission capacity constraint until

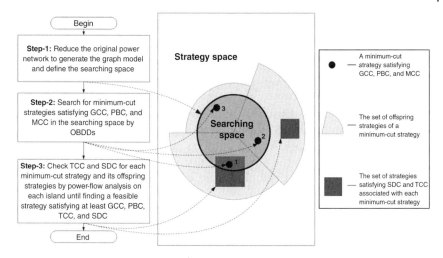

Figure 4.25 Procedure of checking constraints.

finding a separation strategy (called a feasible separation strategy) that at least satisfies all constraints except the minimum-cut constraint.

Step 1: Reduce the original power network using the network reduction and decomposition measures introduced in Sections 4.6.3 and 4.6.4 to generate the graph model of the network, which may comprise a number of subgraphs for parallel processing if decomposition is performed. This stage reduces the original huge strategy space to a searching space of a proper size, illustrated respectively by the large rectangle and the smaller circle in Figure 4.25.

Step 2: Search for minimum-cut strategies satisfying the generation coherency constraint, power balance constraint, and minimum-cut constraint using OBDDs. The strategies found are illustrated by the three black dots numbered 1–3 in Figure 4.25.

Step 3: Check the small disruption constraint and transmission capacity constraint for power-flow calculation results on islands formed by each minimum-cut strategy and its offspring strategies (as indicated by the circular sector containing each black dot) by following the checking process shown in Figure 4.24 until finding a feasible separation strategy satisfying all constraints except the minimum-cut constraint. Note that all offspring strategies of a minimum-cut strategy satisfy the generation coherency constraint and power balance constraint but not the minimum-cut constraint, and feasible separation strategies from these offspring strategies have to additionally satisfy the small disruption constraint and transmission capacity constraint. In Figure 4.25, the set of such feasible separation strategies associated with a minimum-cut strategy is represented by a gray square, of which there are two, which illustrates a scenario that the minimum-cut strategy No. 1 itself satisfies all five constraints; strategy No. 2 itself does not satisfy either the small disruption constraint or the transmission capacity constraint but some of its offspring strategies do; and neither strategy No. 3 nor any of its offspring strategies satisfy both of these two constraints.

In the following, we discuss in what time frame each stage should be performed in a practical CSS scheme. Step 1 only needs to know the network topology, control areas, and likely profiles on generation dispatch and power flows in order to reduce the power network and generate its graph model. Therefore, it can be performed offline, for example, hourly, a day ahead, or even in the planning stage.

Step 2 needs to know the accurate power-flow condition including generations and loads to meet the power balance constraint. If slow coherency analysis is performed to predict generator grouping, only a linearized model of the system is enough without the need to know what specific contingency may happen. In that case, Step 2 and Step 3 only need to be performed whenever the current power-flow solution becomes available, for example, every 3–5 minutes from the online state estimator. If precise real power balance in any island formed is not requested, or in other words, the tolerance ε_p of real power imbalance is moderately large, Step 2 as well as the subsequent Step 3 may be performed offline right after Step 1. In both cases, a strategy table can be established and maintained either offline or online for a number of representative power-flow conditions and contingency scenarios. Each strategy in the table is a feasible separation strategy satisfying at least the generation coherency, power balance, and transmission capacity constraints and is appropriate for one specific pattern of generator grouping and the loading conditions close to one of the concerned representative power-flow conditions. Which strategy should be performed in real time needs to be determined by real-time measurements on the actual out-of-step condition.

For some power systems with more uncertainties in generator coherency and dynamics, the pattern of generator grouping may have to be judged by real-time measurements on the actual system dynamics or even postdisturbance system condition. Then Step 2 and Step 3 must be performed in real time. If Step 1 effectively simplifies the power network, more likely the computations in real time can be finished within 1 second or even faster.

References

1 Sun, K., Zheng, D., and Lu, Q. (2003). Splitting strategies for islanding operation of large-scale power systems using OBDD-based methods. *IEEE Transactions on Power Systems* 18 (2): 912–923.

2 Zhao, Q., Sun, K., Zheng, D. et al. (2003). A study of system splitting strategies for island operation of power system: a two-phase method based on OBDDs. *IEEE Transactions on Power Systems* 18 (4): 1556–1565.

3 Sun, K., Zheng, D., and Lu, Q. (2006). Searching for feasible splitting strategies of controlled system islanding. *IEE Proceedings Generation, Transmission & Distribution* 153 (1): 89–98.

4 Sun, K., Zheng, D., and Lu, Q. (Feb 2005). A simulation study of OBDD-based proper splitting strategies for power systems under consideration of transient stability. *IEEE Transactions on Power Systems* 20 (1): 389–399.

5 Li, X. and Zhao, Q. (2007). Parallel implementation of OBDD-based splitting surface search for power system. *IEEE Transactions on Power Systems* 22 (4): 1558–1593.

6 Jin, M., Sidhu, T.S., and Sun, K. (2007). A new system splitting scheme based on the unified stability control framework. *IEEE Transactions on Power Systems* 22 (1): 433–441.

7 Xue, Y., Van Custem, T., and Ribbens-Pavella, M. (1988). A simple direct method for fast transient stability assessment of large power systems. *IEEE Transactions on Power Systems* 3 (2): 400–412.

8 Pavella, M., Ernst, D., and Ruiz-Vega, D. (2000). *Transient Stability of Power Systems: A Unified Approach to Assessment and Control.* Norwell, MA: Kluwer.

9 Sun, K., Sidhu, T. S., and Jin, M., (2005). "Online pre-analysis and real-time matching for controlled splitting of large-scale power networks," *IEEE International Conference on Future Power Systems, Amsterdam, Nov. 16–18, 2005.*

10 You, H., Vittal, V., and Yang, Z. (2003). Self-healing in power systems: an approach using islanding and rate of frequency decline-based load shedding. *IEEE Transactions on Power Systems* 18: 174–181.

11 You, H., Vittal, V., and Wang, X. (2004). Slow coherency-based islanding. *IEEE Transactions on Power Systems* 19 (1): 483–491.

12 Wang, X. and Vittal, V. (2004). "System islanding using minimal cutsets with minimum net flow," *IEEE PES Power Systems Conference and Exposition, Oct. 10–13, 2004.*

13 Yang, B. and Vittal, V. (2006). Slow-coherency-based controlled islanding: a demonstration of the approach on the August 14, 2003 blackout scenario. *IEEE Transactions on Power Systems* 21 (4): 1840–1847.

14 Xu, G. and Vittal, V. (2010). Slow coherency based cutset determination algorithm for large power systems. *IEEE Transactions on Power Systems* 25 (2): 877–884.

15 Xu, G., Vittal, V., Meklin, A., and Thalman, J.E. (2011). Controlled islanding demonstrations on the WECC system. *IEEE Transactions on Power Systems* 26 (1): 334–343.

16 Yang, B., Vittal, V., Heydt, G. T. et al. (2007). "A novel slow coherency based graph theoretic islanding strategy," *IEEE PES General Meeting, 24–28 June, 2007.*

17 Ding, L., Gonzalez-Longatt, F.M., Wall, P., and Terzija, V. (2013). Two-step spectral clustering controlled islanding algorithm. *IEEE Transactions on Power Systems* 28 (1): 75–84.

18 Ding, L., Wall, P., and Terzija, V. (2014). Constrained spectral clustering based controlled islanding. *International Journal of Electrical Power & Energy Systems* 63: 687–694.

19 Sánchez-García, R.J., Fennelly, M., Norris, S. et al. (2014). Hierarchical spectral clustering of power grids. *IEEE Transactions on Power Systems* 29 (5): 2229–2237.

20 Song, H., Wu, J., and Wu, K. (2014). A wide-area measurement systems-based adaptive strategy for controlled islanding in bulk power systems. *Energies* 7: 2631–2657.

21 Quirós-Tortós, J., Sánchez-García, R., Brodzki, J. et al. (2015). Constrained spectral clustering-based methodology for intentional controlled islanding of large-scale power systems. *IET Generation Transmission and Distribution* 9 (1): 31–42.

22 Ding, L., Guo, Y., and Wall, P. (2017). Performance and suitability assessment of controlled islanding methods for online WAMPAC application. *International Journal of Electrical Power & Energy Systems* 84: 252–260.

23 Li, J., Liu, C.-C., and Schneider, K.P. (2010). Controlled partitioning of a power network considering real and reactive power balance. *IEEE Transactions on Smart Grid* 1 (3): 261–269.

24 Liu, W., Liu, L., and Cartes, D.A. (2007). Binary particle swarm optimization based defensive islanding of large scale power systems. *International Journal of Computer Science and Applications* 4: 69–83.

25 Abdelaziz, A.Y., El-Khattam, W., and Tageldin, M. (2014). Application of angle-modulated particle swarm optimization technique in power system controlled separation WAP. *International Journal of Intelligent Systems and Applications in Engineering* 2 (3): 51–57.

26 Aghamohammadi, M.R. and Shahmohammadi, A. (2012). Intentional islanding using a new algorithm based on ant search mechanism. *International Journal of Electrical Power & Energy Systems* 35: 138–147.

27 Tang, F., Zhou, H., Wu, Q. et al. (2015). A Tabu search algorithm for the power system islanding problem. *Energies* 8: 11315–11341.

28 Jabari, F., Seyedi, H., and Ravadanegh, S.N. (2015). Large-scale power system controlled islanding based on backward elimination method and primary maximum expansion areas considering static voltage stability. *International Journal of Electrical Power & Energy Systems* 67: 368–380.

29 Wu, Y., Tang, Y., Han, B., and Ni, M. (2015). A topology analysis and genetic algorithm combined approach for power network intentional islanding. *International Journal of Electrical Power & Energy Systems* 71: 174–183.

30 Trodden, P.A., Bukhsh, W.A., Grothey, A., and McKinnon, K.I.M. (2013). MILP formulation for controlled islanding of power networks. *International Journal of Electrical Power & Energy Systems* 45: 501–508.

31 Trodden, P.A., Bukhsh, W.A., Grothey, A., and McKinnon, K.I.M. (2014). Optimization-based islanding of power networks using piecewise linear AC power flow. *IEEE Transactions on Power Systems* 29 (3): 1212–1220.

32 Ding, T., Sun, K., Huang, C. et al. (2015). Mixed-integer linear programming-based splitting strategies for power system islanding operation considering network connectivity. *IEEE Systems Journal* 12 (1): doi: 10.1109/JSYST.2015.2493880.

33 Ding, T., Sun, K., Yang, Q. et al. (in press). Mixed integer second order cone relaxation with dynamic simulation for proper power system islanding operations. *IEEE Journal on Emerging and Selected Topics in Circuits and Systems* accepted.

34 Wang, C.G., Zhang, B.H., Hao, Z.G. et al. (2010). A novel real time searching method for power system splitting boundary. *IEEE Transactions on Power Systems* 25 (4): 1902–1909.

35 Raak, F., Susuki, Y., and Hikihara, T. (2016). Data-driven partitioning of power networks via Koopman mode analysis. *IEEE Transactions on Power Systems* 31 (4): 2799–2808.

36 Adibi, M.M., Kafka, R.J., Maram, S., and Mili, L.M. (2006). On power system controlled separation. *IEEE Transactions on Power Systems* 21 (4): 1894–1902.

37 Diao, R., Vittal, V., Sun, K. et al. (2009). "Decision tree assisted controlled islanding for preventing cascading events," *IEEE PES Power Systems Conference and Exposition, Seattle, WA, March 15–18, 2009*.

38 Senroy, N., Heydt, G.T., and Vittal, V. (2006). Decision tree assisted controlled islanding. *IEEE Transactions on Power Systems* 21 (4): 1790–1797.

39 Ahsan, M.Q., Chowdhury, A.H., Ahmed, S.S. et al. (2012). Technique to develop auto load shedding and islanding scheme to prevent power system blackout. *IEEE Transactions on Power Systems* 27 (1): 198–205.

40 Sun, K., Hur, K., and Zhang, P. (2011). A new unified scheme for controlled power system separation using synchronized phasor measurements. *IEEE Transactions on Power Systems* 26 (3): 1544–1554.

41 Chow, J.H., Galarza, R., Accari, P. et al. (1995). Inertial and slow coherency aggregation algorithms for power system dynamic model reduction. *IEEE Transactions on Power Systems* 10 (2): 680–685.

42 You, H., Vittal, V., and Wang, X. (2004). Slow coherency-based islanding. *IEEE Transactions on Power Systems* 19 (1): 483–491.

43 Automatic Underfrequency Load Shedding, NERC Standard PRC-006-2.

44 Thomas J. Schaefer, The complexity of satisfiability problems, Proceedings of the 10th Annual ACM Symposium on Theory of Computing. San Diego, CA. pp. 216–226, 1978.

45 Papadimitriou, C.H. (1982). *Combinatorial Optimization: Algorithms and Complexity*. Englewood Cliffs, NJ: Prentice-Hall.

46 Stoer, M. and Wagner, F. (1997). A simple min-cut algorithm. *Journal of the ACM* 44 (4): 585–591.

47 Dahlhausl, E., Johnson, D. S., Papadimitriou, C. H. (1992). "The complexity of multiway cuts (extended abstract)," *Proceedings of the 24th Annual ACM Symposium on Theory of Computing, Victoria, British Columbia, Canada, May 4–6, 1992.*

48 Kim, K.H. (1982). *Boolean Matrix Theory and Applications.* New York: Marcel Dekker.

49 Bryant, R.E. (1986). Graph-based algorithms for Boolean function manipulation. *IEEE Transactions on Computers* C-35: 677–691.

5

Online Decision Support for Controlled System Separation

5.1 Online Decision on the Separation Strategy

Real-time wide-area synchrophasor measurements can be used to identify the actual pattern of generator grouping, to predict an out-of-step condition, and to decide when controlled system separation (CSS) should be performed. This section presents one approach for online decision on the separation strategy:

- A strategy table is established offline and maintained online, including feasible separation strategies appropriate for one specific pattern of generator grouping and the current power-flow condition.
- Generator grouping is identified online using wide-area synchrophasor measurements to select a matched separation strategy from the table. Accordingly, the separation boundary and points are determined.
- Separation timing is decided in real time when an out-of-step condition is predicted on the separation boundary.

The synchrophasor-based wide-area measurement system (WAMS) is important in online decision support for CSS. This chapter will introduce the WAMS-based techniques for identification of generator grouping according to inter-area oscillations between elementary coherent groups (ECGs). When generators are going out of step, asynchronism usually occurs first on the oscillation between two ECGs, which is often related to a dominant inter-area oscillation mode. The out-of-step prediction and separation timing decision in this chapter focus on 2-cut separation, that is, formation of two islands.

Failures occurring between two ECGs can be reflected from a real-time change in the mode shape of their inter-area oscillation. The frequencies of inter-area oscillations are mainly in the range of 0.1–0.9 Hz.

Power System Control Under Cascading Failures: Understanding, Mitigation, and System Restoration, First Edition. Kai Sun, Yunhe Hou, Wei Sun and Junjian Qi.

Oscillation frequencies faster than 0.9 Hz are usually of local oscillation modes. Instability caused by a local mode is more likely controlled by a local protective action rather than CSS.

In the rest of this chapter, a measurement-based spectral analysis method is presented for determining 2-cut separation boundaries. Then, its limitations on power systems are discussed because of the unavoidable drifting nature of the oscillation frequency with an electromechanical mode. Also, the concept of the "Frequency-Amplitude Curve" is introduced to characterize such frequency drifting. A phase-locked loop-based method is introduced for more accurate identification of generator grouping. Then, methods for out-of-step prediction are introduced. Finally, a WAMS-based three-stage scheme for online decision support of CSS will be presented and demonstrated.

5.1.1 Spectral Analysis-Based Method

This method has two steps: (i) identification of a few inter-area modes and (ii) prediction of the separation boundary.

Step 1: Identify a few inter-area modes that are dominant in terms of the amplitude of oscillation.

Let N_G be the number of ECGs. Synchrophasors are installed at the terminal buses of large generating units in each ECG. Assume $\delta_1,\ldots,\delta_{N_G}$ to be the average phase angles of ECGs calculated from synchrophasor data over a sliding time window, T.

Calculate angle difference $\delta_{ij} = \delta_i - \delta_j$ between any two adjacent ECGs. Apply fast Fourier transform (FFT) to each δ_{ij} to identify the frequencies of several (e.g., two or three) strongest oscillation modes in terms of their magnitudes. If the same oscillation mode is observed at different locations of the system, its central frequency can be estimated by averaging its frequencies from FFT results on different pairs of ECGs that identify the mode as one of the strongest modes. Finally, a few dominant modes with distinct frequencies are identified.

Then, apply the modal analysis technique in Trudnowski [1] to judge whether a dominant mode is an inter-area or local mode. The technique is briefly introduced as follows. Introduce squared-coherency function γ_{ij}^2 as defined in Eq. (5.1) about δ_i and δ_j, where ω is the angular frequency of the mode, $S_{ij}(\omega)$ is the cross-spectral density (CSD) of δ_i and δ_j, $S_{ii}(\omega)$ and $S_{jj}(\omega)$ are respectively their power spectral densities (PSD), $\mathcal{F}(\cdot)$ denotes FFT, and $E\{\,\}$ calculates the expectation.

$$\gamma_{ij}^2(\omega) = \frac{\left|S_{ij}(\omega)\right|^2}{S_{ii}(\omega)S_{jj}(\omega)} = \frac{E\left\{\mathcal{F}^*(\delta_i)\cdot\mathcal{F}(\delta_j)\right\}}{E\left\{\mathcal{F}^*(\delta_i)\cdot\mathcal{F}(\delta_i)\right\}\cdot E\left\{\mathcal{F}^*(\delta_j)\cdot\mathcal{F}(\delta_j)\right\}} \tag{5.1}$$

The CSD and PSD can be estimated using the data of δ_i and δ_j over the latest time window T. If $\gamma_{ij}^2(\omega)$ is close to 1 (e.g. > 0.7) for some i and j, this means that the mode involves significantly two ECGs i and j and hence the mode is considered an inter-area mode rather than a local mode; otherwise, the mode can be filtered out. Order the identified inter-area modes by magnitude, and select the top-N_M as dominant inter-area modes with angular frequencies $\omega_1,\ldots,\omega_{N_M}$ corresponding to frequencies $f_1,\ldots,f_{N_M}$. Here, N_M is a number from 2 to 5.

Step 2: The next step is to predict a separation boundary regarding each dominant inter-area mode.

For any two ECGs i and j, the phase angle of CSD S_{ij} is denoted by β_{ij} and can be calculated to estimate the mode shape at any ω. If $|\beta_{ij}(\omega)|$ is close to $0°$, two ECGs swing together about ω; if $|\beta_{ij}(\omega)|$ is close to $180°$, they swing against each other. To distinguish these two conditions, define two thresholds $\beta_1 < 90°$ and $90° < \beta_2 < 180°$ for β_{ij}. For example, $\beta_1 = 60°$ and $\beta_2 = 120°$. Accordingly, the following algorithm is adopted to infer potential separation boundaries:

1) Calculate $\beta_{ij}(\omega_n)$ for each dominant inter-area mode ω_n ($n = 1, \ldots, N_M$). If $|\beta_{ij}(\omega_n)| < \beta_1$, two ECGs are treated as "swinging together about ω_n"; if $|\beta_{ij}(\omega_n)| > \beta_2$, they are treated as "swinging against each other about ω_n"; otherwise, the relationship is considered "undetermined about ω_n."
2) If a 2-cut partition among all ECGs can be determined from this analysis, it indicates a probable separation boundary about ω_n; otherwise, no boundary is associated with ω_n.
3) Order all determined boundaries by the magnitudes of the associated modes, and select the top N_I as probable separation boundaries to monitor in real time for out-of-step conditions.

Each probable separation boundary k ($k = 1, \ldots, N_I$) divides $\delta_1,\ldots,\delta_{N_G}$ into two groups. Calculate the average angle of each group over the window T, and calculate their differences. To support out-of-step prediction on that separation boundary, other modal properties of the oscillation across the boundary can also be estimated from the data over the window T, for example, the damping ratio using Prony analysis [2] or analytical wavelet transform [3].

5.1.2 Frequency-Amplitude Characteristics of Electromechanical Oscillation

5.1.2.1 Limitations of the Spectral Analysis-Based Method
The accuracy of this spectral analysis-based method is limited due to the nature of FFT algorithm: it has spectral leakage when fed with a finite length of signal, and it only considers a finite number of discrete frequency components due to computational complexity. Thus, the estimated frequency and phase of

a signal of interest are influenced heavily by nearby frequency components. Consequently, significant errors also exist in the other modal parameters. That disadvantage becomes obvious when several oscillation modes have close frequencies. Besides, in a power system, the signals captured by synchrophasors often show floating modal parameters, especially for frequencies and phases in transient periods following disturbances, which is because a power system is essentially a nonlinear dynamical system. More specifically, the real-time oscillation frequency of an electromechanical mode may decrease with the increase of the oscillation amplitude. Such a phenomenon is mathematically explained in Wang and Sun [4], and is illustrated on a single-machine-infinite-bus (SMIB) shown in Figure 5.1 with the model given by (5.2).

$$\ddot{\delta} = \frac{\omega_0}{2H}\left(P_m - P_e\right) - \frac{D}{2H}\dot{\delta}$$
$$P_m = P_{max}\sin\delta_s$$
$$P_e = P_{max}\sin\delta$$

(5.2)

In Figure 5.1 and model, δ is the rotor angle, ω_0 is the synchronous frequency; E', P_m, and P_e represent the electromotive force (EMF), mechanic power, and electric power, respectively; H and D represent the inertia and damping constants of the machine, respectively; X is the total reactance between the EMF and the infinite bus, and the resistance in between is ignored. For simplicity, all parameters except for δ are assumed to remain constant. Define $P_{max} = E' \cdot 1/X$ as the theoretical maximum power transfer to the infinite bus and let δ_s denote the rotor angle at the stable equilibrium point (SEP). Model (5.2) can be transformed into a nonlinear second order differential equation:

$$\ddot{\delta} + \frac{D}{2H}\dot{\delta} + \frac{\omega_0}{2H}P_{max}\left(\sin\delta - \sin\delta_s\right) = 0$$

(5.3)

This equation describes a nonlinear oscillator having one oscillation mode, whose natural frequency f_n is defined as the oscillation frequency of a harmonic oscillator system that linearizes (5.3) at the SEP. f_n can be calculated by

$$f_n = \frac{1}{2\pi}\sqrt{\frac{P_{max}\omega_0}{2H}\cos\delta_s} \, .$$

(5.4)

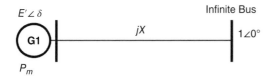

Figure 5.1 Single-machine-infinite-bus system (SMIB).

Note that the actual oscillation frequency of the power system following a disturbance (e.g., starting from an initial state different from the SEP) is not equal to the natural frequency because of the nonlinear nature of this power system.

Nonlinear differential Eq. (5.3) has no accurate closed-form solution. Its trajectory starting from an initial state having $\delta(0)$ and $\dot{\delta}(0)$ can be solved by numerical integration as an initial value problem of the differential equation. For example, consider parameters $H = 3$s, $D = 0$, $X = 1$ p.u., $E' = 1.7$ p.u., $\delta_s = 15°$, and $\omega_0 = 2\pi \times 60 \, \text{rad s}^{-1}$. Then, P_{max} and P_m are calculated to be 1.7 and 0.44 p.u., respectively. There are two equilibrium points: the SEP at $\delta_s = 15°$ and the unstable equilibrium point at $\delta = 180° - 15° = 165°$. The natural frequency of the only oscillation mode is 1.62 Hz from (5.4).

Consider two scenarios shown in Figure 5.2, which are numerical solutions of the system starting from two different initial states:

- Small-disturbance scenario: the trajectory starting from $\delta(0) = 25°$ and $\dot{\delta}(0) = 0$ has nearly linear system dynamics, that is, the trajectory of a harmonic oscillator.
- Large-disturbance scenario: the trajectory starting from $\delta(0) = 157°$ and $\dot{\delta}(0) = 0$ has significant nonlinearities in its swings.

A comparison in Figure 5.2 indicates that this nonlinear system exhibits slower swings and more significant asymmetry in the waveform when the disturbance

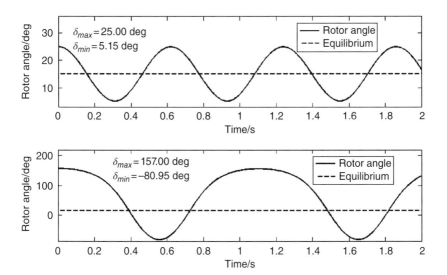

Figure 5.2 Rotor angles of an SMIB system under small (upper) and large (lower) disturbances.

of the system becomes bigger. The following are more detailed observations about nonlinearities of the oscillation:

- First, for the small-disturbance scenario, the period of oscillation is 0.62 s (i.e., the time between adjacent two maxima or two minima), or in other words, the oscillation frequency is 1.61 Hz, which is very close to the natural frequency 1.62 Hz. However, for the large-disturbance scenario, the period increases to 1.09 s and the frequency decreases to 0.92 Hz. Hence, the oscillation frequency decreases from the natural frequency with the increase of oscillation amplitude.
- Second, the waveform and amplitudes of the oscillation are not symmetric about the SEP. For example, the positive and negative amplitudes of the rotor angle following the small disturbance are 10° and 9.85°, respectively, which are fairly close but still different; however, for the large-disturbance scenario, that difference increases to 46°.
- Third, the two halves of one cycle of the oscillation are not symmetric either. For the large-disturbance scenario, the time length of any upper half of cycle is about 0.76 s while the length of any lower half is only about 0.33 s.

This SMIB system ignores oscillation damping. In a realistic power system, because of the existence of frequency-dependent loads and other damping factors, any power system oscillation will sooner or later be damped from large amplitudes to small amplitudes. This means that following a large disturbance, the waveform about an oscillation mode may exhibit swings slower than its natural frequency at the beginning and then gradually increase its frequency, approaching its natural frequency. This fact discovers the drifting oscillation frequency and asymmetric waveform about any power system oscillation.

Figure 5.3 illustrates a gradually damped rotor angle oscillation waveform under a large disturbance. In the figure, "TW1," "TW2," "TW3," and "TW4" are the oscillation periods estimated between two adjacent maxima or minima, which clearly indicate gradually increased oscillation frequency with the decrease of oscillation amplitude.

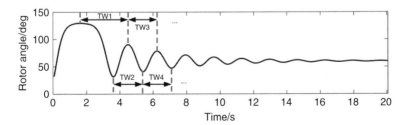

Figure 5.3 Gradually damped rotor angle oscillation under a large disturbance.

Therefore, any spectral analysis technique that is designed for outputs of linear harmonic oscillators or for sinusoidal signals having constant natural frequencies must introduce errors in the estimated modal properties including the oscillation frequency, damping, and even the mode shape.

5.1.2.2 Frequency-Amplitude Curve

For an SMIB system without damping, its oscillation frequency during any half of a swing cycle is derived mathematically in Wang and Sun [4] as a function of the maximum and minimum rotor angles given in (5.5), which assumes the conservation of mechanical energy with the generator and needs to solve two elliptic integrals of the first kind.

$$f\left(\delta_{max},\delta_{\min}\right)=\sqrt{\frac{\omega_0 P_{\max}}{4H}}\times\left[\int_0^{\delta_{max}}\frac{d\delta}{\mu\left(\delta_s+\delta,\delta_s+\delta_{\max}\right)}+\int_{\delta_{min}}^0\frac{d\delta}{\mu\left(\delta_s+\delta,\delta_s+\delta_{\min}\right)}\right]^{-1}$$

where $\mu\left(x,y\right)=\sqrt{\cos x-\cos y+\left(x-y\right)\sin\delta_s}$

$$(5.5)$$

Accordingly, the concept of frequency-amplitude curve (or for short, F-A curve) is defined to characterize how the oscillation frequency (OF) may change with the oscillation amplitude (OA) about an electromechanical mode, as illustrated in Figure 5.4. Here, the OA can be either $\delta_{\max}$ or $|\delta_{\min}|$, or the average.

The F-A curve is a comparative concept of the well-known power-voltage (P-V) curve for voltage stability analysis, which projects the trace of the continuously changing SEP until saddle-node bifurcation from a high-dimensional state space onto a two-dimensional power-voltage plane. The F-A curve actually projects the system trajectory from a high-dimensional state space within the stability boundary onto a two-dimensional OA-OF plane, and hence can be used for analyses on nonlinear power system oscillation and angle stability. As illustrated in Figure 5.4, the F-A curve has the following properties [4, 5]:

- The curve intersects with the OF and OA axes respectively at the SEP point $(0, f_n)$ and a "nose point" or "knee point" $(\pi\text{-}2\delta_s, 0)$: at the SEP, the OA is zero

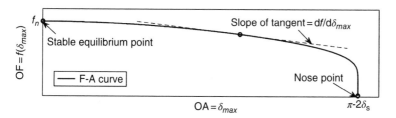

Figure 5.4 F-A curve on an electromechanical mode.

and the OF equals the natural frequency f_n; at the nose point, $\pi\text{-}2\delta_s$ gives the largest allowed OA for any stable oscillation, or in other words, it is actually the stability limit of the angle, while the waveform is not oscillatory anymore so as to lose the definition of oscillation frequency.

- Oscillation frequency monotonically decreases with the increase of the oscillation amplitude and the curve has a slope equal to $-\dfrac{1}{3\pi^2}\sqrt{\dfrac{P_{\max}\omega_0}{2H\ \cos\delta_s}}$ at the SEP.

It is pointed out that the nose point is the projection of part of the stability boundary, which contains the controlling unstable equilibrium point (CUEP). Projected onto the two-dimensional OA-OF plane, an F-A curve can help visualize how the disturbed system state moves back to its SEP over time in a straightforward way, and its real-time margin to the stability boundary is the distance to the nose point. Wang and Sun [4] also demonstrate that for each dominant electromechanical mode with a multimachine power system, an F-A curve can also be estimated from measurements on how the oscillation frequency varies with the amplitude about that mode. The resulting F-A curve has a similar shape to that in Figure 5.4 for an SMIB system.

Coming back to the online decision of generator grouping for CSS, we should consider the drifting oscillation frequency about a dominant inter-area mode with the change of its oscillation amplitude because the power system before CSS has been significantly stressed and exhibits more nonlinearities in its oscillations. Knowing the F-A characteristic of the mode is important to avoid treating it as multiple modes in the spectrum with multiple frequencies. In the following, a real-time algorithm is presented for tracking the modal properties of each dominant mode in real time using synchrophasor measurements so that generator grouping can be more reliably identified.

5.1.3 Phase-Locked Loop-Based Method

5.1.3.1 Flow Chart of the PLL-Based Method

This section introduces the phase-locked loop (PLL)–based algorithm proposed in Sun et al. [6] for identification of dominant inter-area modes and their shapes for a power system. Compared to the spectral-analysis-based method, this method is better in tracking drifting modal properties. It utilizes PLL as a closed-loop feedback control system to automatically adjust a locally generated signal to optimally fit the targeted signal, that is, the real-time measurement from a synchrophasor, so that the modal properties including coherency and mode shape on the targeted signal can be directly obtained from the generated signal.

A PLL [7] is a type of closed-loop feedback control system that automatically adjusts specific modal properties, for example frequency and phase, of a signal rebuilt locally to lock on those of an input signal. Due to its feedback nature,

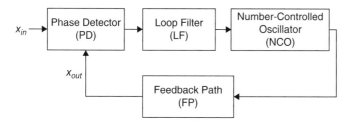

Figure 5.5 PLL algorithm.

the rebuilt signal can optimally fit the input signal. Thus, the frequency and phase can be dynamically estimated at a high accuracy. PLL has been successfully applied to several fields about power system signal analysis; for example, fault detection and harmonic estimation.

The structure of a discrete PLL is illustrated in Figure 5.5, where x_{in} and x_{out} are input and rebuilt signals, respectively, with phases θ^{ref} and θ. Typically, the closed loop contains four blocks: the phase detector (PD), loop filter (LF), number-controlled oscillator (NCO), and feedback path (FP). The phase detector compares two signals and produces an error signal proportional to their phase difference. The error signal goes through the loop filter (usually a low-pass filter) to drive a number-controlled oscillator to create an output phase θ going back through the feedback path. Despite the floating of the input signal, such a negative feedback loop makes the output phase θ dynamically locked to the input phase θ^{ref}.

In a typical implementation of PLL, the phase detector may simply be a digital multiplier, the loop filter usually adopts a proportional-integral controller and the number-controlled oscillator is essentially an integrator converting frequency into phase. Figure 5.6 illustrates a simple linearized Z-domain model for PLL, where $LF(z)$ and $NCO(z)$ are defined by (5.6) and (5.7). In (5.6), K_P and K_I are, respectively, the gains of the proportional and integral paths influencing the tracking performance.

$$LF(z) = K_P + \frac{K_I z^{-1}}{1 - z^{-1}} \tag{5.6}$$

$$NCO(z) = \frac{z^{-1}}{1 - z^{-1}} \tag{5.7}$$

The PLL-based approach takes three steps as shown in Figure 5.7:

Step 1: Mode detection
Step 2: Signal tracking
Step 3: Mode shape analysis

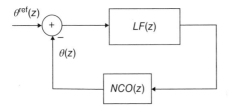

Figure 5.6 A simple PLL algorithm in the *Z*-domain.

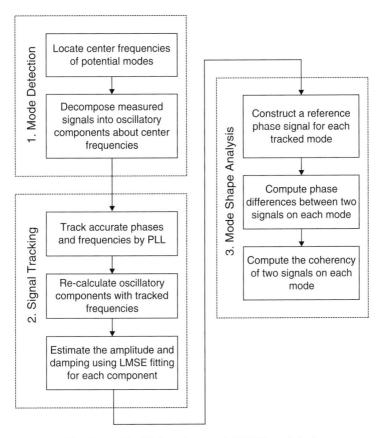

Figure 5.7 Flowchart of the PLL-based approach. "LMSE" stands for "least mean square error".

5.1.3.2 Mode Detection

Assume that N_G ECGs of the system are respectively monitored by synchrophasors (at least one representative generator of each group is monitored). The *i*th ECG has a time series of measurements denoted by $x_i(1)$ to $x_i(L)$ at the interval of Δt (e.g., 1/30 s for a phasor measurement unit at the 30 Hz sampling rate). The length of the time window is $T = L \times \Delta t$.

The first step is to identify low-frequency oscillation modes in each signal x_i measured by a synchrophasor. According to knowledge on the power system, a number (denoted by M) of targeted oscillation modes can be preselected. The initial estimates on their frequencies are denoted by $f_1^0, \ldots, f_M^0$ with expected standard deviations $\sigma_1^0, \ldots, \sigma_M^0$, respectively. As defined in Eq. (5.8), M complex Gaussian filters are adopted to decompose x_i in the frequency domain into M oscillatory components s_{im} $(i = 1, \ldots, N, m = 1, \ldots, M)$, each with respect to one single center frequency at f_m^0.

$$g_m\left(f\right) = \frac{\exp\left[-\dfrac{\left(f - f_m^0\right)^2}{2\left(\sigma_m^{\,0}\right)^2}\right]}{\sqrt{2\pi}\sigma_m^{\,0}} \quad m = 1, \ldots, M \tag{5.8}$$

s_{im} is calculated by the linear convolution operation

$$s_{im} = x_i \otimes g_m. \tag{5.9}$$

5.1.3.3 Signal Tracking

For each component s_{im} in signal x_i, a PLL is used to accurately track its phase, named θ_{im}. Selection of K_P and K_I should consider the power of s_{im} and can be optimized for the fastest tracking. The proposed PLL-based approach selects K_P and K_I by the following two steps to meet the requirement of real-time applications:

Step 1: Select typical data on x_i over a time window and estimate the power (denoted by A_{im}) of s_{im} by the corresponding peak in the FFT plot on the data.

Step 2: Let $K_{P0} = 1/A_{im}$ and $K_{I0} = 0.01K_{P0}$ as initial values of K_P and K_I. Then, test the tracking performances of the PLL with sample values of K_P and K_I, respectively, in the ranges of $0.1K_{P0}$ to $10K_{P0}$ and $0.1K_{I0}$ to $10K_{I0}$, and then select the best K_P and K_I.

With θ_{im} tracked, the instantaneous frequency f_{im} for component s_{im} in signal x_i can be estimated by

$$f_{im} = \frac{d\theta_{im}}{dt} \quad i = 1, \ldots, N. \tag{5.10}$$

This mode detection and PLL-based signal tracking can also be executed on signal x_i over a time window by the following iterative procedure for more accurate modal estimation, especially if x_i has modes with close frequencies:

Step 1: For a targeted mode m, select $f^0{}_m$ (based on, e.g., FFT of the data), $\sigma^0{}_m$, a small positive number Δ for convergence checking, and a decreasing factor $\lambda \in (0, 1)$ for iteration speed control.

Step 2: Calculate component s_{im} by (5.8) and (5.9) and perform the PLL algorithm to track the frequency as f_{im}.

Step 3: If $|f_{im} - f^0{}_m|/f^0{}_m < \Delta$, then exit; otherwise, let f_{im} and $\lambda\sigma^0{}_m$ be the new $f^0{}_m$ and $\sigma^0{}_m$, and go back to Step 2.

It should be noted that for a signal containing two modes with very close frequencies, its FFT plot may just indicate one peak if the time window is not long enough. In Step 1 of this iterative procedure, their initial frequencies may be selected from different sides of the peak to avoid converging to the same mode.

After the estimated frequencies about mode m are obtained from all N signals, which are $f_{1m}, \ldots, f_{Nm}$, we use their mean value denoted by f_m as the final estimate of the frequency. Then, similar to Eq. (5.9), apply a complex Gaussian filter at f_m with a reduced deviation σ_m, for example, $0.2\sigma^0{}_m$, to signal x_i to obtain y_{im} as a better approximation of the m-th oscillatory component. There is

$$x_i \approx \sum_{m=1,\ldots,M} y_{im}. \tag{5.11}$$

The next step is obtain damping coefficient d_{im} and amplitude a_{im} of component y_{im} together by minimizing either of the following two objective functions

$$c_1\left(a_{im}, d_{im}\right) = \sum_{l=1}^{L} \left\| |y_{im}(l \cdot \Delta t)| - a_{im}e^{-d_{im}\cdot l \cdot \Delta t} \right\|^2 \tag{5.12}$$

$$c_2\left(a_{im}, d_{im}\right) = \sum_{k=1}^{L} \left| \ln|y_{im}(l \cdot \Delta t)| - \left[\ln(a_{im}) - d_{im} \cdot l \cdot \Delta t\right] \right|^2. \tag{5.13}$$

For instance, the second one can be solved as a least mean square error (LMSE) linear fitting problem.

5.1.3.4 Mode Shape Analysis

For a specific center frequency f_m, construct a reference phase signal at each of L time steps of Δt

$$u(f_m, l) = e^{-j2\pi f_m \cdot l \cdot \Delta t}, l = 1, \ldots, L. \tag{5.14}$$

Calculate

$$q_{im}(f_m) = \frac{1}{L}\sum_{l=1}^{L} y_{im}(l \cdot \Delta t) \cdot u(f_m, l). \tag{5.15}$$

$q_{1m}(f_m)$ to $q_{Nm}(f_m)$ together determine the mode shape of the oscillation mode at f_m:

- Magnitude $|q_{im}(f_m)|$ measures the average power of y_{im}'s component at f_m, which, in fact, indicates the participation factor of x_i in oscillation at that frequency.
- $q_{im}(f_m)$'s phase angle (denoted by $\hat{\theta}_{im}$) represents the phase of component y_{im} at f_m over the time window. Thus, $\hat{\theta}_{im} - \hat{\theta}_{jm}$ (i and $j = 1, ..., N$) gives the phase difference between two signals x_i and x_j at f_m.

Each of the M modes could be either a local mode or an inter-area mode. The squared linear coherency index defined in (5.11) can be calculated for any two components y_{im} and y_{jm} (i and $j = 1, ..., N$) measured at different locations. If the coherency index is high (e.g., >0.5), the mode at f_m can be regarded as an inter-area mode over a region covering synchrophasors i and j. This cross-spectrum-based method requires a long enough time window of data for reliable estimation.

The PLL-based approach is able to directly estimate coherency between two signals from their average powers at a specific frequency. As defined in (5.16), a new coherency index $r_{ij} \in [0, 1]$ is calculated for signals x_i and x_j at f_m over the latest T:

$$r_{ij}(f_m) = \frac{|q_{im}(f_m)| \cdot |q_{jm}(f_m)|}{|q_{im}(0)| \cdot |q_{jm}(0)|},$$ (5.16)

where $|q_{im}(0)|$ and $|q_{jm}(0)|$ equal the average powers of y_i and y_j at f_m over the time window, for example

$$q_{im}(0) = \frac{1}{L} \sum_{n=1}^{L} y_{im}(n).$$ (5.17)

If $r_{ij}(f_m)$ is small (e.g., <0.5) for any pair of i and j, the mode at f_m can be regarded as a local mode.

The PLL-based approach can accurately estimate modal parameters for both local and inter-area modes. System operators in control rooms usually pay more attention to inter-area modes. For an inter-area mode m, $\hat{\theta}_{im} - \hat{\theta}_{jm}$ (i and $j = 1, ..., N$) gives the phase difference between two signals x_i and x_j at f_m. Thus, for any pair of signals, the phase difference and the coherency index can be calculated at a specific frequency. All such phase differences and coherency indices together present the mode shape of the oscillation mode at the frequency.

As illustrated in Figure 5.8, let $\hat{\theta}_{1m} = 0°$ and $360°$ (the phase reference) and draw N phasors $r_{1j}(f_m) \angle \hat{\theta}_{jm}$ in a unit circle as a "phase clock" on the mode shape at f_m. The length of each pointer indicates how coherent the corresponding signal is with signal x_1. The N phasors can be divided into a number of clusters by their phase differences. The clustering of N phasors may indicate a

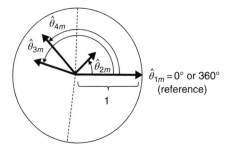

Figure 5.8 "Phase clock" on the mode shape at f_m.

potential out-of-step pattern of generators [10]. For example, the biggest two phase differences between any adjacent phasors in Figure 5.8 are $\left|\hat{\theta}_{4m} - \hat{\theta}_{2m}\right|$ and $\left|360° - \hat{\theta}_{3m}\right|$, which indicate two clusters {1, 2} and {3, 4}.

5.1.4 Timing of Controlled Separation

In real time, the final timing to take the action of CSS is decided based on the real-time absolute angle difference across each probable separation boundary identified online. The angle of each side of one boundary can be calculated by averaging the measured angles from all sychrophasors.

The measurement-based angular stability analysis methods in Sun et al. [8] and Wang and Sun [4] can estimate the stability boundary of the monitored angle difference across a grid interface based only on real-time measurements. The first method constructs an adaptive equivalent system based on the phase portrait over a time window in order to estimate the margin to instability; the second method monitors the real-time oscillation frequency (e.g., tracked by the PLL-based algorithm) with the inter-area mode across a grid interface and estimates the stability margin using the aforementioned F-A curve regarding the mode.

Besides those measurement-based methods for prediction of angle instability across a grid interface, a simpler and easier to implement method is to directly compare the absolute angle difference with a threshold (in 120–180°) determined by offline simulation studies. When the absolute angle difference exceeds that threshold, an out-of-step condition across that boundary is detected or predicted. Then CSS should be performed immediately.

Alternatively, Sun et al. [9] suggest the following extrapolation of the real-time angle difference utilizing its waveforms and modal properties from the latest time window in order to foresee a violation of the threshold before it occurs.

Assume that N_I probable separation boundaries have been identified online. At the current time $t = \tau$, the angle separation $\Delta_k(t)$ on the kth boundary ($k = 1, \ldots, N_I$) is extrapolated for a coming short time window $[\tau, \tau + \Delta t]$. First, decompose $\Delta_k(t)$ into a steady-state signal $A_{k,0}(\tau)$ and damped sinusoidal signals $\alpha_{k,1}(t), \ldots, \alpha_{k,N_M}(t)$, respectively, about N_M dominant inter-area modes with damping $\zeta_{k,n}$ and angular frequency ω_n $(k = 1, \ldots, N_I, n = 1, \ldots, N_M)$, as follows:

$$\Delta_k(t) \approx A_{k,0}(\tau) + \sum_{n=1}^{N_M} \alpha_{k,n}(t), \quad \text{where } \alpha_{k,n}(t) = A_{k,n}(\tau) e^{\frac{-\zeta_{k,n}}{\sqrt{1-\zeta_{k,n}^2}}\omega_n(t-\tau)} \cos(\omega_n t + \varphi_{k,n})$$

(5.18)

$A_{k,0}(\tau)$ can be estimated by averaging $\Delta_k(t)$ over a latest time window T. Amplitudes $A_{k,1}(\tau) \ldots A_{k,N_M}(\tau)$ can be solved from (5.18) together with equations in (5.19), where $\Delta_k(t)$'s derivatives of different orders are estimated using its measurements near time τ.

$$\begin{cases} \dot{\Delta}_k(\tau) = \dot{\alpha}_{k,1}(\tau) + \cdots + \dot{\alpha}_{k,N_M}(\tau) \\ \qquad \cdots \\ \Delta_k^{(2N_M-1)}(\tau) = \alpha_{k,1}^{(2N_M-1)}(\tau) + \cdots + \alpha_{k,N_M}^{(2N_M-1)}(\tau) \end{cases}$$

(5.19)

Define

$$A_k(\tau) = A_{k,0}(\tau) + A_{k,1}(\tau) + \cdots + A_{k,N_M}(\tau),$$

(5.20)

which is an estimate of the theoretically maximum value that Δ_k might reach in an immediate period $T' = 1/\min(f_1, \ldots, f_{N_M})$ and is compared to a preset limit to estimate the margin to instability. If $\alpha_{k,1}(t), \ldots, \alpha_{k,N_M}(t)$ are extracted from $\Delta_k(t)$, for example, by band-pass filters, $A_k(\tau)$ may also be calculated directly by

$$A_k(\tau) = A_{k,0}(\tau) + \sum_{n=1}^{N_M} \sqrt{\alpha_{k,n}^2(\tau) + \left(\frac{\dot{\alpha}_{k,n}(\tau)}{\omega_n} + \frac{\zeta_{k,n}\alpha_{k,n}(\tau)}{\sqrt{1-\zeta_{k,n}^2}}\right)^2}.$$

(5.21)

For the simplest case with $N_M = N_I = 1$, only the first inter-area mode is used to predict angle separation at one probable separation boundary. Ignore subscript k in (5.21). There is

$$A(\tau) = A_0(\tau) + \sqrt{\alpha^2(\tau) + \left(\frac{\dot{\alpha}(\tau)}{\omega} + \frac{\zeta\alpha(\tau)}{\sqrt{1-\zeta^2}}\right)^2},$$

(5.22)

where $\alpha(t) = \Delta(t) - A_0(\tau)$.

In addition, an approximate approach to fast computation of $A_k(\tau)$ is to apply FFT to $\Delta_k(t)$ over $[\tau-T, \tau]$. Denote the FFT magnitudes at $\omega_1, \ldots, \omega_{N_M}$ by $\hat{A}_{k,1}(\tau), \ldots, \hat{A}_{k,N_M}(\tau)$ to approximate $A_{k,1}(\tau), \ldots, A_{k,N_M}(\tau)$, and calculate (5.23) and then $A_k(\tau)$ by (5.21).

$$\alpha_{k,n}(t) \approx \left[\Delta_k(t) - A_{k,0}(\tau)\right] \cdot \hat{A}_{k,n}(\tau) \Big/ \sum_{n=1}^{N_M} \hat{A}_{k,n}(\tau)$$

(5.23)

If the limit (denoted by A_k^{max}) on the angle separation Δ_k across each probable separation boundary is predefined, the stability margin on the k-th boundary is calculated by

$$\text{Margin}_k = \frac{A_k}{A_k^{\text{max}}} \times 100\%. \tag{5.24}$$

An initial guess on that limit can be set at $180° - A_{k,0}$ by treating the two sides, that is, two probable islands, of the boundary as two equivalent generators. Then the limit can be refined by offline analytical or simulation studies.

Sun et al. [10] suggest that both the real-time angle difference together with the oscillation phasing difference ϕ_k across each probable separation boundary should be considered. It concerns the possibility that one probable separation boundary may have much larger phase difference (i.e., closer to 180°) across the boundary but slightly smaller angle difference than another probable separation boundary. In such a case, the former with both a large angle difference and a large phase difference is more likely to be the final separation boundary. Thus, the stability margin on the kth boundary may be revised to

$$\text{Margin}_k = \frac{A_k}{A_k^{\text{max}}} \times \frac{\phi_k}{\phi_k^{\text{max}}} \times 100\%, \tag{5.25}$$

where ϕ_k^{max} is equal to 180° or a preset threshold on the phase difference.

5.2 WAMS-Based Unified Framework for Controlled System Separation

5.2.1 WAMS-Based Three-Stage CSS Scheme

To address the three questions "where," "when," and "how," a unified framework for practical implementation of CSS using the WAMS was proposed in Sun et al. [9], which suggests partitioning all related tasks into three stages as shown in Table 5.1 including the **Offline Analysis** stage, **Online Monitoring** stage, and **Real-Time Control** stage.

Two assumptions are made under this framework on the infrastructures for measurement, communication, and actuation systems to support CSS:

- Synchrophasors should be placed near all main generators and send real-time data streams to the central energy management system (EMS) of the system control center through the WAMS. This is critical for monitoring inter-area oscillations and grouping of generators and accordingly judging the timing of CSS.
- Hardware devices that take the action of CSS are a special type of out-of-step relays, called "separation relays" here, which are coordinated out-of-step protection systems having both blocking and tripping functions. These separation

Table 5.1 Three stages of the proposed WAMS-based unified framework for CSS.

Stages	Main Tasks with the Questions Addressed Indicated
Offline Analysis (annual or day-ahead)	• Optimize the separation points of each potential island to place separation relays (*where* and *how*). • Design and daily update a postseparation control strategy table for each potential island (*how*).
Online Monitoring (every 1 second)	• Monitor inter-area oscillations between generators by WAMS for any potential out-of-step condition (*where* and *when*). • Identify probable separation boundaries made up by offline optimized separation points according to how generators cluster (*where*).
Real-Time Control (milliseconds)	• Trip all relays on the separation boundary where the out-of-step condition is detected or credibly predicted (*when* and *how*). • Perform postseparation control strategies for each island formed (*how*).

relays are placed at offline determined potential separation points according to the studies addressing the question "where," and they are remotely controlled by a CSS program integrated in the central EMS.

- The communication channels from all synchrophasors to the EMS and from the EMS to all separation relays are highly secured and reliable. Once CSS is decided, a right set of separation relays will be tripped simultaneously while the others are blocked in order for desired islands to be formed.

The three questions are addressed in three stages as follows:

- "Where?": the **Offline Analysis** stage reduces the search scope for separation points to a set of potential separation points so as to place separation relays; on that basis, potential separation boundaries are monitored that are made up by some of the separation points in the **Online Monitoring** stage; the final separation boundary is determined in the **Real-Time Control** stage according to the final out-of-step condition detected or credibly predicted.
- "When?": **Online Monitoring** stage provides an early warning of potential system separation by monitoring inter-area oscillations; the **Real-Time Control** stage determines the timing of CSS to trip the right set of separation relays once an out-of-step condition appears.
- "How?": the **Offline Analysis** stage has separation relays placed at optimized locations and constructs a strategy table for postseparation control; the **Real-Time Control** stage performs postseparation control strategies from the table that match the actual islands.

In the following, the tasks in each stage are presented and demonstrated using the WECC (Western Electricity Coordinating Council) 29-generator 179-bus power system.

5.2.2 Offline Analysis Stage

5.2.2.1 Determination of ECGs

In this stage, system planning engineers need to study generator coherency to determine potential out-of-step conditions. In order to give consideration to probable changes in generator coherency under various operating conditions, all generators are divided into a number of ECGs, whose combinations in different ways can cover most scenarios of generator clustering or out-of-step conditions, as discussed in Section 4.5.2. The slow-coherency analysis can be applied to determine these ECGs. Then the ECGs should be tested by simulation studies with credible contingencies. If an ECG is found often losing synchronism, it may further be partitioned into smaller ECGs to reflect the existence of that scenario. Finally, each offline identified ECG is assumed to stay in the same island and is merged to one equivalent generator together with neighboring buses and lines that help connect its generators. Separation points will be determined only between ECGs, while the stability of generators within each ECG is ensured by post-separation control actions.

Figure 5.9 gives the graph model of the WECC 179-bus power system. Assume that four ECGs are identified as indicated by the shade areas, which are merged to four equivalent generators, numbered 1–4, together with neighboring load buses in respective shaded areas. Thus, the original 179-node graph is simplified to a 45-node graph with four generator nodes. The net power output of each equivalent generator equals the total generation minus the total load in the merged area. Thus, the CSS considers out-of-step between 12 possible groups of generators: 1, 2, 3, 4, 1+2, 1+4, 2+3, 3+4, 1+2+3, 1+2+4, 1+3+4, and 2+3+4. The separation points should be able to isolate each of these 12 groups into a single island in order to meet the generation coherency constraint under various out-of-step conditions between the four elementary groups.

5.2.2.2 Determination of Separation Points

For each potential out-of-step scenario among ECGs, the OBDD-based method can be applied to quickly find all separation strategies as sets of separation points that enable satisfactions to constraints, including the generation coherency constraint, power balance constraint, and transmission capacity constraint (the other aforementioned constraints can also be added). Alternatively, separation strategies can be optimized by minimizing the power imbalance while satisfying the other constraints. When the system operating condition changes, the separation points may drift. There is a compromise between the number and performance of the separation points since it may not be economically practical to install separation relays at all possible separation points.

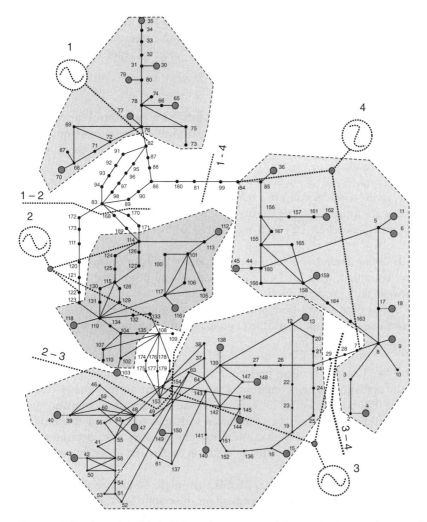

Figure 5.9 Graph model of the WECC 179-bus system with four ECGs respectively merged to equivalent generators.

Considering a range of system conditions with the 179-bus system, assume separation points to be finally selected on the following seven lines of four network interfaces, as indicated by broken lines in Figure 5.9:

- interface 1–2 (lines 83–168, 83–170, and 83–172)
- interface 1–4 (lines 81–99)
- interface 2–3 (lines 142–153 and 153–154)
- interface 3–4 (lines 28–29)

Thus, the portion of the system bounded by any two interfaces may be isolated by separate relays as a potential island whose generation-load imbalance is always within 5% of the system load for a range of representative system conditions.

5.2.2.3 Postseparation Control

Once an island is formed, it may require postseparation control to arrest frequency and voltage excursions and quickly stabilize generators. Adequate real generation reserves and reactive power reserves need to be planned in each potential island.

For every potential island, the amounts and locations of control can be studied offline, giving considerations to a range of operating conditions. Thus, a table of postseparation control strategies corresponding to representative operating conditions can be developed for each potential island by the following steps:

Step 1: Define a number of representative power-flow profiles such as the peak load condition, light load condition, and several other load conditions in between with properly dispatched generation.

Step 2: For each power-flow profile, perform simulation studies on CSS to determine the amounts of control on load and generation in each potential island.

Once the island is formed, perform a strategy from the table that matches best the actual operating condition. Take postseparation real power control with the 179-bus system as an example. Time-domain simulations are performed to disconnect any two interfaces with immediate load shedding or generation rejection of different amounts. The purpose is to determine the right amount of load or generation to adjust in each island in order to maintain frequency within 59.5–60.5 Hz.

Table 5.2 gives all 12 potential islands that might be formed and the postseparation control strategies determined for the system load equal to 60.8 GW as one of representative loading conditions. According to the table, six of 12 potential islands can automatically meet the frequency criteria for postseparation operating conditions. The worst case is island No. 7, which sheds 7.3% of the system load after separation. Similarly, for other loading conditions, postseparation control strategies can be obtained as those in Table 5.2. Finally, a comprehensive strategy table can be built to cover a range of operating conditions for each potential island.

5.2.3 Online Monitoring Stage

In this stage, wide-area synchrophasor measurements from the WAMS are monitored to predict separation boundaries. Failures occurring between two

Table 5.2 Postseparation control strategies for the system load = 60.8 GW.

Number	Potential Island Indicated by the Elementary Group(s) Included	Postseparation Control Strategies
1	1	Not needed
2	2	Shed 3.6% of the system load
3	3	Shed 3.8% of the system load
4	4	Not needed
5	1 + 2	Not needed
6	1 + 4	Not needed
7	2 + 3	Shed 7.3% of the system load
8	3 + 4	Shed 1.5% of the system load
9	1 + 2 + 4	Not needed
10	1 + 2 + 3	Shed 4.1% of the system load
11	2 + 3 + 4	Shed 4.9% of the system load
12	1 + 3 + 4	Not needed

ECGs can be reflected from a real-time change in the mode shape of the oscillation mode across that interface between these two groups. Therefore, this stage online analyzes dominant inter-area oscillation modes for probable separation boundaries. The frequencies of inter-area oscillations are mainly in the range of 0.1–0.9 Hz. Oscillation frequencies faster than 0.9 Hz are usually of local oscillation modes. Instability caused by a local mode is usually controlled by a local protective action rather than CSS. The dominant inter-area modes and their mode shapes can be identified by measurement-based modal analysis.

For illustration purposes, cascading outages of lines 83–172, 83–170, 114–124, 115–130, 83–94, and 83–98 at 40-s intervals in the WECC system are simulated under the 60.8 GW load condition. Generators go out of step at about 201 s. These line outages are all near the so-called California-Oregon Intertie, which was involved in the uncontrolled system separation during the US western blackout event in August 1996 [11].

From simulation results, angle differences between four directly connected ECGs are shown in Figure 5.10a. Directly from the measurements, the steady-state angle differences between ECG 1 and ECG 2 and between ECG 3 and ECG 4 are obviously larger. However, the most probable separation boundary should be predicted from modal analysis on the measurements. To show how inter-area oscillation modes are changed with cascading line outages (Figure 5.10b–e compares the Fourier transform results of four angle differences over time windows of 40–80, 80–120, 120–160, and 160–200 s.

(a)

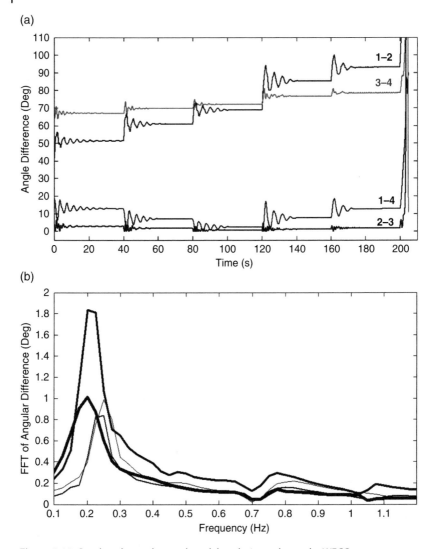

(b)

Figure 5.10 Synchrophasor data and modal analysis results on the WECC system.
(a) Angle differences between four ECGs. (b) Angle difference between ECGs 1 and 2. (c) Angle
difference between ECGs 1 and 4. (d) Angle difference between ECGs 3 and 4.
(e) Angle difference between ECGs 2 and 3. (f) Squared-coherency index for each pair of
ECGs. (g) The angle of CSD for each pair of ECGs.

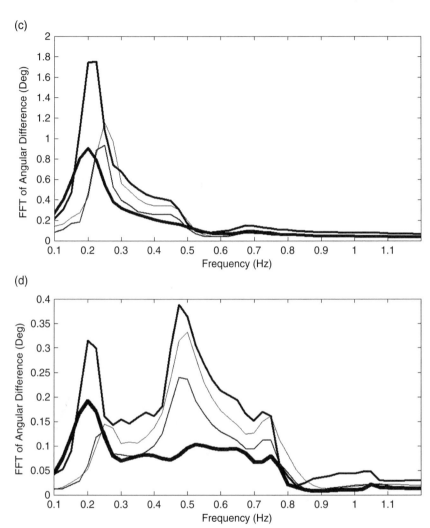

Figure 5.10 (*Continued*)

The results for four time windows are differentiated by increasing line thickness, from which three dominant inter-area modes are at 0.25, 0.5, and 0.75 Hz. If the spectral analysis-based method is applied to estimate the mode shape at $t = 200\,\text{s}$, the squared-coherency index γ_{ij}^2 and angle of CSD for any pair of elementary groups are calculated for the frequency range of 0.1–0.9 Hz

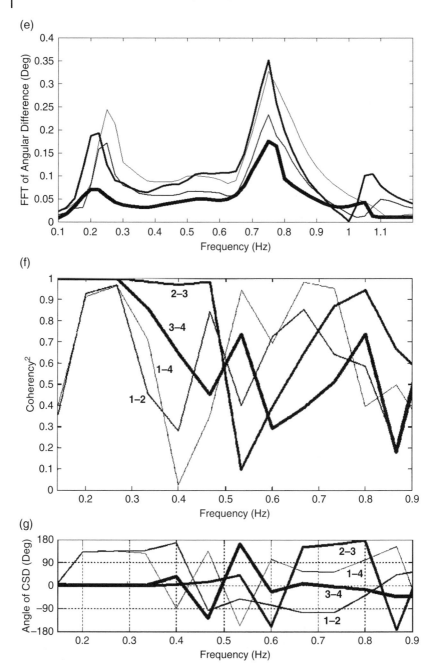

Figure 5.10 (*Continued*)

using measurements over 160–200 s, as shown in Figure 5.10f–g. The following observations can be made:

- ECG 1 swings against ECGs 2 and 4 at 0.25 Hz since the angles of CSD for ECG 1 versus ECG 2 and ECG 1 versus ECG 4 at 0.25 Hz are bigger than 120°,
- ECGs 2, 3, and 4 swing together at 0.25 Hz since the angle of CSD for ECG 2 versus ECG 3 and ECG 3 versus ECG 4 are very close to 0°.

Thus, one probable separation boundary is derived at $t = 200$ s, which disconnects the interface 1–2 (lines 83–168, 83–170, and 83–172) and interface 1–4 (line 81–99) to form two islands in the north and south.

5.2.4 Real-Time Control Stage

The final time to take the action of CSS can be decided based on the real-time absolute angle difference (or its extrapolated value using the method in Section 5.1) across each probable separation boundary identified from the **Online Monitoring** stage. The angle of each side can be calculated by averaging the measured angles from all sychrophasors. A simple and also practical method is to directly compare the absolute angle difference with a threshold (in 120–180°) determined by offline simulation studies. A zero margin calculated from Eq. (5.24) means that an out-of-step condition across that boundary is detected or credibly predicted. That can be a trigger to take the action of CSS immediately.

Alternatively, the timing of separation can be judged by both the real-time angle difference and oscillation phase difference between the two potential islands using (5.25). In this stage, the dominant oscillation modes related to probable separation boundaries may change more significantly. Thus, the PLL-based method performs better than a spectral analysis-based method. Figure 5.11 compares the angle differences and phase differences about two probable separation boundaries: 1 versus 2+3+4 and 1+4 versus 2+3. The phase differences are estimated using the PLL-based method. The boundary separating ECG 1 from ECGs 2–4 has a smaller angle difference than the other boundary separating ECGs 1 and 4 from ECGs 2 and 3. However, the first boundary has a bigger phase difference.

Finally, an out-of-step condition on the boundary separating ECG 1 from ECGs 2–4 is predicted at 201 s with close to zero margin calculated from Eqs. (5.24) or (5.25). Figure 5.12 demonstrates the system response after CSS is performed at 202.5 s, which is 1.5 s after the out-of-step condition is predicted to allow communication latency in sending the CSS action signal. In postseparation control, Strategy 11 in Table 5.2 should be performed to shed 4.9% of the system load in the south island once separation relays at the four separation points are tripped, while no control is needed in the north island according to Strategy 1.

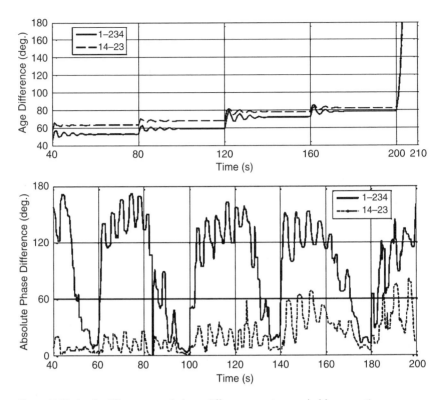

Figure 5.11 Angle differences and phase differences on two probable separation boundaries.

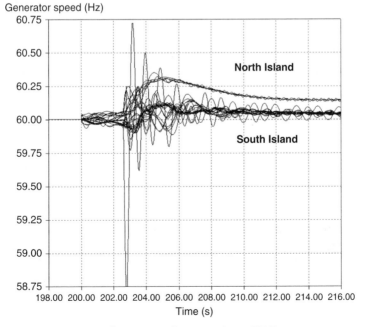

Figure 5.12 Generator frequencies after separation at 202.5 s.

References

1 Trudnowski, D.J. (2008). Estimating electromechanical mode shape from synchrophasor measurements. *IEEE Transactions on Power Systems* 23 (3): 1188–1195.

2 Hauer, J.F. (1991). Application of Prony analysis to the determination of modal content and equivalent models for measured power system response. *IEEE Transactions on Power Delivery* 6 (3): 1062–1068.

3 Hur, K. and Santoso, S. (2009). Estimation of system damping parameters using analytic wavelet transforms. *IEEE Transactions on Power Delivery* 24 (3): 1302–1309.

4 Wang, B. and Sun, K. (2016). Formulation and characterization of power system electromechanical oscillations. *IEEE Transactions on Power Systems* 61 (6): 5082–5093.

5 Wang, B., Su, X., and Sun, K. (2017). Properties of the frequency-amplitude curve. *IEEE Transactions on Power Systems* 32 (1): 826–827.

6 Sun, K., Zhou, Q., and Liu, Y. (2014). A phase locked loop-based approach to real-time modal analysis on synchrophasor measurements. *IEEE Transactions on Smart Grid* 5 (1): 260–269.

7 Best, R.E. (2003). *Phase-Locked Loops Design, Simulation and Applications*, 5e. New York: McGraw-Hill.

8 Sun, K., Lee, S., and Zhang, P. (2011). An adaptive power system equivalent for real-time estimation of stability margin using phase-plane trajectories. *IEEE Transactions on Power Systems* 26: 915–923.

9 Sun, K., Hur, K., and Zhang, P. (2011). A new unified scheme for controlled power system separation using synchronized Phasor measurements. *IEEE Transactions on Power Systems* 26 (3): 1544–1554.

10 Sun, K., Luo, X., and Wong, J. (2012). "Early warning of wide-area angular stability problems using synchrophasors," *IEEE PES General Meeting, 23–26 July 2012, San Diego*.

11 Kosterev, D.N., Taylor, C.W., and Mittelstadt, W.A. (1999). Model validation for the August 10, 1996 WSCC system outage. *IEEE Transactions on Power Systems* 14: 967–979.

6

Constraints of System Restoration

6.1 Physical Constraints During Restoration

A restoration strategy study is performed to verify the feasibility in terms of both steady-state and transient operating conditions.

The steady-state analyses of restoration include

- Balance of generation and load
- Voltage control and steady-state overvoltage analysis
- Capability of black-start units to absorb reactive power produced by charging currents of the transmission system
- For each step of a restoration process, it is necessary to ensure feasibility and compliance with required operational limits
- Verification of the robustness of the tested black-start units, that they are capable of compensating for the potential unavailability of key components

The dynamic analyses of black-start units include the following issues:

- Load-frequency control
- Voltage control
- Large induction motor starting
- Motor starting sequence assessment
- Self-excitation assessment
- System stability
- Transient overvoltage

6.1.1 Generating Unit Start-Up

Utilizing a black-start source (i.e., initial energy sources to crank other unenergized components) to restore generation capacity is one of the most important tasks in power system restoration, since the restoration duration can be significantly prolonged without sufficient energy sources. Therefore, the

Power System Control Under Cascading Failures: Understanding, Mitigation, and System Restoration, First Edition. Kai Sun, Yunhe Hou, Wei Sun and Junjian Qi.
© 2019 John Wiley & Sons Ltd. Published 2019 by John Wiley & Sons Ltd.
Companion website: www.wiley.com/go/sun/cascade

prerequisite of a smooth restoration process is to maintain adequate generation capacity in the preparation phase. In this phase, the entire power grid may be partitioned into several islands, each having black-start sources and non-black-start generating units. Time is so critical that many emergent control actions are performed in this phase. One of the most critical tasks is to restart non-black-start generating units as fast as possible, while maintaining the security and reliability of the entire power grid.

In this section, the general characteristics of various types of generating units are summarized. The requirements of generating unit start-up and major technical challenges are reviewed. The major problem in generating unit start-up, that is the optimal sequence of start-up, is investigated.

The generating units can be categorized into two types:

- Black-start units
- Non-black-start units

The former type requires no off-site power to start, including hydro units, diesel generator sets, aero-derivative gas turbine generator sets, and large gas turbines [1]. The latter type requires off-site power to start, including fossil-fueled units and nuclear units. The start-up problem is to determine the sequence to restart non-black-start units, which use black-start units and other energy sources (like high-voltage direct current (HVDC) links [2], tie lines [3], and energized islands), subject to various steady-state and dynamic constraints.

Constraints considered in generating unit start-up include real power balance, reactive power balance, operational voltage and loading constraint, frequency stability, motor starting assessment [4], self-excitation assessment and cold load effect [1], small signal stability [5], and so on. Some of these constraints are summarized as follows.

6.1.1.1 Active Power Balance and Frequency Control

During the restoration process, it is necessary to maintain the system frequency within allowable limits imposed by turbine resonance, system stability, and the resulting protection settings. This objective is achieved by picking up loads in increments, accommodated by the inertia and response of the restored and synchronized system.

During restoration, operators must consider a prime mover's frequency response to a sudden increase in load. Load pickup in small increments tends to prolong the restoration duration. With large increments, there is a risk of falling into a frequency decline and recurrence of the system outage. The size of load pickup depends on the rate of response of prime movers, which more likely are under manual control at this point. A dynamically calculated guideline, such as allowable load pickup as a percentage of generator capability, would help maintain load and generation balance at an acceptable frequency.

Therefore, the preferred control mode for speed governors associated with black-start unit is constant frequency control. When additional units are added, the preferred control mode for speed governors is droop control.

6.1.1.2   Reactive Power Balance and Overvoltage Control

The generator reactive capability (GRC) curves, furnished by manufacturers and used in operation planning, typically have a greater range than that during the actual operation. Generally, these manufacturer's GRC curves are strictly a function of the synchronous machine design parameters and do not consider plant and system operating conditions as limiting factors. The concern over GRC is warranted by the need for reactive power to provide voltage support for large blocks of power transfer.

During early stages of the restoration process, it is necessary to keep system voltages within the allowable, usually lower than the normal range. This is done in several ways: energizing fewer high-voltage lines, operating generators at minimum voltage levels, deactivating (overriding) switched static capacitors, connecting shunt reactors, adjusting transformer taps, and picking up loads with lagging power factors. Another concern in the energized system is the ability to absorb the cable's charging current during cable reenergization. It should be noted that

- Cables are loaded below their surge impedance loading.
- The MVAr charging current per mile of cable is about 10 times that of an overhead line of the same voltage class.
- About 2 MW of load is picked up per 1 MVAr of charging.

6.1.1.3   Switching Transient Voltage

Energizing equipment during black-start conditions can result in higher over-voltage than during normal operation. During the reintegration phase, it is desirable to energize the section of high-voltage transmission to the maximum value that switching transient voltages would allow. Energizing small sections tends to prolong the restoration process. In energizing a large section, there is a risk of damaging the equipment insulation. Temporary overvoltage can follow switching surges. It may result from several issues:

- Switching circuits that saturate the core of a power transformer when cables and transformers are energized together.
- The harmonic rich transformer inrush currents interact with the harmonic resonances of the power system.
- The resonant frequencies are functions of the series inductance associated with the system's short-circuit strength and shunt capacitances.
- Higher inductances (weak systems) and higher capacitances (long cables) yield lower resonant frequencies and higher chances of temporary overvoltage.

During system restoration, the problem is to determine how to energize a multisection line without exceeding the basic impulse level of the equipment. Some utilities energize section by section one circuit of the double circuit line at a time, and maintain the generator voltage as low as the minimum excitation level allows. This helps reduce the line charging currents and reduce the transient voltage magnitudes. However, the procedure lengthens the restoration. A simple methodology is needed to readily establish guidelines for energizing transmission lines.

6.1.1.4 Self-Excitation

To control the overvoltage along a line or cable due to charging currents, the charging requirements should be large enough to absorb the reactive power. There is a possibility of self-excitation if the charging current is high relative to the size of the generating unit. The result can be an uncontrolled rise in voltage and could result in an equipment failure. Self-excitation can also occur at the load end through inadvertent loss of supply, opening of a transmission line or a cable at the sending end, and leaving the line connected to a large motor or a group of motors. To avoid self-excitation of the generator during restoration, a number of strategies can be used:

- Using the generator with less x_d to energize an extra-high voltage (EHV) transmission line
- Using two or more generators to charge one EHV transmission line
- Changing the impedance of transformer by tap change
- Changing the parameters of EHV transmission line by reactive power compensation

Moreover, characteristics of various types of generating units should be considered in the start-up problem. Hydro units are an important part of black-start capacity. Gas turbines can be restarted immediately when major outages happen. The black-start capacities and ramping rates of gas turbines should be considered. For a thermal unit, the duration of start-up depends on whether the thermal unit is cold-, warm-, or hot-started [6]. The maximum time interval during which a thermal unit cannot be hot-started, and the minimum time interval during which a thermal unit can be safely restarted, should be considered. For reliability concerns, nuclear units can only be restarted after 24 hours of major outages. The off-site power of nuclear stations should be provided reliably to ensure a secure shutdown [7]. Therefore, the timing of generating unit restart typically in the following sequence: hydro units, gas turbines, thermal units, and nuclear units.

To determine an optimal generating unit start-up sequence considering the characteristics of units and comprehensive systemwide constraints, diverse models have been established and solved with various techniques (Table 6.1).

Table 6.1 Features of selected work related to generating unit start-up.

Authors	Solution Techniques	Unit Characteristics	Operational Constraints
C.C. Liu et al., 1993 [8]	Knowledge-based approach	Considered	Considered
A. Ketabi et al., 2001 [9]	Backtracking search	Considered	Not considered
H. Zhong and X.-P. Gu, 2011 [10]	Fuzzy AHP	Considered	Not considered
Q. Liu et al., 2008 [12]	Knapsack model and backtracking	Considered	Not considered
W. Sun et al., 2011 [13]	MILP and violation check	Considered	DC power flow
Y. Hou et al., 2011 [11]	Bilevel optimization	Considered	Full AC power flow

Note. AHP = analytic hierarchy process; MILP = mixed-integer linear programming.

An optimization model to maximize the MWH load served over a given restoration period was proposed in Liu et al. [8] and solved via a knowledge-based system. This model considered comprehensive power plant constraints (such as ramping rate, capacity, minimal output, critical time intervals, etc.) and system operational constraints. The path from black-start units to non-black-start units is determined by a shortest path algorithm to minimize switching actions.

With particular emphasis on satisfying the units' constraints, the backtracking search method was applied in Ketabi et al. [9] to find the optimal unit start-up sequence. The advantage of this is that the search tree generated by this method can be pruned by a feasibility and optimality check compared with brute force searching. Similar to Ketabi et al. [9], the fuzzy analytic hierarchy process was applied to obtain the optimal start-up sequence in Zhong and Gu [10]. The optimization problem for optimal sequence was converted into a quantitative multi-attribute decision-making problem. In both works, the system operational constraints were not addressed in details.

Mathematical optimization methods have been applied to determine the optimal generating unit start-up sequence [11–13]. In Liu et al. [12], the optimization model proposed in Liu et al. [8] was discretized into a sequence of knapsack subproblems. Each subproblem was solved by the backtracking method. By avoiding solving all subproblems simultaneously, this method achieved better efficiency at the cost of global optimality. In Sun et al. [13], the optimal start-up sequence was modeled as a mixed-integer linear programming model, considering comprehensive unit characteristics and DC power flow constraints. If the optimal solution violated some steady-state constraints in the sense of full AC power flow, the corresponding constraints would be added and a recalculation carried out for a revised solution.

In Hou et al. [11], determining the optimal start-up sequence is a major task in Generic Restoration Milestone 1 (GRM1). The proposed methodology is a bilevel optimization algorithm. The upper optimization model is to determine a priority-ranking list of generating unit start-up, considering the units' characteristics, critical time intervals, and user-defined priority. The lower optimization is an optimal power flow model minimizing the ramping time of generating units, considering full AC power flow and other comprehensive operational constraints. This bilevel optimization algorithm is a greedy search with trace-back mechanism, if some subsequence is infeasible. This algorithm is capable of including comprehensive constraints instead of violation checking. It has been implemented as a key module in system restoration navigator [14].

The key features of the aforementioned works are summarized in the following section.

6.1.2 System Sectionalizing and Reconfiguration

After the major outage caused by large disturbances, the power grid may lose part of generation, transmission network, and loads. The topology of the postdisturbance power grid will be substantially different from that of the predisturbance one for two reasons. On one hand, the breakers connected to de-energized buses will be opened, because components are guaranteed to be energized without risk of inadvertently connecting other unplanned components. On the other hand, the surviving power grid will be partitioned into multiple islands to facilitate parallel restoration or to avoid loss of synchronism. To bring the power grid back to the normal operation state, the power grid topology will be altered along the restoration process to rebuild the skeleton of the transmission network and the distribution network. The schemes of both sectionalizing and reconfiguration play an important role in establishing a smooth, secure, and reliable restoration process. These power grid topological changes will be implemented by a series of switching actions meeting certain security criteria.

In the next subsection, the research related to system sectionalizing and reconfiguration will be reviewed. The emphasis will be placed on the modeling and solution methods in system sectionalizing, network reconfiguration, path selection, and redispatch to mitigate switching impact.

6.1.2.1 System Sectionalizing

System sectionalizing is applied to reduce restoration time for both distribution systems and transmission systems. Nonetheless, the effects of system sectionalizing in distribution level are different from those in transmission level. In distribution systems, if a fault happens, it will be located and isolated by sectionalizing switches. As a result, the load can be restored quickly from alternative paths. In transmission systems, after a widespread outage happens, the power grid will be sectionalized into several subsystems. These subsystems can then be restored in a parallel manner and will be reintegrated together (or synchronized) eventually. However, for both distribution and transmission cases, the effectiveness of sectionalizing depends on the number and location of the sectionalizing points.

The work in Nezam Sarmadi et al., Wang et al., and Quiros-Tortos and Terzija [15–17] focused on the optimal sectionalizing strategy of the transmission systems. The general criteria to sectionalizing were outlined in Adibi et al. [18, 19]. In Wang et al. [16], the steady-state stability criteria were considered to sectionalize a power grid into islands. An ordered binary decision diagram (OBDD)-based approach was proposed to establish a sectionalizing scheme satisfying black-start constraints, power balance constraints, and voltage stability constraints. In Nezam Sarmadi et al. [15], the Wide Area Measurement System (WAMS) was used to realize the full observability for all sectionalized islands. In Quiros-Tortos and Terzija [17], a spectral clustering algorithm was proposed to sectionalize a power grid. The *k*-shortest simple paths were calculated to connect black-start units to non-black-start units and loads. The dynamic stability issue in the sectionalizing scheme is still an open question.

For distribution systems, the optimal placement of distributed generator and sectionalizing switches can be obtained by applying heuristic algorithms, such as the genetic algorithm [20] and ant colony optimization algorithm [21].

6.1.2.2 Network Reconfiguration and Path Selection

The control actions in network reconfiguration include energizing skeleton transmission paths, picking up critical loads, cranking generating units, and enabling stable operation of units, close transmission loops, and so on.

The purposes of performing these controls can be categorized into two types: (i) to provide reliable paths delivering energy to non-black-start units and interrupted loads; (ii) to firm up the electrical islands and to enhance reliability by energizing key transmission corridors. Accordingly, the optimality of network reconfiguration can be evaluated in terms of the energy served or, alternatively, in terms of the reliability improvement and transmission efficiency.

For the first purpose, the work in Wang et al. and Gu and Zhong [22, 23] modeled network reconfigurations as optimization models and solved them by the genetic algorithm and the lexicographic method, respectively. In both works, the restored generation capacity was counted in the objective function. The major difference between these two works lay in that the latter treated the network reconfiguration as a multiobjective optimization while the former had a single objective.

For the second purpose, the topological characteristics of the scale-free network of the transmission systems were considered in Liu and Gu [24].

The important loads and transmission lines were identified with scale-free network theory, and the network reconfiguration efficiency was quantified. Then an optimization model was proposed to establish the network reconfiguration strategy and was solved by the discrete particle swarm optimization method. In Escobedo et al. [25], a mixed-integer linear programming model was proposed to determine the optimal transmission switching, aiming to maximize load recovery while satisfying the N-1 and N-2 reliability criteria. This model was solved by a parallel mixed-integer heuristic solver.

Particularly, for establishing paths to non-black-start units or interrupted loads, the path selection algorithms are applied to determine a path with the fewest switching actions subject to comprehensive operational constraints. The restoration performance index of transmission lines was established [26]. The power transfer distribution factor together with the restoration performance index was used to determine a priority list of restoration paths. This proposed method was able to pick up load with lightly loaded lines.

6.1.2.3   Redispatch to Mitigate Switching Impact

The switching action of transmission lines will cause electromagnetic and electromechanic transient processes, which may impose transient overvoltages/overcurrent or mechanical stress on power grid components. Therefore, a dynamic simulation program will be employed to assess the security of switching actions. If the security criteria are not satisfied, remedy actions will be taken to mitigate the switching impact.

In Ye and Liu [27], a mixed-integer nonlinear programming (MINLP) model was proposed to redispatch the power grid so that the standing phase angle of some transmission lines could be reduced. The discrete load increments were used as control measures to implement standing phase angle reduction. This MINLP model was solved by a two-stage decoupled algorithm. This algorithm employed the interior point method to solve a nonlinear model and applied sensitivity analysis to achieve integer feasibility.

In Martins et al. and Viana et al. [28, 29], the transient impacts were caused by a closing transmission loop and energizing long segments of transmission lines, and they were mitigated by optimal redispatch schemes obtained from nonlinear programming (NLP) models. Despite the difference between the electromagnetic transient and electromechanic process, these two pieces of research work were analogous, because predispatch states, postdispatch states, and state transient caused by redispatch actions were simultaneously included in one NLP model. Therefore, the solutions of both NLP models could represent the redispatch actions that guarantee the security of the predispatch states and the postdispatch states. Interior point methods were applied in both pieces of research works.

The key features of the aforementioned work are summarized in Table 6.2.

Table 6.2 Features of selected work related to network reconfiguration.

Authors	Problem Addressed	Model Considerations	Solution Method
S. A. Nezam Sarmadi et al., 2011 [15]	System sectionalizing	PMU placement for full observability	Integer programming
C. Wang et al., 2011 [16]	System sectionalizing	Black start, power balance, voltage	Ordered binary decision diagram method
H. Falaghi et al., 2009 [21]	System sectionalizing	Distribution system with DGs	Ant colony optimization
X. Gu and H. Zhong, 2012 [23]	Transmission line switching	Cooptimization of network, generation, and load pickup	Iterative optimizations of subproblems
Y. Liu and X. Gu, 2007 [24]	Transmission line switching	Importance degree, network reconfiguration efficiency, power flow	Discrete particle swarm optimization
A. R. Escobedo et al., 2014 [25]	Transmission line switching	DC power flow, contingency set	Parallel heuristic mixed-integer programming
C. Wang et al., 2009 [26]	Restoration path selection	Power transfer distribution factor	Calculation with restoration performance index matrices
H. Ye and Y. Liu, 2013 [27]	Standing phase angle reduction	Full AC OPF with discrete load	Interior point method with sensitivity analysis
N. Martins et al., 2008 [28]	Shaft impact reduction	Full AC OPF	Interior point method
E. M. Viana et al., 2013 [29]	Transient overvoltage mitigation	Full AC OPF	Interior point method

Note. PMU = phasor measurement unit; DG = distributed generator; OPF = optimal power flow.

6.1.3 Load Restoration

Power grid customers may lose electricity for two major reasons. The customers' load demand may be shed intentionally by system operators to maintain steady-state power balance, frequency stability, or transient stability. Alternatively, the load demand may be interrupted due to the loss of generation capacity or transmission corridors in case of large disturbances or cascading failures. In both situations, load restoration, that is, restoring service to those interrupted customers at appropriate timing and at a reasonable rate, is one of the most important tasks of the restoration process.

Although the timing to initiate load restoration phase depends on the restoration strategies and the postdisturbance system status, the common features of this phase can be generalized twofold. First, the major generation capacity and the skeleton of transmission systems have been restored. Second, the objective of this phase is to restore load as fully as possible and as fast as reasonable for the sake of security and reliability. Therefore, it becomes a major concern to restore load in relatively large steps in a parallel fashion to minimize the adverse impacts of outages, compared with other restoration phases where loads are merely control means.

However, the security and reliability criteria should be persistently maintained in the seemingly hasty load restoration process. Since the load pickup actions continuously cause disturbances in the yet vulnerable power grids, the steady-state, transient, short-term, and long-term dynamics should be carefully studied to avoid another shutdown or collapse due to inappropriate actions. The technical challenges related to load restoration and associated research advancements are reviewed as follows.

6.1.3.1 Active Power Balance and Frequency Control

During the load restoration process, it is critical to maintain the system frequency within an acceptable range, which is jointly determined by system stability criteria, protective relay settings, and turbine resonance. Nevertheless, the automatic generation control function may be switched off and the flat frequency control in the manual mode will be employed prior to the interconnection of firmed-up electrical islands. Therefore, the frequency deviation is determined by the load behaviors and governor response on the sudden change of load.

To ensure the frequency deviation within the given range, the amount of simultaneous load pickup (or "total restorable level") should be limited according to the system inertial and the total governor response capacity, considering the dynamics of generating units [30] and load itself [31]. In case that frequency deviates from the acceptable range, load shedding and load recovery will be employed to maintain frequency stability [32–33].

6.1.3.2 Load Characteristics Modeling

If the distribution circuits are closed to energize load after a prolonged outage, the load demand will be greater than that during normal operation or before the outage. This phenomenon is termed "cold load pickup," as the load has been off-potential for a long period to reach a "cold" state before being reenergized [34].

The overload effect is related to the types of load [35], weather condition [36], and duration of outage [34]. Although modeling methods of the cold load effect can be found in the research [37–39], this effect is generally difficult to be accurately quantified [34]. To provide additional generation capacity to serve excessive load demand due to the cold load effect, the distributed generators (DGs) have been applied in distribution systems [40].

6.1.3.3 Load Restoration in Distribution Systems

For load restoration in distribution systems, it is assumed that the generation capacity is sufficient. Thus, the major challenge is to figure out the optimal switching sequence to provide electricity to customers. Various solution methods have been developed to solve these problems, including ant algorithm [41], dynamic programming [42], Lagrangian relaxation algorithm [43], genetic algorithm [44], metaheuristic technique [45], interior point with sensibility analysis [46], analytical hierarchy process [47], nondominated sorting genetic algorithm [48], and so on. Special control strategies, such as load curtailment [49] and utilizing microgrids [50], have been proposed to improve the efficiency of load restoration.

6.1.3.4 Load Restoration in Transmission Systems

For load restoration in transmission systems, a larger spectrum of constraints need to be considered [18]. A single variable optimization model to maximize load pickup for a given substation subject to a large set of security constraints has been formulated and solved by a bisectional algorithm [51]. With accurate phasor measurement units data, an analytical method to predict restorable load level has been proposed [52], incorporating polynomial load model and a simplified single generator frequency dynamic model. To overcome the cold load pickup issue, the DGs are introduced in load restoration [53].

The key features of selected work related to load restoration are summarized in Table 6.3.

6.2 Electromagnetic Transients During System Restoration

During black start, the electromagnetic transients are an important consideration, with the power system in transition from one state to another. Electromagnetic transients can significantly impact the safety during system restoration. The importance of the research in this area is well recognized.

At the beginning stage of black start, restoration of generating units is a major task. During this process, self-excitation, which leads to overvoltage, might occur because of parameter resonance. Furthermore, a high-level start-up current for a generating unit may also occur. For the safety of generating units, all these electromagnetic transients should be carefully considered.

Table 6.3 Features of selected work related to load restoration.

Authors	Problem Addressed	Model Considerations	Solution Method
I. Mohanty et al., 2003 [41]	Distribution load restoration	Cold load curve, step-by-step restoration	Ant algorithm
R. Perez-Guerrero et al., 2008 [42]	Distribution load restoration	Mixed-integer linear programming model, no network model	Dynamic programming
R. Perez-Guerrero et al., 2008 [43]	Distribution load restoration	Mixed-integer linear programming model, switching constraint	Lagrangian relaxation
S. Chavali et al., 2002 [44]	Distribution load restoration	Cold load curve, minimize switching time	Genetic algorithm
O. Duque and D. Morinigo, 2006 [45]	Distribution load restoration	Radial network, considering voltage drop and line capacity	Metaheuristic technique based on tabu search
T. T. Borges et al., 2011 [46]	Distribution load restoration	AC OPF with switching	Interior point method and sensibility analysis
H. Qu and Y. Liu, 2012 [51]	Substation load restoration	Maximize substation load pick subject to a large constraint set	Single variable optimization, bisection method
W. Liu et al., 2013 [52]	Transmission load restoration	Maximize restorable load level with integer programming	Integer programming

During reestablishment of the overhead line or the underground cable, three kinds of overvoltage may occur. They are power frequency overvoltage, switching overvoltage, and resonance overvoltage.

In addition, the harmonic resonance overvoltage originates from the switching operation and nonlinear feature of the equipment. The sources of harmonic currents are explained next. First, transformer saturation occurs and produces harmonic currents when transformers become overexcited due to power frequency overvoltage. Next, the magnetizing inrush current caused by energizing transformers also leads to harmonic currents. Finally, after a series of switching operations on transmission lines, the no-load transformers might be placed into the system. Then it may generate a large magnetizing inrush current that can damage transformers and other electrical equipment.

Load pickup is the major task at the last stage of system restoration. The transient characteristics of the load with different types should be understood. System restoration can be threatened by a variety of factors and conditions. The transient phenomena during a black start will be discussed.

6.2.1 Generator Self-Excitation

During black start, the system employs generators as the start-up power to energize no-load lines. It might lead to generator self-excitation problems, which cause overvoltage at the terminal and damage the insulation of generators.

6.2.1.1 Self-Excitation Criteria During Black Start

There are a variety of methods for self-excitation study. Time-domain simulation is accurate but time consuming. During the black start, some parameters are hard to be estimated within a short period of time. By establishing the characteristic equation, self-excitation could be judged according to the existence of positive real roots or positive complex roots. The physical process and phenomena of self-excitation are combined to analyze the black start's self-excitation by using the impedance method. It only requires available and simple parameters. Although they are not very precise, they could meet the requirements for self-excitation study during black start.

6.2.1.2 Suppression of Self-Excitation

To suppress self-excitation and restrain the resonance, the following methods can be employed.

- Using modern automatic control excitation devices, which can eliminate the simultaneous self-excitation.
- A larger resistance could be connected to the stator winding and cut off during the normal operation, so as to prevent large power losses.
- The closing operation should be conducted on the system side with a large capacity, not on the isolated generator side.
- Increasing the number of generators so that the sum of x_d and x_q is less than x_c. The self-excitation condition is hence not satisfied. However, this method for suppressing self-excitation in the system recovery may not be applicable during black start, due to the limited number of black-start generators.
- In the EHV grid, the shunt reactors x_l can be installed on the line to compensate for capacitance x_c so that the sum is greater than that of x_d and x_q, and hence the self-excitation condition can be changed.

6.2.2 Switching Overvoltage

Switching overvoltage can be produced by closing an unloaded line, opening an isolating switch, or interrupting low currents in inductive or capacitive circuits where the possibility of restrikes exists. Opening operations intended for interrupting a circuit current are responsible for switching overvoltages. Routine operations are capable of producing overvoltage effects by altering the system configuration. In the black start, the overvoltage issue should be paid more attention, as it may damage equipment insulation and affect the recovery sequence because of the frequent line operations.

Although overvoltages cannot be completely avoided, their impacts can be minimized. Generally, the occurrence and magnitude of overvoltage can be limited by utilizing appropriate measures, such as series or parallel compensation, closing resistors, surge suppression, metal oxide varistors, or snubbers containing a combination of resistors and capacitors.

6.2.2.1 Closing of a Line

A cable being energized from a transformer is often implemented in a black start. Take a simplest case of a switching operation for example, as shown in Figure 6.1a. The cable and transformer are represented by the capacitance and leakage inductance, respectively. The equivalent circuit is shown in Figure 6.1b. The transient response is determined by the combined impedance of the transformer feeding the system and the total surge impedance of the connected lines.

The total closing overvoltage is given by the sum of the power frequency source overvoltage and the transient overvoltage generated at the line. The overvoltage factor for the source is given by the following equation.

$$K_{sf} = \frac{1}{\cos 2\pi f \sqrt{LCl} - \dfrac{X_s}{Z} \sin 2\pi f \sqrt{LCl}}, \tag{6.1}$$

where f denotes the power frequency; L and C are the positive sequence inductance and capacitance per length of line, respectively; l is the line length; and X_s and Z denote the short circuit reactance of the source and surge impedance of the line.

(a)

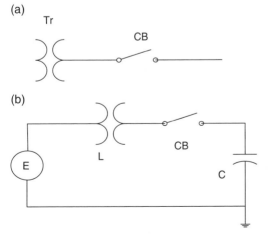

Tr

CB

(b)

E

L

CB

C

Figure 6.1 Closing a line.
(a) Switching operation.
(b) Equivalent circuit.

From this expression (6.1), it can be concluded that the overvoltage factor is proportional to the following issues:

- Instantaneous voltage differences between the source voltage and line voltage as the contact of the circuit breaker closes
- Damping impedance of the lines connected at the source side of the circuit
- Terminal impedance of the unloaded line being energized

Note that the amplitude of the overvoltage will reduce with the increasing size of the system. In addition, the superposition of individual responses with different frequency results in the reduction of the factor.

6.2.2.2   Energizing Unloaded Transformers

During system restoration, energization of unloaded transformers through a long line will increase the likelihood of overvoltage. When closing the transformer, the discharge path for the current trapped along the line is established through the magnetizing impedance of the transformer. The transformer presents a high impedance and thus the discharge is very slow as long as the transformer is not saturated. However, it may enter the saturation and its impedance drops quickly, leading to an increasing current and rapid discharge of the cable.

Energization of an unloaded transformer can cause voltage values to be several times higher than the primary voltage that can be generated. It will cause a failure of the transformer. The critical length of the cable can be determined by the travel time of resonant frequency of the line and transformer. As a result, changing the length of the line would be the simplest alternative to decrease the effects of switching transients. Another approach is to install capacitors connected to the secondary terminal of the transformer.

6.2.2.3   Limiting Overvoltage

Switching overvoltage includes the power frequency overvoltage and the transient overvoltage. Limiting the magnitude of the former is usually sufficient to reduce the total overvoltage within an acceptable range.

Appropriate measures can be adopted to constrain the overvoltage magnitude by limiting the power frequency overvoltage as follows:

- Providing polarity controlled closing
- Adding closing and/or opening resistors across the circuit breaker contacts
- Providing a method combing polarity control
- Adding parallel compensation
- Reducing the supply side reactance

On the other hand, the transient overvoltage factor can be controlled by

- Synchronizing closing, accomplished either by closing at a voltage zero of the supply side or matching the polarity of the line and the supply side
- Using a preinsertion resistor

6.2.2.4 Approach for Evaluating Transient Voltage

As far as energizing transmission lines is concerned, three factors are taken into account: length of line to be energized, adequacy of online generation, and size of underlying load [53].

If a large section of lines are energized, there is a risk of damaging the equipment insulation. However, energizing small sections tends to prolong the restoration process.

The utilization of inadequate sources for energizing line could lead to a higher transient voltage than what the piece of equipment can withstand. The loads at both end of the line tend to reduce the sustained and transient voltage. It is important for system operators to know the minimum size of load pickup to avoid transient overvoltages. Detailed methods such as the Electromagnetic Transients Program (EMTP) tools can meet the need.

6.2.2.5 Open-Ended Line Model

When lines are energized by different sizes of generator and line dead loads, the sustained and transient voltages on the sending and receiving ends could be determined for energizing lines with different voltage and lengths. The approach could be applied at the early stages of power system restoration.

The open-ended line model is shown in Figure 6.2. The generator is connected to a high-voltage line, represented by the π type equivalent circuit with series and parallel impedances Z_3 and $\frac{1}{2}Z_2$. The step-up transformer impedance is denoted as Z_1. To evaluate switching voltage, the generator subtransient reactance X_d'' is lumped with Z_1.

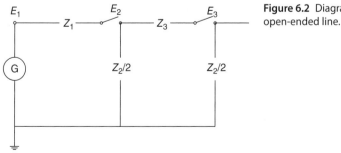

Figure 6.2 Diagram of open-ended line.

The following two equations relate sending-end voltage E_2 and receiving-end voltage E_3 with generator voltage E_1.

$$\begin{cases} E_2 = (N_1/D_1)E_1 \\ E_3 = (N_2/D_1)E_1 \end{cases}$$ (6.2)

where $N_1 = 2Z_2(2Z_2 + Z_3)$; $N_2 = 2Z_2(2Z_2 + Z_3)$; $D_1 = (Z_1 + 2Z_2)(2Z_2 + Z_3) + (2Z_2 Z_1)$.

To evaluate switching transient voltages, the coupling between phases and pole disparity in energizing lines is neglected in the following model:

$$E_1 - E_2 = R_1 I_1 + L_1 dI_1/dt$$
$$E_2 = \frac{2}{C} \int I_2 dt$$ (6.3)
$$E_2 - E_3 = R_3 I_3 + L_3 dI_3/dt.$$

Using $E_1(t) = E_1 \cdot \cos(\omega t)$, the transient solutions for $E_2(t)$ and $E_3(t)$ could be written as

$$\begin{cases} E_2(t) = E_1(t) \sum \exp(-\alpha_i t)\cos(\beta_i t + \delta_i) \\ E_3(t) = E_1(t) \sum \exp(-\alpha_j t)\cos(\beta_j t + \delta_j) \end{cases}$$ (6.4)

where i or j = 1, 2, 3.

In the π equivalent circuit, E_2 and E_3 are related to E_1 by the sum of a 60-cycle sinusoid, and two high-order harmonics decay with time.

6.2.2.6 Load-Ended Line Model

Regarding the load-ended line, it is convenient to represent the line by the T type equivalent circuit as shown in Figure 6.3. The step-down transformer and loads are lumped together and represented as Z_4. The sending-end voltage E_2

Figure 6.3 Diagram of load-ended model.

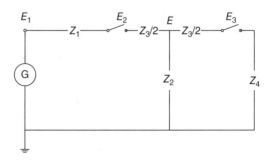

and receiving-end voltage E_3 are related to generator voltage E_1 by the following two equations:

$$\begin{cases} E_2 = (N_1/D_1)E_1 \\ E_3 = (N_2/D_1)E_1 \end{cases} \tag{6.5}$$

where $N_1 = (Z_2 + Z_3/2)(Z_4 + Z_3/2) + (Z_2 Z_3/2); N_2 = Z_2 Z_4; D_1 = N_1 + Z_1(Z_2 + Z_4 + Z_3/2)$.

In the load-ended model, the following parameters are used to evaluate the switching overvoltage:

$$\begin{aligned} & E_1 - E_2 = R_1 I_1 + L_1 dI_1/dt \\ & E = \frac{1}{C}\int I_2 dt \\ & E - E_3 = \frac{1}{2}\left(R_3 I_3 + L_3 dI_3/dt\right) \\ & E_3 = R_4 I_4 + L_4 dI_4/dt. \end{aligned} \tag{6.6}$$

As the case in open-ended line, the transient solutions for $E_2(t)$ and $E_3(t)$ are given here:

$$\begin{cases} E_2(t) = E_1(t)\left[\exp(-a_1 t + b_1) + \sum\exp(-\alpha_i t)\cos(\beta_i t + \delta_i)\right] \\ E_3(t) = E_1(t)\left[\exp(-a_2 t + b_2) + \sum\exp(-\alpha_j t)\cos(\beta_j t + \delta_j)\right] \end{cases} \tag{6.7}$$

In the T equivalent circuit, E_2 and E_3 are related to E_1 by the sum of a 60-cycle sinusoid, a decaying function, and one high-order harmonic that also decays with time.

The models described here could be used to determine the transient overvoltage during the black start or other situations. It has been shown that this approach could accelerate the restoration process. The transient overvoltage could be avoided by considerations of

- Length of high-voltage line to be energized
- Adequacy of energizing sources
- Adequacy of the loads at either end of the line

6.2.3 Resonant Overvoltage in the Case of Energizing No-Load Transformer

At the early stage of system restoration, a large inrush current may be incurred as a result of parallel resonance between line capacitance and transformer reactance when the no-load transformer is put into operation. This inrush current enlarges harmonic distortion and generates a high harmonic resonant overvoltage. At the same time, due to the nonlinear characteristic of iron core in the transformer, mutual inductor, and arc suppression coil, the iron core component will be saturated to endanger system restoration because a continuous ferromagnetic resonance overvoltage with a high magnitude will be activated by the AC power supply. To prevent the equipment from an overvoltage hazard, research on the harmonic resonance overvoltage and ferromagnetic overvoltage is necessary.

6.2.3.1 Harmonic Resonance Overvoltage

During system restoration, if a no-load or light-loaded transformer is put into operation, a large inrush current including a large amount of higher harmonics will be incurred due to the iron core saturation. This current becomes the major harmonic source. When the resistance and line capacitance produce resonance near a harmonic frequency, the sudden increase of the system resistance may result in a higher voltage by the inrush current of the transformer. The rising terminal voltage deteriorates the saturation of the iron icon, leading to a further increase of harmonic current. Since the system damping is smaller at the initial stage of system restoration, harmonic resonance tends to occur. The overvoltage lasting time is so long that it will incur relay tripping.

Harmonic resonance is caused by harmonics, so the major approach to suppress harmonic resonance is to reduce harmonics. The appropriate measures to reduce harmonics on or near the harmonic source are described as follows:

- Install an AC filter devices
  High pass filters and single-tuned filters can be installed near the harmonic source to absorb the harmonic current. This is an effective approach to reduce harmonics by combining power factor compensation and voltage adjustment.
- Change the harmonic source configuration
  The devices with harmonic complementarity should be distributed properly so that the resonance mode with a large amount of harmonics can be changed. This method has particular requirements on the configuration of the devices.
- Install a static Var compensator
  The capacitive part of a static compensator, such as *TCR*, *TCT*, or *SR*, could be designed as the filter to reduce harmonics. Moreover, it has other functions such as confining voltage fluctuation and flicking, and power factor compensation with comprehensive technical and economic benefits. However, the investment is large and a special design is needed.

6.2.3.2 Ferroresonance Overvoltage

Magnetic saturation of the iron core introduces nonlinear inductance, which may lead to ferroresonance. During a black start with a no-load transformer,

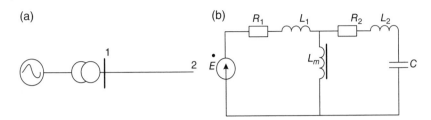

Figure 6.4 Generator and transformer connected with long no-load line. (a) A typical system with ferromagnetic overvoltage. (b) Equivalent circuit.

potential transformers, and capacitors, the impact current will saturate the iron core and generate ferroresonance under certain conditions.

The induction of ferromagnetic components under alternating magnetic flux will change with the even times as the angular frequency 2ω, 4ω, 6ω. As a result, higher harmonics with 2ω, 3ω, etc., are generated.

A typical system with ferromagnetic overvoltage is shown in Figure 6.4a, in which the generator and transformer are connected to a long no-load line. Figure 6.4b depicts the equivalent circuit with lumped parameters.

L_1 and L_2 are the linear inductance, including the leakage inductance and line inductance. L_m is the magnetizing inductance of transformers. C is the line-to-ground capacitance. R_1 and R_2 are equivalent resistances for power supply and the line. This circuit may produce odd and even at times high-frequency resonance.

In the equivalent circuit here, L_1 and L_2 are much smaller than L_m. The resonant frequency for the linear part of the circuit is

$$\omega_0 = \frac{1}{\sqrt{(L_1 + L_2)C}}. \tag{6.8}$$

Due to the nonlinear feature of L_m, odd harmonics are included, in addition to the current with the fundamental component. If ω_0 is equal or close to a higher harmonic angular frequency $n\omega$, there may be $n\omega$ resonance.

As for the odd harmonics, Figure 6.5 is the equivalent circuit of an n-order harmonic. The current source in Figure 6.5a represents the odd high-frequency current components caused by the nonlinear magnetizing characteristics of the transformer. The analysis only focuses on the n-order harmonic without considering the power frequency source. After the equivalent transformation between the voltage and current sources, Figure 6.5b is obtained, and the voltage source may be written as

$$\dot{E} = -\dot{I}(R_1 + jn\omega L_1). \tag{6.9}$$

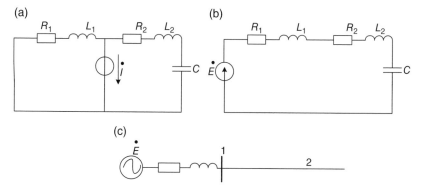

Figure 6.5 Odd high-frequency resonance analysis. (a) Odd high-frequency current components. (b) N-order harmonic equivalent diagram.

The n-order harmonic equivalent diagram corresponding to Figure 4.9b is shown in Figure 6.5c.

To sum up, the reason for higher harmonic resonance generated in Figure 6.5 includes the general ferromagnetic resonance condition, and some specific conditions listed next.

- The input impedance for the head of the line should be capacitive.
- The resonance angular frequency for the linear part must be close to $n\omega$.

Generally, in a real system, the second, third, and fifth order harmonic resonances are more likely to occur. The winding of the transformer adopts the delta connection, which prevents the third harmonic resonance. As for the fifth harmonic and higher harmonics, the equivalent circuit loss and the corona loss will increase with frequency rising; thus, the resonance can be suppressed.

To prevent the occurrence of ferromagnetic resonance, three measures can be taken:

- Changing parameters for the inductance and capacitance to prevent the resonance condition.
- Consuming the resonance energy to eliminate the resonance.
- Using grounding mode in power system design, or using temporary switching action in operation.

6.2.4 Impact of Magnetizing Inrush Current on Transformer

During the normal operation of the transformer, the magnetizing current is small, usually only 1–3% of the rated current and even less than 1% for large-capacity transformers. The transformer magnetizing current depends on the magnetizing inductance as well as the saturation degree of the transformers. Under normal circumstances, the transformer will not saturate, so the influence of inrush current on the differential protection is often negligible. When the transformer is connected to the system with no load under the external voltage recovery after fault clearing, the transformer voltage increases from zero or a very small value to the operating voltage instantaneously. During the transient of the voltage rising, the transformer will be severely saturated, and this results in large magnetizing inrush currents. Its maximum can reach four to eight times of the rated current. During the early stage of system restoration, with the closing of the self-start-up unit and a series of no-load line, no-load transformer energizing is inevitable. Once a large inrush current is incurred, it will hinder safety and stability of the power grid.

6.2.4.1 Characteristics and Identification Methods of Magnetizing Inrush Current

The iron core saturation is the main reason for the occurrence of the magnetizing inrush current generation. The magnetizing circuit is essentially an electrical circuit with iron core winding. The primary side of transformers can be equivalent to nonlinear inductance. Under normal circumstances, the core is not saturated and the relative permeability is high. Magnetizing inductance of transformer windings is large; hence, the magnetizing current is small. During the transient process, the core will reach saturation level with the relative permeability close to one. The inductance of transformer windings reduces sharply, and thus a large inrush current will occur.

The transformer inrush current and fault current are quite different. If it cannot be distinguished with the fault current, the differential protection may act incorrectly. The false action would affect the restoration process and recovery time. At present, the methods identifying the inrush current include the second harmonic restraint, intermittent angle, waveform symmetry, the magnetic properties, and the equivalent circuit.

6.2.4.2 Magnetizing Inrush Current Suppression

The inrush current can be suppressed by two techniques. One is via the external control adopted to constrain the inrush current. The other is aimed at changing the internal structure such as the virtual air gap. The specific methods are listed next.

- Control the three-phase switch angle (e.g., time) for closing
  If the peak value of the external AC voltage reaches the maximum at the closing moment, no inrush current will be generated in the transformer. Controlling the three-phase closing angle could be performed to reduce inrush current. When the closing angle is 90°, the magnetic flux generated inside the transformer is close to zero. Moreover, the corresponding inrush current is small. However, in practice, the error time for closing still needs to be studied.

- Install grounding resistance
  The unbalanced three-phase inrush current may exist during the closing of a no-load transformer. A grounding resistance could be applied to connect with the neutral point of three-phase transformer so as to withstand this imbalance current. As a result, the inrush current would decline. In addition, the grounding resistance can also weaken the voltage applied on the transformer core to prevent the saturation. This approach may be used in combination with the method before and the magnitude of inrush current can be suppressed to 40%. However, further research should be conducted, such as grounding resistance and delay time selection.

- Change windings distribution of transformer
  The generation of the inrush current is accompanied by core saturation with its magnetic permeability closed to vacuum permeability. Under this condition, the primary side of the transformer can be deemed an air core coil without the iron core. Therefore, the magnetic flux is extended to areas outside the core. If the transformer primary windings is moved from the internal to external, the corresponding cross-sectional area is enlarged. The inductance of the air coil is proportional to the cross-sectional area and inversely proportional to the inrush current. Hence, the inrush current will be reduced. The transformer primary or secondary windings distribution can be changed to increase the transient or surge equivalent inductance to suppress the inrush current. However, the application of this method must change the structure of the transformer, which will lead to other problems and thus limits its development.

6.2.5 Voltage and Frequency Analysis in Picking up Load

After a system outage, the restoration process consists of three stages: the generator black start, the network reconstruction, and the load restoration. To prevent overvoltage of the line, part of the load will be recovered during line charging. However, load restoration at this moment is not the final objective. When the network is reconstructed and the system has certain capability for power supply, the large amount of load needs to be recovered. Since the transmission network at this time is not strong enough, the excess load connected to the grid will result in sharp frequency and voltage drop. On the other hand, the small amount of load recovery will extend the recovery time. As a consequence, the maximum load should be recovered without violating any operational constraints.

6.2.5.1 Auxiliary Load Recovery in Black Start

If no overvoltage or self-excitation issues arise after the charge of the no-load long line, the next step will concentrate on the start-up of the auxiliary load in the plants so as to start large-capacity units. The gas turbines or hydraulic turbines with smaller capability are usually chosen as the black-start power supply. The loads for those units are pumps, fans, induction motors, and so on. During start-up, the induction motor is in locked-rotor state, so the start-up current is considerably larger compared with the rated current, which will impact system frequency and voltage greatly. In addition, the weak power supply and network structure will deteriorate the situation and lead to extremely low voltage and frequency. As a result, the generator may be tripped and the system may collapse again. Therefore, to ensure the operational feasibility of the black-start sequence at the initial state, it is necessary to study issues regarding voltage and frequency stability during the start-up of large-capacity generators.

6.2.5.2 Induction Motor Model

Due to the large start-up current existing in the large induction motor, the flux in the motor will be saturated, so the model considering the saturation should be adopted. The iron loss r_c attributed to the load and efficiency factor η is modified by ignoring the reactance x_1. The equivalent circuit is shown in Figure 6.6: r_s and x_s are the stator resistances and stator leakage reactance; x_m is the excitation reactance; r_1 and r_2 are the rotor resistances of the internal and external cage, respectively; x_1 and x_2 are leakage reactances; s is slip. The estimation formulas for those electrical parameters are as follows:

$$\begin{cases} r_s = \cos\phi\left[1-\eta'/(1-s)\right] \\ r_r = s\eta'/\left[(1-s)\cos\phi\right] \\ x_m = \eta'/\left[(1-s)\sin\phi\right] \\ T_{rat} = T_{st}/T = \dfrac{T_{st}\cdot\omega_s(1-s)}{\cos\phi\cdot\eta'} \\ r_{st} = r_r \cdot T_{rat}\left(I_s/I_{st}\right)^2\cos^2\phi/s \\ r_1 = r_{st}\left(1+m^2\right)-r_r m^2 \\ r_2 = r_1 r_r/(r_1-r_r) \\ x_2 = (r_1+r_2)/m, \end{cases} \tag{6.10}$$

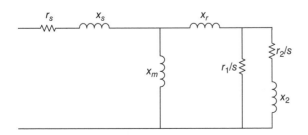

Figure 6.6 Equivalent circuit of an induction motor.

where cos ϕ is the power factor at full load, η' is the revised efficiency factor, T_{st} and I_{st} are the starting torque and current with the rated voltage, and m is the design ratio (usually $m = 1$). Without taking into account the saturation, the motor parameters can be derived from the data given by the nameplates. Taking account of saturation, the voltage and current should be increased in the case of the reduced-voltage start. The voltage could be set to 0.8 $p.\,u.$ of the starting voltage.

6.2.5.3 Influence of Motor Start-Up on Black Start

During the start-up of a generator with a high capacity, the characteristics with large start current and low power factor will lead to the terminal voltage drop. However, no serious drop emerges in the system frequency, as shown in Figure 6.7. For the black-start process, the system frequency decrease should be kept in 0.5 Hz compared to that of the prefault. If more than 0.5 Hz, the black-start scheme should be rescheduled.

In addition, the voltage drop caused by the induction motor is concerned with the short-circuit capability provided by the system, that is, the larger capability corresponds to a smaller voltage reduction. At the initial stage of the black start, only several units are operating, which provides a small short-circuit capacity and thus causes the voltage drop significantly.

Electromagnetic torque of induction motor is:

$$T = \frac{1}{\Omega} \frac{m_1 U^2 \dfrac{R_2}{s}}{\left(R_1 + \dfrac{R_2}{s}\right)^2 + \left(X_{1\sigma} + X_{2\sigma}\right)^2},$$

(6.11)

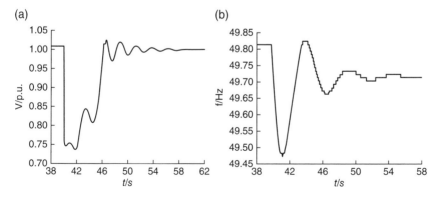

Figure 6.7 Curves for voltage and frequency. (a) Motor terminal voltage. (b) System frequency.

in which T is the electromagnetic torque, Ω is the rotor mechanical angular speed, U is motor terminal voltage, s is slip, R_1, $X_{1\sigma}$ are the resistance and reactance of the stator side, and R_2, $X_{2\sigma}$ are the resistance and reactance of the rotor side. In Figure 6.7, the motor torque and the voltage show the quadratic relationship. The terminal voltage drop will cause a greater decline of the torque. Once the start torque of the induction motor is less than the damping torque, the motor start will fail, which leads to black-start restoration failure.

6.2.5.4 Cold Load Pickup

Load capacity in a short period after pickup is often greater than the steady-state load, that is, the phenomenon of cold load recovery, which is mainly caused by losing the asynchrony of the constant temperature-controlled load.

When power supplies are restored, the system's constant temperature-controlled loads such as electric heaters, refrigerators, and other appliances will be started at the same time, and this results in a sudden increase in load power in a short time. The amount of load demand can be up to 10 times the steady-state load. Owing to large fluctuations of system frequency in the recovery, a rapid increase in load power is likely to bring about the system overload and frequency instability. In colder climates, the cold load will be a more serious problem.

At the late stage of the load pickup, a large number of temperature control devices are active, making load increase up to 4–5 times of that of prefault. The overload may last several hours. With the outage time increasing, this feature will become more obvious. In general, the load can be divided into three categories: temperature-controlled loads, fixed loads, and manually controlled loads. During system recovery, these three load components show different start-up characteristics. For temperature load, its total capacity increases a

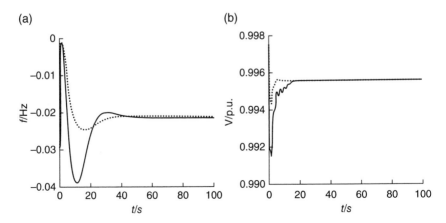

Figure 6.8 Response curves. (a) Typical response characteristics of the bus frequency. (b) Typical bus voltage waveform.

great deal compared to that of prefault, due to the loss of diversity. So the start-up for the temperature-controlled load is a transient process from beginning of start-up. The fixed load changes slightly during the fault.

During the restoration and load pick-up process, the automatic controlled load will also be started at the same time. The active and reactive power are often much larger than the value under normal conditions, so the system frequency and voltage fluctuation are relatively large in the restoration process, and system voltage and frequency stability are therefore affected. Figure 6.8 is the system response curve with and without the cold load pickup characteristics.

Considering the cold load recovery characteristics, the system frequency and voltage fluctuate significantly, due to the sudden increase of the load power. To avoid overloading, the characteristics of the cold load must be taken into account and the cold load recovery model should be established.

References

1 Feltes, J.W. and Grande-Moran, C. (2008). Black start studies for system restoration, in *Proceedings of the 2008 IEEE Power and Energy Society General Meeting, July 2008*, pp. 1–8.

2 Bahrman, M. and Bjorklund, P.E. (2014). The new black start: system restoration with help from voltage-sourced converters. *IEEE Power Energ. Mag.* 12 (1): 44–53.

3 Lieu, K.-L., Liu, C.-C., and Chu, R.F. (1995). Tie line utilization during power system restoration. *IEEE Trans. Power Syst.* 10 (1): 192–199.

4 Tz, R.D.S. and Mason, G.A. (1984). Blackstart utilization of remote combustion turbines, analytical analysis and field test. *IEEE Trans. Power Syst.* PAS-103 (8): 2186–2191.

5 Zhou, B., Chan, K.W., and Yu, T. (2009). Application of eigen-analysis method in auxiliary decision of black start scheme, in *Proceedings of the 8th International Conference on Advances in Power System Control, Operation and Management (APSCOM 2009), November 2009*, pp. 1–6.

6 de Mello, F.P. and Westcott, J.C. (1994). Steam plant startup and control in system restoration. *IEEE Trans. Power Syst.* 9 (1): 93–101.

7 Adibi, M.M., Adsunski, G., Jenkins, R., and Gill, P. (1995). Nuclear plant requirements during power system restoration. *IEEE Trans. Power Syst.* 10 (3): 1486–1491.

8 Liu, C.-C., Liou, K.-L., Chu, R.F., and Holen, A.T. (1993). Generation capability dispatch for bulk power system restoration: a knowledge-based approach. *IEEE Trans. Power Syst.* 8 (1): 316–325.

9 Ketabi, A., Asmar, H., Ranjbar, A.M., and Feuillet, R. (2001). An approach for optimal units start-up during bulk power system restoration, in *Proceedings of the 2001 Large Engineering Systems Conference on Power Engineering (LESCOPE '01) 2001*, pp. 190–194.

10 Zhong, H.-R. and Gu, X.-P. Determination of optimal unit start-up sequences based on fuzzy AHP in power system restoration, in *Proceedings of the 2011 4th International Conference on Electric Utility Deregulation and Restructuring and Power Technologies (DRPT), 2011*, pp. 1541–1545.

11 Hou, Y., Liu, C.-C., Sun, K. et al. (2011). Computation of milestones for decision support during system restoration. *IEEE Trans. Power Syst.* 26 (3): 1399–1409.

12 Liu, Q., Shi, L., Zhou, M. et al. (2008). A new solution to generators start-up sequence during power system restoration, in *Proceedings of the 2008 Third International Conference on Electric Utility Deregulation and Restructuring and Power Technologies (DRPT 2008), April 2008*, pp. 2845–2849.

13 Sun, W., Liu, C.-C., and Zhang, L. (Aug. 2011). Optimal generator start-up strategy for bulk power system restoration. *IEEE Trans. Power Syst.* 26 (3): 1357–1366.

14 Liu, S., Podmore, R., and Hou, Y. (2012). System restoration navigator: a decision support tool for system restoration, in *Proceedings of the 2012 IEEE Power and Energy Society General Meeting, July 2012*, pp. 1–5.

15 Nezam Sarmadi, S.A., Dobakhshari, A.S., Azizi, S., and Ranjbar, A.M. (2011). A sectionalizing method in power system restoration based on WAMS. *IEEE Trans. Smart Grid* 2 (1): 190–197.

16 Wang, C., Vittal, V., and Sun, K. (2011). OBDD-based sectionalizing strategies for parallel power system restoration. *IEEE Trans. Power Syst.* 26 (3): 1426–1433.

17 Quiros-Tortos, J. and Terzija, V. (2013). A graph theory based new approach for power system restoration, in *Proceedings of the 2013 IEEE Grenoble PowerTech (POWERTECH), 2013*, pp. 1–6.

18 Adibi, M., Clelland, P., Fink, L. et al. (May 1987). Power system restoration – a task force report. *IEEE Trans. Power Syst.* 2 (2): 271–277.

19 Adibi, M.M. and Fink, L.H. (Sep.-Oct. 2006). Overcoming restoration challenges associated with major power system disturbances – restoration from cascading failures. *IEEE Power Energ. Mag.* 4 (5): 68–77.

20 Nematollahi, M. and Tadayon, M. (2013). Optimal sectionalizing switches and DG placement considering critical system condition, in *Proceedings of the 2013 21st Iranian Conference on Electrical Engineering (ICEE), 2013*, pp. 1–6.

21 Falaghi, H., Haghifam, M., and Singh, C. (Jan. 2009). Ant colony optimization-based method for placement of sectionalizing switches in distribution networks using a fuzzy multiobjective approach. *IEEE Trans. Power Delivery* 24 (1): 268–276.

22 Wang, C., Liu, Y., Qu, H., and Yuan, Z. (2009). Genetic algorithm based restoration scheme for power system skeleton, in *Proceedings of the 2009 Fifth International Conference on Natural Computation (ICNC '09), August 2009*, pp. 651–655.

23 Gu, X. and Zhong, H. (2012). Optimisation of network reconfiguration based on a two-layer unit-restarting framework for power system restoration. *IET Gener. Transm. Distrib.* 6 (7): 693–700.

24 Liu, Y. and Gu, X. (2007). Skeleton-network reconfiguration based on topological characteristics of scale-free networks and discrete particle swarm optimization. *IEEE Trans. Power Syst.* 22 (3): 1267–1274.

25 Escobedo, A.R., Moreno-Centeno, E., and Hedman, K.W. (Mar. 2014). Topology control for load shed recovery. *IEEE Trans. Power Syst.* 29 (2): 908–916.

26 Wang, C., Vittal, V., Kolluri, V.S., and Mandal, S. (Aug. 2010). PTDF-based automatic restoration path selection. *IEEE Trans. Power Syst.* 25 (3): 1686–1695.

27 Ye, H. and Liu, Y. (Dec. 2013). A new method for standing phase angle reduction in system restoration by incorporating load pickup as a control means. *Int. J. Electr. Power Energy Syst.* 53 (0): 664–674.

28 Martins, N., De Oliveira, E.J., Moreira, W.C. et al. (May 2008). Redispatch to reduce rotor shaft impacts upon transmission loop closure. *IEEE Trans. Power Syst.* 23 (2): 592–600.

29 Viana, E.M., de Oliveira, E.J., Martins, N. et al. (Feb. 2013). An optimal power flow function to aid restoration studies of long transmission segments. *IEEE Trans. Power Syst.* 28 (1): 121–129.

30 Jadid, S. and Salami, A. (2004). Accurate model of hydroelectric power plant for load pickup during power system restoration, in *Proceedings of the 2004 IEEE Region 10 Conference (TENCON 2004), November 2004*, Vol. 3, pp. 307–310.

31 Ketabi, A., Ranjbar, A.M., and Feuillet, R. (2000). A new method for dynamic calculation of load steps during power system restoration, in *Proceedings of the 2000 Canadian Conference on Electrical and Computer Engineering, 2000*, Vol. 1, pp. 158–162.

32 Chuvychin, V., Gurov, N., and Rubcov, S. (2005). Adaptive underfrequency load shedding and underfrequency load restoration system, in *Proceedings of the 2005 IEEE Russia Power Tech., June 2005*, pp. 1–6.

33 Chuvychin, V.N., Gurov, N.S., Venkata, S.S., and Brown, R.E. (1996). An adaptive approach to load shedding and spinning reserve control during underfrequency conditions. *IEEE Trans. Power Syst.* 11 (4): 1805–1810.

34 IEEE Power system relaying committee, Cold load pickup issues, May 2008 [Online]. Available: http://resourcecenter.ieee-pes.org/pes/product/technical-publications/PESTR0038.

35 Agneholm, E. and Daalder, J. (2000). Cold load pick-up of residential load. *IEE Proc. Generat. Transm. Distrib.* 147 (1): 44–50.

36 McDonald, J.E. and Bruning, A.M. (1979). Cold load pickup. *IEEE Trans. Power Syst.* PAS-98 (4): 1384–1386.

37 Mirza, O.H. (1997). Usage of CLPU curve to deal with the cold load pickup problem. *IEEE Trans. Power Delivery* 12 (2): 660–667.

38 Ihara, S. and Schweppe, F.C. (1981). Physically based modeling of cold load pickup. *IEEE Trans. Power Syst.* PAS-100 (9): 4142–4150.

39 Gibo, N., Kikuma, T., Takenaka, K., Hatano, R., and Yamamoto, N. (2009). Estimation method of load type and capacity by using starting load characteristics, in *Proceedings of the 2009 IEEE/PESPower Systems Conference and Exposition (PSCE '09), March 2009*, pp. 1–6.

40 Kumar, V., Kumar, H.C.R., Gupta, I., and Gupta, H.O. (2010). DG integrated approach for service restoration under cold load pickup. *IEEE Trans. Power Delivery* 25 (1): 398–406.

41 Mohanty, I. Kalita, J. Das, S. Pahwa, A. and Buehler, E. (2003). Ant algorithms for the optimal restoration of distribution feeders during cold load pickup, in *Proceedings of the 2003 IEEE Swarm Intelligence Symposium (SIS '03), April 2003*, pp. 132–137.

42 Perez-Guerrero, R., Heydt, G.T., Jack, N.J. et al. (2008). Optimal restoration of distribution systems using dynamic programming. *IEEE Trans. Power Delivery* 23 (3): 1589–1596.

43 Perez-Guerrero, R.E. and Heydt, G.T. (2008). Distribution system restoration via subgradient-based Lagrangian relaxation. *IEEE Trans. Power Syst.* 23 (3): 1162–1169.

44 Chavali, S., Pahwa, A., and Das, S. (2002). A genetic algorithm approach for optimal distribution feeder restoration during cold load pickup, in *Proceedings of the 2002 Congress on Evolutionary Computation (CEC '02), 2002*, Vol. 2, pp. 1816 1819.

45 Duque, O. and Morinigo, D. (2006). Load restoration in electric distribution networks using a metaheuristic technique, in *Proceedings of the 2006 IEEE Mediterranean Electrotechnical Conference (MELECON), May 2006*, pp. 1040–1043.

46 Borges, T.T., Carneiro, S., Garcia, P.A.N. et al. (2011). Distribution systems restoration using the Interior Point Method and sensibility analysis, in *Proceedings of the 2011 IEEE Power and Energy Society General Meeting, July 2011*, pp. 1–4.

47 Chen, W.-H. (2010). Quantitative decision-making model for distribution system restoration. *IEEE Trans. Power Syst.* 25 (1): 313–321.

48 Kumar, Y., Das, B., and Sharma, J. (2008). Multiobjective, multiconstraint service restoration of electric power distribution system with priority customers. *IEEE Trans. Power Delivery* 23 (1): 261–270.

49 Kleinberg, M.R., Miu, K., and Chiang, H.-D. (2011). Improving service restoration of power distribution systems through load curtailment of in-service customers. *IEEE Trans. Power Syst.* 26 (3): 1110–1117.

50 Li, J., Ma, X.-Y., Liu, C.-C., and Schneider, K.P. (2014). Distribution system restoration with microgrids using spanning tree search. *IEEE Trans. Power Syst.* 29 (6): 3021–3029.

51 Qu, H. and Liu, Y. (2012). Maximizing restorable load amount for specific substation during system restoration. *Int. J. Electr. Power Energy Syst.* 43 (1): 1213–1220.

52 Liu, W., Lin, Z., Wen, F., and Ledwich, G. (2013). A wide area monitoring system based load restoration method. *IEEE Trans. Power Syst.* 28 (2): 2025–2034.

53 El-Zonkoly, A.M. (2012). Power system single step restoration incorporating cold load pickup aided by distributed generation. *Int. J. Electr. Power Energy Syst.* 35 (1): 186–193.

7

Restoration Methodology and Implementation Algorithms

Power system restoration is highly challenging as it covers a wide spectrum of challenges [1, 2] and involves various independent entities [3]. Despite that different entities have specific major concerns and obligations to restore their equipment, control actions should be coordinated to maintain security and reliability of the power system. To ensure reliability during restoration, various regulatory authorities and independent system operators or regional transmission organizations (ISO/RTO) have prepared guidelines and restoration plans with highlights on the importance of coordination.

Generally, a procedure for system restoration has three stages, that is preparation, system restoration, and load restoration [4, 5]. According to these general stages of restoration, system restoration strategies can be categorized into five types, that is, build-upward, build-downward, build-inward, build-outward, and build-together. Descriptions of these strategies can be found in Fink et al. [5]. To establish a restoration plan, the technical feasibility under both steady-state and transient operating conditions needs to be checked [6]. Technical constraints include active power balance and frequency control, reactive power balance and overvoltage control, switching transient voltage [7], self-excitation, cold load pickup [8], system stability, protective systems, and load control [9]. This chapter addresses the restoration of generating units and loads.

7.1 Algorithms for Generating Unit Start-Up

7.1.1 A General Bilevel Framework [10]

The objective is to provide cranking power and restart available non-black-start units as quickly as possible. Critical loads should be picked up as well. For constructing the generating units' start-up sequence, to maintain the stability

Power System Control Under Cascading Failures: Understanding, Mitigation, and System Restoration, First Edition. Kai Sun, Yunhe Hou, Wei Sun and Junjian Qi.
© 2019 John Wiley & Sons Ltd. Published 2019 by John Wiley & Sons Ltd.
Companion website: www.wiley.com/go/sun/cascade

of generating units and system voltage, some loads may be restored. To achieve this objective, the following restoration actions should be included:

- Starting up generating units with black-start capability
- Energizing paths
- Cranking non-black-start units/picking up critical loads

Generally speaking, the following criteria have to be considered when developing an algorithm:

- Efficiency
  The impact of a system blackout largely depends on the duration of the blackout. In order to take remedy actions within the shortest time, proposed algorithms have to be as fast and as effective as possible.
- Generality
  The target is to develop a generic restoration methodology that reflects common characteristics of different systems.
- Adaptability
  In the real world, the characteristics of power systems are complex and vary from system to system. The proposed methods should be adaptable for different systems showing different characteristics. The system characteristic itself should not affect the effectiveness of proposed algorithms.
- Openness
  Power systems will not stay unchanged; they undergo adjustment and expansion. Therefore, the designed toolbox should not only provide decision restoration support for current system but also adapt to the system in the future.
- Flexibility
  Even for the same power system, the system parameters and operating conditions may be changed. Constraints may also be changed to meet the requirement of different scenarios. The algorithms should be flexible enough to accommodate changes in system conditions.

7.1.1.1 A Generic Model of Generating Units

The feasible regions of generating units depend not only on constraints of the system but also on the physical constraints. As a generic decision support system, characteristics of different generating units should be incorporated. A generic model of a generating unit is shown in Figure 7.1. The parameters are described in Table 7.1. Based on the physical constraints of generating units, besides the parameters shared by all generating units, such as capacity, ramping rate, and minimal output, other parameters are needed for different types of generating units. For example, for fossil units, start-up power requirement and critical maximal interval are necessary; for super-critical-once-through (SCOT) units, start-up power requirement and critical minimum interval should be considered; for black-start units, such as hydro units, start-up power

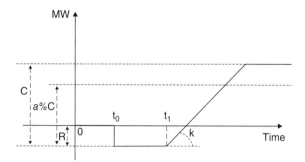

Figure 7.1 Generic model of generating units.

Table 7.1 Parameters of generating units.

Cap.	Start-up Requirement	Ramping Rate	Min. Output	Cranking to Paralleling	Critical Max. Interval	Critical Min. Interval
C	R	R	$\alpha\%$	T_1	T_2	T_3

requirement and time from cranking to parallel are zero. To simplify the model, a critical load is described as a generating unit with positive start-up power requirement and zero ramping rate. Therefore, for a generating unit restarted at time t_0, the following equations hold:

$$T_3 \leq t_0 \leq T_2 \tag{7.1}$$

$$T_1 = t_1 - t_0. \tag{7.2}$$

Generally, the generator output at time t may be written as

$$P(t) = \min\left\{k \cdot \max\left[t - (t_0 + T_1), 0\right], C\right\} - R \cdot U(t - t_0), \tag{7.3}$$

where $U(.)$ is a unit step function, defined as

$$U(t) = \begin{cases} 1 & t \geq 0 \\ 0 & t < 0 \end{cases} \tag{7.4}$$

The output of the unit step function is shown in Figure 7.2.

7.1.1.2 Algorithms for Restoring Generating Units

Let x_i denote the generating unit or critical load restarted at stage S, and θ_S be the set of all restarted generating units and critical loads at stage S. Let $f_S(x_i, \theta_S)$

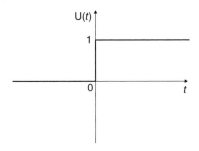

Figure 7.2 The unit step function.

be the shortest time to crank all generating units or critical loads after stage S. The recursive computation can be denoted as

$$f_S\left(x_i,\theta_S\right)=\min_{x_i\in\theta_S}\left\{\Delta t_{x_i-x_j}+f_{S+1}\left(x_j,\theta_{S+1}\right)\right\} \tag{7.5}$$

where x_j is the last generating unit restarted (so x_j must be in θ_{S+1}). $\Delta t_{x_i-x_j}$ is the time to crank the generating unit (or critical load) x_j from θ_{S+1}. At each stage S, the technical constraints are represented as

$$PF\left(\Omega_{E(S)},\mathbf{P}_{G(S)},\mathbf{Q}_{G(S)},\mathbf{P}_{CL(S)},\mathbf{Q}_{CL(S)},\mathbf{P}_{DL(S)},\mathbf{Q}_{DL(S)}\right)=0 \tag{7.6}$$

$$\mathbf{P}_\Pi\in\mathbf{FR}_P\left(\Pi\right),\mathbf{Q}_\Pi\in\mathbf{FR}_Q\left(\Pi\right),\Pi=\mathbf{G}(S),\mathbf{CL}(S),\text{ and }\mathbf{DL}(S) \tag{7.7}$$

$$\underline{V_B}\le V_B\le\overline{V_B},\qquad\forall B\in\Omega_{E(S)} \tag{7.8}$$

$$\underline{P_L}\le P_L\le\overline{P_L},\qquad\forall L\in\Omega_{E(S)}, \tag{7.9}$$

where energized block set $\Omega_{E(S)}$ includes all the energized buses and lines at stage S, and $\mathbf{P}_{G(S)}$, $\mathbf{Q}_{G(S)}$, $\mathbf{P}_{CL(S)}$, $\mathbf{Q}_{CL(S)}$, $\mathbf{P}_{DL(S)}$, and $\mathbf{Q}_{DL(S)}$ are vectors of real power of generating units, reactive power of generating units, real power of critical loads, reactive power of critical loads, real power of dispatchable loads, and reactive power of dispatchable loads, respectively. $PF(\cdot)$ is the power flow equations. $\mathbf{FR}_P(\Pi)$ and $\mathbf{FR}_Q(\Pi)$ denote feasible regions of real power and reactive power of the set Π. $\mathbf{G}(S)$, $\mathbf{CL}(S)$, and $\mathbf{DL}(S)$ are sets of generating units, critical loads, and dispatchable loads at stage S, respectively. Π represents any one of these three sets. $\mathbf{P}_\Pi$ and $\mathbf{Q}_\Pi$ are real power and reactive power that belong to set Π, respectively. V_B is the voltage at bus B, and $\underline{V_B},\overline{V_B}$ are the corresponding lower and upper limits. P_L is the real power flow on line L, and $\underline{P_L},\overline{P_L}$ are the corresponding lower and upper limits.

In this model, Eq. (7.6) represents the power flow equations at each stage of restoration; Eq. (7.7) shows that the real power and reactive power of each generating unit, critical load, and dispatchable load should stay within the feasible regions at each stage; and Eqs. (7.8) and (7.9) indicate that the voltage at each bus and power flow through each line should stay within limits. More constraints may be involved in this algorithm.

To solve this complex multistage optimization problem, a method with two interacting subproblems is proposed. In these two subproblems, an energized block of the system, that is $\Omega_{E(S)}$, is determined by the primary problem, while the operating point for the block, that is, $P_{G(S)}$, $Q_{G(S)}$, $P_{CL(S)}$, $Q_{CL(S)}$, $P_{DL(S)}$, and $Q_{DL(S)}$, is specified by the secondary problem. This is illustrated in Figure 7.3. With this framework, practical constraints related to system operation are incorporated.

The objective of the primary problem is to minimize the duration of the process. In this problem, restoring each generating unit is formulated as a stage. Till the final stage, all available generating units will be cranked. A tree with all possible sequences for restoration is employed to model this process, and dynamic programming is used to solve this problem. Two key algorithms are included.

1) Finding a sequence of cranking available generating units or of picking up critical loads to shorten the implemented time. Because this problem is a combinational problem, the combinatorial explosion is one major challenge. To ensure that the tree is within an acceptable scale, only generating units that are close to the energized block will be employed in the next stages.
2) Finding the shortest path for cranking non-black-start generating units or for picking up critical loads. Because the system is lightly loaded at this stage, the voltage may be over limits when cranking non-black-start

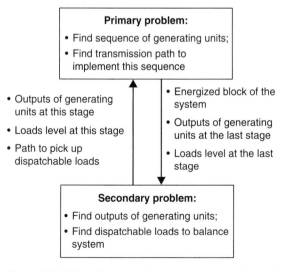

Figure 7.3 A bilevel framework for restoring generating units.

generating units or picking up critical loads. To find a path for cranking a non-black-start unit or a critical load, the charging current of each line is employed as a weight to avoid steady-state overvoltage. A path with the least charging current is selected.

To satisfy the operating constraints, outputs of generating units are adjusted and some dispatchable loads may be picked up, considering the ramping rate of each generating unit and the time for establishing paths. To achieve this objective, the secondary problem is designed.

The objective of the secondary problem is to determine the outputs of restarted generating units and pick up dispatchable loads. To satisfy the operating constraints, outputs of generating units are adjusted and some dispatchable loads may be picked up, while considering the ramping rate of each generating unit and the time for establishing paths. To achieve this objective, the secondary problem is designed. Two tasks are involved in the secondary problem:

1) Determining outputs of restarted generating units: Considering characteristics of generating units, the region of available adjustment in a time interval is limited. The objective of this problem is to determine an operating point for each restarted unit with minimal adjustment of all restarted generating units. In other words, it is to find an acceptable operating point for the entire system within the minimal time, considering the ramping rates of generating units.

2) Picking up available dispatchable loads for power balance: The power system should be instantaneously balanced. If adjusting generating units' outputs cannot eliminate the violations, some dispatchable loads may be restored.

In summary, the primary problem determines an energized block of the system, while the secondary problem specifies the operating point of the block.

7.1.2 Algorithms for the Primary Problem

Two major tasks of the primary problem are

- to find a sequence of generating units and
- to find the transmission lines to implement the sequence.

To find a reasonable system structure subject to system constraints, three basic steps should be considered:

- structuring decision trees to identify the sequence of generating units,
- searching a neighboring set of an energized block, and
- finding transmission paths for cranking generating units.

7.1.2.1 Structuring Decision Trees to Identify the Sequence of Generating Units

With the topology of a system, a decision tree can be constructed based on the available generating units in a neighboring set of an energized block. By using the tree, the paths for cranking the generating units can be established. To satisfy the system constraints, each state in the tree will be checked in the secondary problem. At the same time, the time for cranking the generating units is also achieved from the solution of the secondary problem.

After obtaining the decision tree, the next task is to search the best sequence of cranking the generating units. For this task, the standard dynamic programming is used. The time to crank the next generating unit is represented by the length of each arc. For this optimization problem with multiple stages, let x_i represent the unit restarted in the stage k, and θ_k be the set of all restarted generating units in the stage k. Let $f_k(x_i,\theta_k)$ be the shortest distance to the last stage, or the shortest time for cranking all generating units after the stage k. The recursive expression is shown as

$$f_k\left(x_i,\theta_k\right)=\min\left\{\Delta t_{x_i-x_j}+f_{k+1}\left(x_j,\theta_{k+1}\right)\right\}, \tag{7.10}$$

where x_j is the last generating unit restarted (so x_j must be in θ_{k+1}). $\Delta t_{x_i-x_j}$ is the time for cranking the generating unit x_j from the generating unit x_i. This time depends on the characteristic of generators. Based on this idea, the sequence for cranking all generating units in minimal time can be achieved. The recursive Eq. (7.10) can be solved by dynamic programming. However, for a realistic power grid with a number of non-black-start (NBS) units, the computational burden of dynamic programming is huge. Therefore, some greedy search methods with traceback mechanism may be applied.

7.1.2.2 Searching a Neighboring Set of an Energized Block

As the task is to solve a combinational optimization problem within a limited amount of computing time, it is critical to limit the number of states at each stage. Based on the restoration process in industry practice, an algorithm is proposed. In this algorithm, only the generating units that can be cranked within a given number of breakers operation will be used as the states for the next stage. Actually, in industry practice, to restart the system efficiently and reliably, generating units near the energized block will be cranked first, and then the distant generating units are cranked by the extended energized block.

A key issue of this algorithm is to find the neighboring set of an energized block. The straightforward search algorithm might be time consuming, especially for a large-scale power system. A detailed algorithm based on the transformation of the connection matrix is proposed to find the neighboring set of an energized block. It is described as follows:

For a graph, the connection matrix is defined as

$$CM = \left[d_{ij}\right] = \begin{cases} 1 & i = j \text{ or } i \,\&\, j \text{ are connected directly} \\ 0 & i \,\&\, j \text{ are NOT connected directly} \end{cases} \quad (7.11)$$

All the buses connected with bus i can be found by non-zero elements in row i or column i (the connection matrix is symmetrical). By searching the non-zero elements of all rows or columns related to the block, the neighboring set is obtained.

To find all the buses near bus i within distance n, it is necessary to use the graph search algorithm, which is to calculate the following distance matrix:

$$\mathbf{D} = \mathbf{C}^n \quad (7.12)$$

Therefore, all the buses near the bus i within distance n are the column numbers of the non-zero entries in the i-th row. The following steps are used.

- Step 1: Depth = D, $k = 1$.
- Step 2: Generate the connection matrix with Eq. (7.11), where non-zero elements represent those buses that are involved in the block or connected with the block directly.
- Step 3: For the block $\mathbf{\Omega}_K$ at step k, establish the transformation matrix as

$$\mathbf{TM}(k) = \left[t_{ij}^k\right] = \begin{cases} 1 & i = j \text{ or line}_{i-j} \in \mathbf{\Omega}_K \\ 0 & \text{else} \end{cases}, \quad (7.13)$$

where elements of $\mathbf{TM}(k)$ indicate lines belong to $\mathbf{\Omega}_K$.
- Step 4: $\mathbf{CM}(k+1) \leftarrow \mathbf{CM}(k) \cdot \mathbf{TM}(k)$.
  According to the definition of $\mathbf{TM}(k)$, $\mathbf{CM}(k) \cdot \mathbf{TM}(k)$ adds up all columns within the energized block. Therefore, each column represents the bus within this block with the same elements. Furthermore, according to Eq. (7.11), non-zero elements represent those buses that are involved in the block ($d_{ij}^k > 1$) or connected with the block directly ($d_{ij}^k = 1$). By detecting non-zero elements of $\mathbf{TM}(k + 1)$, the buses within the block $\mathbf{\Omega}_K$ or connected with it directly will be found. Mathematically, it can be written as follows:
- Step 5: $\forall j \in \mathbf{\Omega}_K$, find set $\mathbf{I}_{K+1} = \left\{i \middle| d_{ij}^k \neq 0 \text{ and } i \notin \mathbf{\Omega}_K \right\}$, where $\mathbf{I}_{K+1}$ is the set with buses connected with block $\mathbf{\Omega}_K$ directly. Since all columns within the block are the same, the buses connected with the block directly can be detected from an arbitrary column.
- Step 6: $\mathbf{\Omega}_{K+1} \leftarrow \mathbf{\Omega}_K \cup \mathbf{I}_{K+1}$
  This step updates the buses involved in this block.
- Step 7: $k \leftarrow k + 1$. If $k \leq D$, go to Step 3; else output $\mathbf{\Omega}_{K+1}$.
  The idea of this algorithm is that non-zero elements of column j indicate buses connected with bus j directly. By generating $\mathbf{TM}$ D times and evaluating $\mathbf{CM} \cdot \mathbf{TM}$ D times, all buses connected with the block within D lines will be found.

Figure 7.4 Example for neighborhood searching.

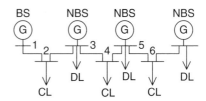

For example, for a system shown in Figure 7.4, the connection matrix is

$$\mathbf{C} = \begin{bmatrix} 1 & 1 & 0 & 0 & 0 & 0 & 0 \\ 1 & 1 & 1 & 0 & 0 & 0 & 0 \\ 0 & 1 & 1 & 1 & 0 & 0 & 0 \\ 0 & 0 & 1 & 1 & 1 & 0 & 0 \\ 0 & 0 & 0 & 1 & 1 & 1 & 0 \\ 0 & 0 & 0 & 0 & 1 & 1 & 1 \\ 0 & 0 & 0 & 0 & 0 & 1 & 1 \end{bmatrix}. \tag{7.14}$$

If one needs to find all the buses near bus 1 within a distance of 2, the distance matrix is

$$\mathbf{D} = \mathbf{C}^2 = \begin{bmatrix} 1 & 1 & 0 & 0 & 0 & 0 & 0 \\ 1 & 1 & 1 & 0 & 0 & 0 & 0 \\ 0 & 1 & 1 & 1 & 0 & 0 & 0 \\ 0 & 0 & 1 & 1 & 1 & 0 & 0 \\ 0 & 0 & 0 & 1 & 1 & 1 & 0 \\ 0 & 0 & 0 & 0 & 1 & 1 & 1 \\ 0 & 0 & 0 & 0 & 0 & 1 & 1 \end{bmatrix}^2 = \begin{bmatrix} 2 & 2 & 1 & 0 & 0 & 0 & 0 \\ 2 & 3 & 2 & 1 & 0 & 0 & 0 \\ 1 & 2 & 3 & 2 & 1 & 0 & 0 \\ 0 & 1 & 2 & 3 & 2 & 1 & 0 \\ 0 & 0 & 1 & 2 & 3 & 2 & 1 \\ 0 & 0 & 0 & 1 & 2 & 3 & 2 \\ 0 & 0 & 0 & 0 & 1 & 2 & 2 \end{bmatrix}. \tag{7.15}$$

Thus, all the buses near bus 1 within distance 2 are bus 1, bus 2, and bus 3. Generally, the unenergized buses, which can be energized by n branches from the energized blocks, are determined from the matrix $\mathbf{C}^n$.

7.1.2.3 Finding Transmission Paths for Cranking Generating Units

When finding transmission paths for cranking generating units, it is necessary to avoid excessive charging currents (or steady-state overvoltage). For this purpose, the charging current of each line is employed as a weighting coefficient to find a path to crank an NBS unit. Dijkstra's algorithm is employed to find the shortest path for cranking an NBS unit. With this algorithm, a path that has the least charging current, can be found. This path to crank an NBS unit is likely to avoid the violation of the voltage constraints.

Considering that transformers usually have small charging currents, a path with transformers might be selected by the algorithm. However, the likelihood of ferroresonance of transformers may be raised with an increasing number of transformers on a path. To avoid this problem, the best way is to combine charging currents of lines and operating times of breakers as weights of a path. In the proposed method, a path with transformers has a lower priority.

Because the classical Dijkstra's algorithm can only find the shortest path from a bus to another bus, an extended algorithm, which can find the shortest path from a block to a bus, is proposed. Figure 7.5 shows the flowchart of finding the shortest path from an energized block to a bus. Dijkstra's algorithm is implemented as follows.

Specifically, the execution process of the algorithm is shown here. Let the node at which restoration is starting be called the initial node. Let the distance of node Y be the distance from the initial node to Y. Dijkstra's algorithm will assign some initial distance values and will try to improve them step by step. Let the set **S** stand for the nodes that have been calculated for the shortest route and set **U** stand for the nodes that have not been calculated for the shortest route.

- Step 1: Assign to every node a tentative distance value: set it to zero for our initial node and to infinity for all other nodes.
- Step 2: Mark all nodes unvisited. Set the initial node as current, which is put into set **S**. Create a set of the unvisited nodes called the unvisited set, consisting of all the nodes, namely set **U**.

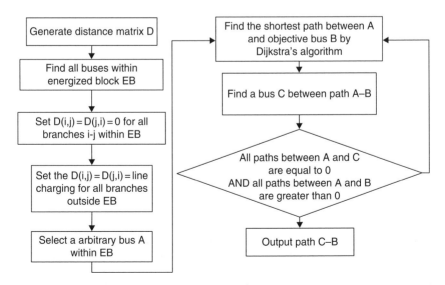

Figure 7.5 Flow chart for finding the shortest path from an energized block to a bus.

Figure 7.6 An illustrative example of Dijkstra's algorithm.

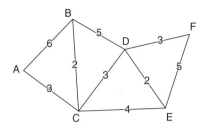

- Step 3: For the current node, consider all of its unvisited neighbors and calculate their tentative distances. For example, if the current node A is marked with a distance of 6, and the edge connecting it with a neighbor B has length 2, then the distance to B (through A) will be $6 + 2 = 8$. If this distance is less than the previously recorded tentative distance of B, then overwrite that distance. Even though a neighbor has been examined, it is not marked as "visited" at this time, and it remains in the unvisited set.
- Step 4: After considering all of the neighbors of the current node, mark the current node as visited and remove it from the unvisited set **U** to the set **S**. A visited node will never be checked again.
- Step 5: If the destination node has been marked visited (when planning a route between two specific nodes) or if the smallest tentative distance between the nodes in the unvisited set is infinity (when planning a complete traversal, this occurs when there is no connection between the initial node and remaining unvisited nodes), then stop. The algorithm has finished.
- Step 6: Select the unvisited node that is marked with the smallest tentative distance and set it as the new "current node," then go back to Step 3.

For example, for a given graph shown in Figure 7.6, the steps of Dijkstra's algorithm are listed in Table 7.2.

7.1.3 Algorithms for the Second Problem

The objective of the second problem is to find the outputs of restarted NBS generating units and the load level of dispatchable loads to meet the system constraints. When the constraints cannot be met in the block, which is identified by the primary problem, picking up new dispatchable loads, as remedial actions, will be determined.

An optimal power flow (OPF) algorithm with the system constraints is used to find the level of dispatchable loads and the acceptable outputs of generating units. The minimal adjustments of all generating units are employed as the objective function. Considering the ramping rate of each generator, the acceptable operating point of the system will be determined within the shortest time by using this algorithm. Two key subalgorithms are included.

Table 7.2 Solution steps for the illustrated example.

Step	Set S	Set U
1	Initial point: A, A ∈ S; Current shortest route: A → A = 0; Start from A	(B,C,D,E,F) ∈ U; A → B = 6, A → C = 3, A → other nodes = ∞; Shortest route: A → C = 3
2	Current node C: (A,C) ∈ S; Current shortest route: A → A = 0, A → C = 3; Start from A → C	(B,D,E,F) ∈ U; A → C → B = 5 (shorter than previous route A → B = 6), so the weight of B is A → C → B = 5, A → C → D = 6, A → C → E = 7, A → C → other nodes = ∞; Shortest route: A → C → B = 5
3	Current node B: (A,C,B) ∈ S; Current shortest route: A → A = 0, A → C = 3, A → C → B = 5; Start from A → C → B	(D,E,F) ∈ U; A → C → B → D − 10 (longer than previous route A → C → D = 6), so the weight of D is A → C → D = 6, A → C → B other nodes = ∞; Shortest route: A → C → D = 6
4	Current node D: (A,C,B,D) ∈ S; Current shortest route: A → A = 0, A → C = 3, A → C → B = 5, A → C → D = 6; Start from A → C → D	(E,F) ∈ U; A → C → D → E = 8 (longer than previous route A → C → E = 7), so the weight of E is A → C → E = 7, A → C → D → F = 9; Shortest route: A → C → E = 7
5	Current node E: (A,C,B,D,E) ∈ S; Current shortest route: A → A = 0, A → C = 3, A → C → B = 5, A → C → D = 6, A → C → E = 7; Start from A → C → E	(F) ∈ U; A → C → E → F = 12 (longer than previous route A → C → D → F = 9), so the weight of F is A → C → D → F = 9; Shortest route: A → C → D → F = 9
6	Current node F: (A,C,B,D,E,F) ∈ S; Current shortest route: A → A = 0, A → C = 3, A → C → B = 5, A → C → D = 6, A → C → E = 7, A → C → D → F = 9; Start from A → C → D	U = Φ, algorithm has finished

7.1.3.1 Finding an Acceptable Operating Point of the System

In order to ensure the minimum duration of restarting generating units and to satisfy all constraints, an OPF problem is used to describe this problem. The same model with zero ramping rates is used to describe critical loads. By using an OPF algorithm with minimizing adjustments of generating units, the shortest duration for implementation of the state, provided by the primary problem, is thus achieved. To find a solution within a reasonable computation time, a vectorial interior point method is used. The algorithm for discrete or continuous dispatchable load model is designed to balance the system.

$$\min \sum_{P_{Gi}^{(S)} \in \mathbf{P}_{G(S)}} \left(P_{Gi}^{(S)} - P_{Gi}^{(S-1)} \right)^2 \tag{7.16}$$

subject to

$$\underline{P_{Gi}^{(S)}} \le P_{Gi}^{(S)} \le \overline{P_{Gi}^{(S)}}, \ \forall Gi \in \mathbf{G}(\mathbf{S}) \tag{7.17}$$

$$PF\left(\mathbf{\Omega}_{\mathbf{E}(\mathbf{S})}, \mathbf{P}_{\mathbf{G}(\mathbf{S})}, \mathbf{Q}_{\mathbf{G}(\mathbf{S})}, \mathbf{P}_{\mathbf{CL}(\mathbf{S})}, \mathbf{Q}_{\mathbf{CL}(\mathbf{S})}, \mathbf{P}_{\mathbf{DL}(\mathbf{S})}, \mathbf{Q}_{\mathbf{DL}(\mathbf{S})}\right) = 0 \tag{7.18}$$

$$\mathbf{P}_\Pi \in \mathbf{FR}_\mathbf{P}\left(\Pi\right), \mathbf{Q}_\Pi \in \mathbf{FR}_\mathbf{Q}\left(\Pi\right), \Pi = \mathbf{G}(\mathbf{S}), \mathbf{CL}(\mathbf{S}), \text{ and } \mathbf{DL}(\mathbf{S}) \tag{7.19}$$

$$\underline{V_B} \le V_B \le \overline{V_B}, \qquad \forall B \in \mathbf{\Omega}_{\mathbf{E}(\mathbf{S})} \tag{7.20}$$

$$\underline{P_L} \le P_L \le \overline{P_L}, \qquad \forall L \in \mathbf{\Omega}_{\mathbf{E}(\mathbf{S})}, \tag{7.21}$$

where $P_{Gi}^{(S)}$ and $P_{Gi}^{(S-1)}$ are the output of generating unit Gi at stage S and S-1, respectively, and $\overline{P_{Gi}^{(S)}}$ and $\underline{P_{Gi}^{(S)}}$ are the upper and lower limit of the output of generating unit G_i at stage S. Other symbols are the same as the corresponding ones in Eqs. (7.6)–(7.9).

In this problem, Eq. (7.17) is the limit of each generating unit, and it is determined by the upper limit and ramping rate of a generating unit; Eq. (7.18) represents power flow equations for the energized block at stage S; Eq. (7.19) describes physical constraints of each generating unit, critical load, and dispatchable load; and Eqs. (7.20) and (7.21) are limits of voltage and power flow through each line at stage S. It should be noted that all constraints at stage S-1 are already met in the last step; only the constraints for stage S are involved here.

The problem described by Eqs. (7.16)–(7.21) is an OPF problem. To find a solution with a reasonable computing time, a vectorial interior point method is employed [11]. By the proposed algorithm, to balance the system at each stage some dispatchable loads might be picked up.

Generally, the OPF algorithm is only implemented on a given network, that is, only components that have been connected are involved in this computation. In this work, if the OPF defined by Eqs. (7.16)–(7.21) is divergent, a proactive strategy is employed to find dispatchable loads to improve controllability of the system at each step during system restoration.

7.1.3.2 Finding Dispatchable Loads to Balance the System

Only the dispatchable loads, which have been connected with the system, will be involved in the OPF algorithm. However, dispatchable loads may not be available via the transmission paths, which are used to crank generating units. In this case, the passive strategy of picking up dispatchable loads should be modified to an initiative one. By the algorithm to find the neighboring set of an energized block, which is introduced in the primary problem, the dispatchable loads within a limited depth are found. These dispatchable loads will be picked up one by one until the OPF converges. If the OPF cannot converge considering all available dispatchable loads, the state will be removed from the decision tree, which is achieved by the primary problem.

To ensure that a restoration plan can be found, two enhancements may be involved in the basic flowchart:

- Traceback
- Passive search

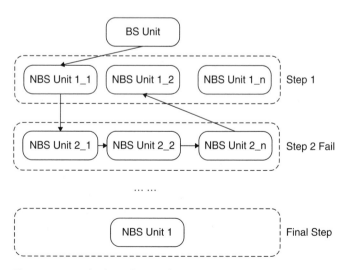

Figure 7.7 Traceback mechanism for restoring generating units.

The key process of traceback is described as follows: when none of the optional generating units can be cranked, the restoration process returns to the previous step to crank other optional generating units. When all generating units have been cranked or the last optional generating unit of the first step has been tracked, the calculation will terminate. The process is shown in Figure 7.7.

Passive search, which takes effort to find dispatchable loads and pick up these loads when needed, will be activated if the feasible steady-state operating point is not identified. The following steps are involved:

- Step 1: Find the neighboring set of the energized block.
- Step 2: Identify the dispatchable loads within the neighboring set.
- Step 3: Find the shortest path to crank the dispatchable loads.
- Step 4: Solve OPF problem defined by Eqs. (7.16)–(7.21) with new dispatchable loads and the associated transmission path. If the OPF problem can find a solution, output the solution; or else delete this state on the decision tree.

7.2 Algorithms for Load Restoration

After major generating units and the transmission skeleton have been restored, restoring service to customers based on priorities is the major concern of the load restoration phase. Despite the fact that load restoration for distribution systems is widely addressed, the design and implementation of a methodology to establish load restoration strategies for transmission systems is highly challenging. The difficulties may lie in the following aspects:

1) Load restoration requests coordinated control of restoration participants. For example, the transmission owner (TO) specifies the magnitude and location of load pickup for distribution owners (DOs) [12]. Generation owners (GOs) manage the capacity resource as directed by ISO to guard against contingencies [13]. Therefore, such coordination is accomplished on a systemwide basis. Mathematically, the computation for the optimal load restoration plan combines a nonconvex (due to power flow constraints) and combinatorial nature (considering discrete load increments of feeders), and is computationally expensive.
2) Security constraints need to be considered in various time scales [14]. A tradeoff should be made to keep the computational complexity tractable while providing feasible restoration strategies.

In view of these technical challenges, this part introduces a methodology and associated algorithms to aid restoration participants in maximizing load pickup based on load priorities while maintaining frequency security, adequate reserve levels, and meeting comprehensive steady-state constraints. As the entire restoration process is highly complicated, the following assumptions are made for a moderate research scope:

1) The communication infrastructure works properly and enables a smooth collaboration between restoration participants.
2) The model of the entire power system and individual components can be established in a high confidence level with the historical data and system operators' experience.
3) The power system has been restored with sufficient strength, although some components may be unavailable, or put into operation gradually due to physical or technical constraints.
4) The sequence to restart generating units and to switch transmission is given information. It is not implied here that restorations are divided into independent problems. Instead, the major concern in load restoration is to coordinate the load pickup with generation and transmission resources at hand, realizing that generation and transmission constraints should be met. The determination of generating unit start-up sequence and transmission switching can be found in references focusing on these topics (e.g., [15–17]).
5) The automatic generation control may be deferred prior to interconnection of TO islands [13, 18]. The frequency is regulated within a small range around the nominal value. Thus, approximate real power balances should be maintained.

Industry practices in load restoration suggest that the amount, timing, and location of the load being restored should be controlled by restoration participants coordinately [12, 19, 20]. To mitigate unknown impacts, participants wait for voltage and frequency to stabilize prior to picking up the next block of load increments [12, 19].

The load restoration process is therefore modeled as a two-stage process. First, the total restorable load level at each stage is determined by the frequency response capacity of the synchronized generating units. Second, each generating unit's output at each stage is restricted by intertemporal operational constraints, such as start-up requirement and minimal technical output.

The methodology to establish the load restoration strategy consists of the following subtasks:

1) Estimate output bounds of generating units and the total restorable load level.
2) Formulate a model to compute the maximum load pickup for DOs at each stage.
3) Estimate the duration of each stage. These subtasks will be described in the following chapter.

For sake of convenient reference, the symbols used in this section are listed as follows.

Sets

Ω_L^m	Set of unserved loads at stage m.
Ω_{Lk}^m	Set of unserved loads on bus k at stage m.
Ω_G^m	Set of synchronized generating units at stage m.
$\Omega_{Br}^m, \Omega_{Bu}^m$	Set of energized branches and buses at stage m.

Parameters

$\bar{P}_{Gi}^m, \underline{P}_{Gi}^m$	Upper and lower bound of real power output of generating unit i at stage m.
$\bar{Q}_{Gi}^m, \underline{Q}_{Gi}^m$	Upper and lower bound of reactive power output of generating unit i at stage m.
$\bar{S}_i, \underline{S}_i$	Upper and lower bound of apparent power flow of branch i.
G_{kl}^m, B_{kl}^m	Real and imaginary part of admittance matrix element between bus k and bus l at stage m.
P_{L0i}^m, Q_{L0i}^m	Restored load demand at the nominal voltage on bus i at stage m.
$\Delta P_{L0i}^m, \Delta Q_{L0i}^m$	Load demand of load increment i at the nominal voltage at stage m.
I_i, β_i	Priority weighting and cold load effect coefficient of load increment i.
R_i, ρ_i, C_i	Ramping rate, load pickup factor, and capacity of generating unit i.
$\bar{\tau}$	Maximum ramping time for synchronous reserve of generating units.
$\bar{P}_L^m$	Maximum restorable load level at stage m.
a_i, b_i, c_i	Voltage-dependent coefficients of load i.
$\bar{V}_i, \underline{V}_i$	Upper and lower bound of voltage of bus i.

Decision Variables

P_{Gi}^m, Q_{Gi}^m	Real and reactive power set points of generating unit i at stage m.
P_{Li}^m, Q_{Li}^m	Restored load demand on bus i at stage m.
τ_i^m	Effective ramping time of generating unit i for spinning reserve at stage m.
K_i^m	Effective responsive reserve of generating unit i at stage m.
u_i^m	On/off state of load increment i at stage m. (1 = restored at this stage, 0 = remains interrupted)
$\Delta P_{Li}^m, \Delta Q_{Li}^m$	Load demand of load increment i at stage m.
V_i^m, θ_i^m	Voltage and phase angle of bus i at stage m.
S_i^m	Apparent power flow of branch i at stage m.

7.2.1 Estimate Operational Region Bound

The generating unit's steady-state model described in [10] is used to estimate the output of each generating unit. As shown in Figure 7.1, for the generating unit i, M_i is the start-up requirement, α_i is the minimum technical output ratio, and t_{Pi} is the remaining time from being cranked to parallel operation.

For synchronized generating units, the output range is restricted by capacities and minimal technical outputs. For unparalleled generating units, the outputs are fixed as start-up requirements. Therefore, $\bar{P}_{Gi}^{m}$ and $\underline{P}_{Gi}^{m}$ are determined by

$$\bar{P}_{Gi}^{m} = \min\left\{ P_{Gi}^{m-1}, D\left(t^{m-1} + \gamma - t_{Pi}\right)C_i \right\} \tag{7.22}$$

$$\underline{P}_{Gi}^{m} = \max\left\{ P_{Gi}^{m-1}, D\left(P_{Gi}^{m-1} - \alpha_i C_i\right) D\left(t^{m-1} + \gamma - t_{Pi}\right)\alpha_i C_i \right\}, \tag{7.23}$$

where t^{m-1} is the end time of the previous stage, γ is the allowed ramping time for unparalleled generating units, and

$$D(x) = \begin{cases} 1 & x \geq 0, \\ 0 & x < 0. \end{cases} \tag{7.24}$$

Note that if $t^{m-1} + \gamma - t_{Pl} > 0$, the associated generating unit will be paralleled to the power system at the intermediate of the stage and must be reloaded to stabilize the operation. Equation (7.23) enforces this unit only ramp up in the next stages until it reaches the minimal technical output.

The load pickup should not be so large that the frequency deviation triggers the protective relays to trip generating units or loads. The restorable load level can be determined using generator and governor models. In this section, it is determined in terms of the load pickup factors and capacities of synchronized generating units as follows:

$$\bar{P}_{L}^{m} = \sum_{i \in \Omega_G^m} \rho_i C_i, \qquad \Omega_G^m = \left\{ i \mid t_{Pi} = 0 \right\} \tag{7.25}$$

Note that the load pickup factor ρ_i is a system-specific parameter. For example, PJM Interconnection, L.L.C., uses 5% for fossil steam units, 15% for hydro units, and 25% for combustion turbines [11]. Midcontinent Independent System Operator, Inc. (MISO), uses $\rho_i = 5\%$ for all units [18].

7.2.2 Formulate MINLR Model to Maximize Load Pickup

The mixed integer nonlinear load restoration (MINLR) model is formulated here to maximize the load pickup subject to various constraints.

1) *Objective function*: The objective of each stage is to maximize the load pickup for DOs through load priorities as

$$\max \sum_{i \in \Omega_L^m} u_i^m I_i \Delta P_{L0i}^m. \tag{7.26}$$

2) *Power balance constraints*: At the end of each stage, a steady state should be achieved with an approximate active/reactive power balance as in Eqs. (7.27) and (7.28). The voltage-dependent load characteristics (i.e., ZIP model: constant impedance (Z), constant current (I), constant power (P)) are expressed in a quadratic form as in Eqs. (7.29) and (7.30). For ease of notation, the load increments restored prior to stage m on the same bus are modeled using aggregated coefficients in Eq. (7.29).

$$
P_{Gk}^m - P_{Lk}^m - V_k^m \sum_{l \in \Omega_{Bu}^m} V_l^m \left(G_{kl}^m \cos\theta_{kl}^m + B_{kl}^m \sin\theta_{kl}^m \right)
$$
$$
- \sum_{r \in \Omega_{Lk}^m} u_r^m \Delta P_{Lr}^m = 0, \qquad \forall k \in \Omega_{Bu}^m \tag{7.27}
$$

$$
Q_{Gk}^m - Q_{Lk}^m - V_k^m \sum_{l \in \Omega_{Bu}^m} V_l^m \left(G_{kl}^m \sin\theta_{kl}^m - B_{kl}^m \cos\theta_{kl}^m \right)
$$
$$
- \sum_{r \in \Omega_{Lk}^m} u_r^m \Delta Q_{Lr}^m = 0, \qquad \forall k \in \Omega_{Bu}^m \tag{7.28}
$$

$$
\begin{cases} P_{Lk}^m = \left(a_k \left(V_k^m\right)^2 + b_k V_k^m + c_k \right) P_{L0k}^m \\ Q_{Lk}^m = \left(a_k \left(V_k^m\right)^2 + b_k V_k^m + c_k \right) Q_{L0k}^m \end{cases}, \qquad \forall k \in \Omega_{Bu}^m \tag{7.29}
$$

$$
\begin{cases} \Delta P_{Li}^m = \left(a_i \left(V_i^m\right)^2 + b_i V_i^m + c_i \right) \Delta P_{L0i}^m \\ \Delta Q_{Li}^m = \left(a_i \left(V_i^m\right)^2 + b_i V_i^m + c_i \right) \Delta Q_{L0i}^m \end{cases}, \qquad \forall i \in \Omega_{Lk}^m, \forall k \in \Omega_{Bu}^m \tag{7.30}
$$

3) *Operational constraints*: The steady state at the end of each stage should meet operational constraints as follows:

$$
\begin{cases} \underline{P}_{Gi}^m \le P_{Gi}^m \le \overline{P}_{Gi}^m \\ \underline{Q}_{Gi}^m \le Q_{Gi}^m \le \overline{Q}_{Gi}^m \end{cases}, \qquad \forall i \in \Omega_G^m \tag{7.31}
$$

$$
\underline{S}_i \le S_i^m \le \overline{S}_i, \qquad \forall i \in \Omega_{Br}^m \tag{7.32}
$$

$$
\underline{V}_i \le V_i^m \le \overline{V}_i, \qquad \forall i \in \Omega_{Bu}^m. \tag{7.33}
$$

4) *Spinning (synchronous) reserve constraints*: Spinning reserves in power system restoration may be either online generation capacity that can be loaded within a given period $\bar{\tau}$ (typically 10 min, see e.g., [11, 21]), or loads that can

be shed manually in this period. This reserve must be enough to survive the largest loss of energy contingency. Conservatively, this reserve consists of only generators so that no load is shed in loss of energy contingencies. These constraints are as follows:

$$
\begin{aligned}
&P_{Gi}^m + \tau_i^m R_i \leq C_i, \\
&0 \leq \tau_i^m \leq \bar{\tau}, \\
&P_{Gi}^m \leq \sum_{j \in \Omega_G^m, j \neq i} \left(\tau_j^m R_j \right), \qquad \forall i \in \Omega_G^m.
\end{aligned}
\tag{7.34}
$$

The spinning reserve level is defined as the reserve associated with the largest loss of energy contingency according to

$$
\min \left\{ \sum_{j \in \Omega_G^m, j \neq i} \left(\tau_j^m R_j \right) - P_{Gi}^m, \quad i \in \Omega_G^m \right\}.
\tag{7.35}
$$

5) *Responsive (dynamic) reserve constraints*: Responsive reserves may consist of reserves on automatic governor response (up to the total reserve), and system load with underfrequency load shedding (UFLS). The load with UFLS is not considered in this reserve, so that the power system which relies merely on the governor response to restore frequency should withstand any credible contingency that causes frequency decay to occur.

$$
\begin{aligned}
&K_i^m \leq \rho_i C_i \\
&K_i^m \leq C_i - P_{Gi}^m \\
&P_{Gi}^m \leq \sum_{j \in \Omega_G^m, j \neq i} K_j^m \qquad \forall i \in \Omega_G^m
\end{aligned}
\tag{7.36}
$$

Similarly, the responsive reserve level is defined by

$$
\min \left\{ \sum_{j \in \Omega_G^m, j \neq i} K_j^m - P_{Gi}^m, \quad i \in \Omega_G^m \right\}.
\tag{7.37}
$$

6) *Frequency security constraints*: The total load pickup should be limited up to the total restorable level. Considering the cold load effect, which may last from seconds to minutes, the total load pickup is restricted due to the overload compared with normal conditions.

Estimation of the cold load effect after an extended outage is complicated [22]. In this section, it is assumed that the load level over normal conditions is described by a cold load effect coefficient β_i. This coefficient stands for the maximum load demand over normal condition upon the reclosure of feeder breakers. Thus, the total load pickup is restricted by

$$\sum_{i\in\Omega_L^m}\beta_i u_i^m \Delta P_{L0i}^m \leq \bar{P}_L^m.\tag{7.38}$$

Note that the identification of load model parameters as β_i will be the subject of future research independent of this model. The procedure of the proposed methodology is as follows:

- Step 1: Let $t = 0$, and $m = 1$. Identify the initial model of the power system, including P_{Gi}^0, Q_{Gi}^0, t_{Pi}, and the admittance matrix.
- Step 2: Calculate $\underline{P}_{Gi}^m$, $\bar{P}_{Gi}^m$, and $\bar{P}_L^m$ using Eqs. (7.22)–(7.25).
- Step 3: Establish paths to load buses, close transmission loops if necessary, update Ω_{Br}^m, Ω_{Bu}^m, and the admittance matrix.
- Step 4: Build the MINLR model (Eqs. (7.26)–(7.38)) and solve it.
- Step 5: Calculate Δt^m with Eq. (7.39). Let $t = t + \Delta t^m$ and $t_{Pi} = [t_{Pi} - \Delta t^m]_+$, update Ω_G^m, Ω_L^m and Ω_{Lk}^m.
- Step 6: If all load increments are restored, stop; otherwise set $m = m + 1$, and go to Step 2.

The duration of each stage is primarily determined by the time for the following dispatch actions and processes:

1) Establish communications between restoration participants
2) Reenergize the path to the load buses by closing breakers and switches.
3) Close breakers of feeders to reenergize load increments.
4) Ensure generating units ramp to the power set points.

The methods for estimating the duration for 2) and 3) can be found in Adibi and Milanicz, and Mota et al. [23, 24]. Assuming the generation operators carry out the adjustments of generators in a parallel manner, the duration of 4) at stage m is the maximum ramping time of the generating units given by

$$\Delta t^m = \max\left\{[t_{Pi} - t_{m-1}]_+ + \left|P_{Gi}^m - P_{Gi}^{m-1}\right|/k_i|, \forall i\right\},\tag{7.39}$$

where $[x]_+ = \max\{x, 0\}$.

7.2.3 Branch-and-Cut Solver: Design and Justification

The construction of an efficient branch-and-cut (B&C) solver for MINLR models is described in this section. The start point of the proposed B&C solver is as follows.

Assumption (Solvability of the relaxed MINLR model): Relaxation of u as continuous variables in [0,1] will lead to a solvable nonlinear programming model. Note that relaxation of MINLR converts it into an OPF model with continuous dispatchable loads under ZIP models. It is feasible to find *a local optimal solution* using interior point methods. Therefore, the B&C algorithm is applicable to solve MINLR models.

7.2.3.1 General Branch-and-Cut Algorithm

The B&C algorithm is based on the branch-and-bound (B&B) algorithm. The basic idea of B&B is to solve the continuous relaxation of a given integer programming model and to branch on a selected integer variable (if the optimal solution is not integer feasible). For a maximization model, a relaxed model will provide an upper bound (U) for the objective function. If a node is integer feasible, it provides a lower bound (L). Update L if a bigger (i.e., better) lower bound is found. The branching on some nodes can be stopped if no significant improvement can be found as $|L\text{-}U|/|U| < \varepsilon$, where ε is the termination criterion, and $|L\text{-}U|/|U|$ is the gap.

The B&C method is an extension of B&B by including cutting planes. The cutting planes achieve a tighter feasible set as well as preprocess subproblems with heuristics [25].

7.2.3.2 Proof of Applicability of Cutting Planes

This subsection presents the proof of the applicability of cutting planes targeting on Eq. (7.38) to solve MINLR models.

The decision variables of MINLR models are collectively denoted as x. Accordingly, Eq. (7.38) can be written as $\boldsymbol{\Omega}_{\text{Feq}} = \{x \mid w^{\text{T}}x < v, x_i \in \{0,1\}$ for all $i \in \boldsymbol{\Omega}_{\text{I}}\}$, where $\boldsymbol{\Omega}_{\text{I}}$ is the set of binary variables.

The relaxation (or convexification) of $\boldsymbol{\Omega}_{\text{Feq}}$ is

$$\text{conv}\left(\boldsymbol{\Omega}_{\text{Feq}}\right) = \left\{x \mid, w^{\text{T}}x < v \mid, x_i \in [0,1] \text{ for all } i \in \boldsymbol{\Omega}_{\text{I}}\right\}.$$

Using this result, the MINLR model (7.26)–(7.38) is recast as

$$\left(\text{MINLR}\right) \quad \begin{array}{c} \max \ c^{\text{T}}x \\ \text{s.t.} \quad x \in \boldsymbol{\Omega}_{\text{MINLR}} \end{array},$$

where $\boldsymbol{\Omega}_{\text{MINLR}} = \{x \mid x \in \boldsymbol{\Omega}_x \cap \boldsymbol{\Omega}_{\text{Feq}}\}$ and $\boldsymbol{\Omega}_x$ stands for a nonconvex set defined by Eqs. (7.6)–(7.13).

The relaxation of MINLR in the B&B framework is as follows:

$$\left(\text{MINLR-r}\right) \quad \begin{array}{c} \max \ c^{\text{T}}x \\ \text{s.t.} \quad x \in \boldsymbol{\Omega}_{\text{MINLR-r}} \end{array},$$

where $\boldsymbol{\Omega}_{\text{MINLR-r}} = \{x \mid x \in \boldsymbol{\Omega}_x \cap \text{conv}(\boldsymbol{\Omega}_{\text{Feq}})\}$ and $\boldsymbol{\Omega}_{\text{MINLR}} \subseteq \boldsymbol{\Omega}_{\text{MINLR-r}}$.

Definition (valid inequalities, [25], pp. 114, 117–121): $Ax \le d$ is valid for $\boldsymbol{\Omega}$ if $\forall x_0 \in \boldsymbol{\Omega}$ satisfies $Ax_0 \le d$.

Applicability of cutting planes: If $Ax \le d$ is valid for $\boldsymbol{\Omega}_{\text{Feq}}$, it is valid to tighten $\boldsymbol{\Omega}_{\text{MINLR-r}}$.

Proof: For $\forall x_0 \in \boldsymbol{\Omega}_{\text{MINLR}}$, $x_0 \in \boldsymbol{\Omega}_{\text{Feq}}$. As $Ax \le d$ is valid for $\boldsymbol{\Omega}_{\text{Feq}}$, one has $Ax_0 \le d$.

By contrast, for $\forall x' \in (\mathbf{\Omega}_{\text{MINLR-r}} - \mathbf{\Omega}_{\text{MINLR}})$, $x' \notin \mathbf{\Omega}_{\text{Feq}}$, thus $Ax' > d$. In other words, $Ax \leq d$ cuts off part of $(\mathbf{\Omega}_{\text{MINLR-r}} - \mathbf{\Omega}_{\text{MINLR}})$, which may lead to a tighter relaxation.

7.2.3.3 Incorporation of Cutting Planes for MINLR

The following cutting planes are identified as efficient and valid for (7.38). Note that after each branching, Eq. (7.38) is converted into

$$\sum_j \hat{u}_j \beta_j \Delta P_{\text{L0}j}^m \leq \overline{P}_L^m - \sum_i u_i \beta_i \Delta P_{\text{L0}i}^m, \tag{7.40}$$

where $\hat{u}_j \in (0,1)$ and $u_i \in \{0,1\}$.

- Fixing variable cut

  Fixing variables is a presolve technique for reducing the number of integer variables [26]. Intuitively, after branching on u_i, if the model apparently becomes infeasible on setting some other u_j to one, u_j should be fixed as zero from the current node and its descendants. In other words, the proposed solver mathematically adds equalities for Eq. (7.40) using

  $$\hat{u}_j = 0, \forall j \in F_1$$
  $$F_1 = \left\{ j \mid, \beta_j \Delta P_{\text{L0}j}^m > \overline{P}_L^m - \sum u_i \beta_i \Delta P_{\text{L0}i}^m \mid, u_i \in \{0,1\} \right\}. \tag{7.41}$$

- Gomory rounding cut ([25], p. 119)

  The valid inequality for Eq. (7.41) is given by

  $$\sum_j \hat{u}_j \left\lfloor \vartheta \beta_j \Delta P_{\text{L0}j}^m \right\rfloor \leq \left\lfloor \vartheta \left(\overline{P}_L^m - \sum_i u_i \beta_i \Delta P_{\text{L0}i}^m \right) \right\rfloor, \tag{7.42}$$

  where $\vartheta > 0$, and $\lfloor \bullet \rfloor$ is the biggest integer less than the input argument. Typically, one can set

  $$\vartheta = 1 / \min \left\{ \beta_j \Delta P_{\text{L0}j}^m \right\}. \tag{7.43}$$

  After solving the MINLR model, the solution is given with $\hat{u}_j^*$. A new Gomory rounding cut can be generated to cut this integer-infeasible solution by finding $\vartheta^* > 0$ such that

  $$\sum_j \hat{u}_j^* \left\lfloor \vartheta^* \beta_j \Delta P_{\text{L0}j}^m \right\rfloor > \left\lfloor \vartheta^* \left(\overline{P}_L^m - \sum_i u_i \beta_i \Delta P_{\text{L0}i}^m \right) \right\rfloor. \tag{7.44}$$

- Knapsack cover cut (KC) ([25], pp. 147–151)

  The first step in generating the knapsack cover cut of Eq. (7.41) is to find a set $C \subseteq N$ such that

  $$\sum_{k \in C} \beta_k \Delta P_{\text{L0}k}^m > \overline{P}_L^m - \sum_i u_i \beta_i \Delta P_{\text{L0}i}^m. \tag{7.45}$$

Then the knapsack cover cut is given by

$$\sum_{k \in C} \hat{u}_k < |C| - 1, \tag{7.46}$$

where $|C|$ is the number of elements in C.

7.2.4 Selection of Branching Methods

The branching method is another key factor affecting the scale of B&C trees. There are widely used general-purpose branching methods (see e.g., [27]). The following three branching methods are used for comparison.

1) Maximum fractional branching

This branching method selects the relaxed integer variable closest to 0.5, mathematically selecting the variable through

$$\max_j \left\{ \min \left(1 - \hat{u}_j, \hat{u}_j \right) \right\}, \tag{7.47}$$

where $\hat{u}_j$ belongs to the optimal solution of the parent nodes.

2) Pseudo-cost branching

This branching method branches on the variable that most changes the objective function. For the MINLR model, the branching variable is selected by

$$\max_j \left\{ \min(\Delta P_{L0j}^m I_j \left(1 - \hat{u}_j \right), \Delta P_{L0j}^m I_j \hat{u}_j) \right\}. \tag{7.48}$$

3) Biggest-load-increment branching (BLB)

A problem-specific branching method is proposed here for the MINLR model. The proposed solver branches on the variable that is associated with the biggest (or most significant with priority ranking) interrupted load increment. The proposed solver selects the variable by

$$\max_j \left\{ \Delta P_{L0j}^m I_j | j \in F_2 \right\}, \ F_2 = \left\{ j | \hat{u}_j \notin \{0,1\} \right\}. \tag{7.49}$$

7.3 Case Studies

7.3.1 Illustrative Example for Restoring Generating Units

This example performs a step-by-step computation. Each step will crank one NBS unit or one critical load. The major computational task within each step includes

1) find the unenergized bus set within a given depth;
2) find a list of NBS units and critical loads to be restored;
3) select one NBS unit or critical load to be restored;
4) establish a path to the selected component; and
5) find a feasible operating point to crank the selected component with MW/MVar balance constraints. The following steps present this logic.

- **Step 0:** BS = 73107, Search Depth = 2. Find the unenergized bus set within this range. The bus set is {73103, 73104, 73106, 73105}. See Figure 7.8.

  Within this bus set, three NBS units (i.e., 73106, 73105, and 73103) and one critical load (i.e., 73104) are found. By checking feasibility, only the NBS on bus 73106 (start-up requirement 20 MW) and 73105 (start-up requirement 10 MW) can be cranked in the next step (see Table 7.1). By the system restoration navigator (SRN) logic, the NBS with less start-up requirement and faster ramping capacity will be cranked with higher priority. The next step will try to crank an NBS unit on bus 73105. The time of this step is 60 minutes (since the crank to parallel of black start (BS) is 60 minutes).

- **Step 1:** Crank NBS unit on bus 73105. Find the path to connect bus 73105 by Dijkstra's algorithm. The time for building the path (73107-73106-73105, 73107-73103) is 5 minutes (see Table 7.3). The energized components are shown in Figure 7.9.

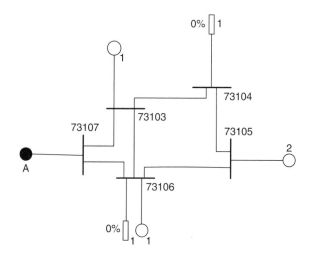

Figure 7.8 Searching for unenergized buses at Step 0.

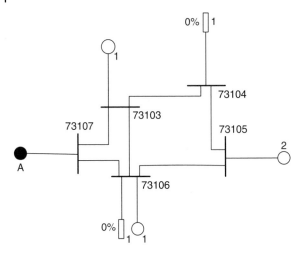

Figure 7.9 System topology and state of components at Step 1.

Table 7.3 Component selection at Step 1.

Bus	Type	Feasible	Ramping (MW min⁻¹)	Start-Up Requirement (MW)
73103	NBS	No		
73104	Critical load	No		
73105	NBS	Yes	1.6	10
73106	NBS	Yes	1.6	20

Build the OPF model as follows,

$$\min\left(P^1_{G.73107} - P^0_{G.73107}\right)^2$$

subject to: power flow and operational constraints

$\left(\text{NBS on bus 73105 is modeled as fixed load demand with start-up requirement}\right)$,

$$(7.50)$$

where $P^1_{G.73107}$ is the output of BS on bus 73107 at Step 1, $P^0_{G.73107}$ is the output of BS on bus 73107 at Step 0. Solve this OPF model. The ramping of BS unit to crank NBS on bus 73105 is 3 minutes ($10/3.3\,\text{MW min}^{-1}$). The total time duration of this step is 5 minutes (ramping time of BS is within the time for building path). At the end of this step, the time is 65 $(60 + 5)$ minutes.

- **Step 2:** Search Depth = 2. Within this neighborhood, two NBS units (i.e., 73106 and 73103) and one critical load (i.e., 73104) are found. The critical load is regarded as a ramping rate equal to zero. Based on the SRN logic, the critical load will be cranked after the NBS units (NBSU) are cranked (unless a higher user-defined priority than NBSU is set for critical loads). NBSU on 73103 has a larger ramping rate ($3\,\mathrm{MW\,min^{-1}}$) than that on 73106 ($1.6\,\mathrm{MW\,min^{-1}}$). Therefore, SRN will crank NBSU on 73103 at this step (see Table 7.4). The energized components are shown in Figure 7.10.

Note that SRN treats NBSU and critical loads consistently. Namely, critical loads are regarded as zero-ramping-capacity NBSU equivalently. Based on the start-up priority rule, critical loads are placed in the tail of the optional NBSU list in this step. On the other hand, dispatchable loads will be picked up, if necessary (e.g., OPF fails to find a feasible power flow solution), in cranking one NBSU or one critical load.

Build the OPF model as follows,

$$\min\left(P^2_{G.73107} - P^1_{G.73107}\right)^2 + \left(P^2_{G.73105} - P^1_{G.73105}\right)^2$$

subject to: power flow and operational constraints

$$\left(\text{NBS on bus 73103 is modeled as fixed load demand with start-up requirement}\right)$$

$$(7.51)$$

Table 7.4 Component selection at Step 2.

Bus	Type	Feasible	Ramping (MW min⁻¹)	Start-Up Requirement (MW)
73103	NBS	Yes	3.0	10
73104	Critical load	Yes	0	15
73106	NBS	Yes	1.6	20

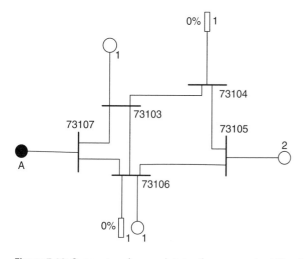

Figure 7.10 System topology and state of components at Step 2.

Solve this OPF model. The ramping time for BS to crank this unit is 3.04 minutes. At the end of this step, the time is 68.04 (65 + 3.04) minutes.

- **Step 3:** Search Depth = 2. Within this neighborhood, one NBSU (i.e., 73106) and one critical load (i.e., 73104) are found. SRN will crank NBSU on 73106 at this step (see Table 7.5). The energized components are shown in Figure 7.11. Build the OPF model as follows:

$$\min\left(P_{G.73107}^3 - P_{G.73107}^2\right)^2 + \left(P_{G.73105}^3 - P_{G.73105}^2\right)^2 + \left(P_{G.73103}^3 - P_{G.73103}^2\right)^2$$

subject to: power flow and operational constraints

$$\left(\text{NBS on bus 73106 is modeled as fixed load demand with start-up requirement}\right).$$

(7.52)

Solve this OPF model. The ramping time for BS to crank all online units is 60 minutes. At the end of this step, the time is 128.04 (68.04 + 60) minutes.

- **Step 4:** The critical load on 73104 will be cranked at this step. The path 73103–73104 is established. The time for building this path is 5 minutes. The energized components are shown in Figure 7.12.

Table 7.5 Component selection at step 3.

Bus	Type	Feasible	Ramping (MW min⁻¹)	Start-Up Requirement (MW)
73104	Critical load	Yes	0	15
73106	NBS	Yes	1.6	20

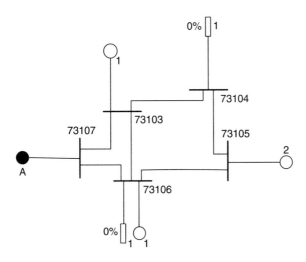

Figure 7.11 System topology and state of components at Step 3.

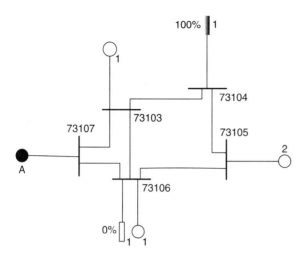

Figure 7.12 System topology and state of components at Step 4.

Build the OPF model as follows:

$$\min\left(P^4_{G.73107} - P^3_{G.73107} \right)^2 + \left(P^4_{G.73105} - P^3_{G.73105} \right)^2 + \left(P^4_{G.73103} - P^3_{G.73103} \right)^2 + \left(P^4_{G.73106} - P^3_{G.73106} \right)^2$$

subject to: power flow and operational constraints

(Critical load on bus 73104 is modeled as fixed load demand with 100% demand).

$$(7.53)$$

Solve the OPF model to pick up the critical load (100% demand). At the end of this step, the time is 133.04 (128.04 + 5) minutes.

7.3.2 Optimal Load Restoration Strategies for RTS 24-Bus System

The first example to illustrate the proposed methodology is the RTS 24-bus system with 17 load buses. Assume that the load in each load bus is fed equally by 10 feeders (i.e., consists of 10 identical load increments). The initial state of the load restoration stage is given by the algorithm in Hou et al. [10]. This case shows that the computational complexity of the MINLR model is extremely high, even for tiny power systems.

This system includes 29 transmission lines, five transformers, one reactor, and one synchronous condenser. The hydro units on bus 22 serve as the black-start resources. Other generating units are as non-black-start units. The condenser is modeled as a reactive source. The system data are in IEEE [28].

To clearly demonstrate the proposed methodology, it is assumed that each generation bus consists of three identical generating units. When the loss of energy contingencies is considered, the generation bus will lose one-third of

Table 7.6 Aggregate characteristics of generators.

Bus	Time to Parallel (minutes)	α_i (%)	Initial Active Power (p.u.)	Initial Reactive Power (p.u.)	R_i (p.u./h)	C_i (p.u.)
22	0	0	1.020	0.108	3.0	3.00
18	0	30	0.349	−0.116	1.5	4.00
16	0	50	0.270	0.184	1.0	1.55
1	0	50	0.063	0.087	1.0	1.92
2	10	50	−0.150	−0.110	1.0	1.92
15	14	50	−0.170	−0.130	1.0	2.15
13	29	60	−0.250	−0.190	1.5	5.91
21	14	30	−0.280	−0.210	1.0	4.00
7	51	50	−0.300	−0.225	1.0	3.00
23	80	50	−0.500	−0.380	1.5	6.60

the total capacity. The aggregate parameters and the initial states of the units, as well as the energized branches in the initial state, are shown in Table 7.6 and Figure 7.13, respectively.

The computational settings are as follows. The bus voltages are restricted between 0.95 and 1.05 p.u. To simplify the case study, the reactive bounds of generating units are independent of the active bounds, as in IEEE [28]. Let $I_i = 1, \varepsilon = 0.05, \beta_i = 1, a_i = 0, b_i = 0, c_i = 1.0$, and $\rho_i = 5\%$.

The first stage is shown as an example to investigate the computational complexity of the MINLR model. As shown in Table 7.7, the number of nodes in the B&C tree is large without cutting planes. The cut planes significantly reduce the computation complexity, enabling a reasonable CPU (central processing unit) time.

The complete restoration plan is achieved by using the methodology proposed in Section 7.2. This plan consists of 23 stages. The estimated duration of the restoration process is 333.20 minutes. The total number of nodes in B&C trees reaches 886. The CPU time to compute this plan is 66 seconds. The real power output of generation buses, reserve levels, and the restored load level of each load bus are shown in Figures 7.14a, 7.15a, and 7.16, respectively. If not considering reserve constraints, the real power outputs and the reserve levels are as in Figures 7.14b, and 7.15b, respectively.

As in Figure 7.14, if the reserve constraints are not considered, the outputs of generating units become uneven. Although load restoration duration without reserve constraints will drop to 326 minutes, this restoration process cannot survive the loss of energy contingency due to insufficient (negative) reserve, as in Figure 7.15b. By contrast, with reserve constraints, the responsive reserve

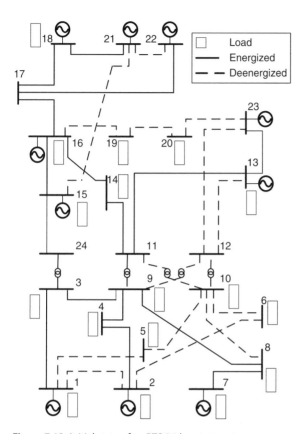

Figure 7.13 Initial state of an RTS 24-bus test system.

Table 7.7 Computational complexity of the first stage in load restoration.

Solution Strategy	Number of Nodes in B&C Tree
No cuts + MFB	4505
No cuts + PCB	4445
GC + KC + PCB	341
GC + KC + BLB	341
FC + BLB	5

Note. PCB = pseudo-cost branching; MFB = maximum fractional branching; GC = Gomory rounding cut; KC = knapsack cover cut; FC = fixing variable cut; BLB = biggest-load-increment branching.

(a)

(b)

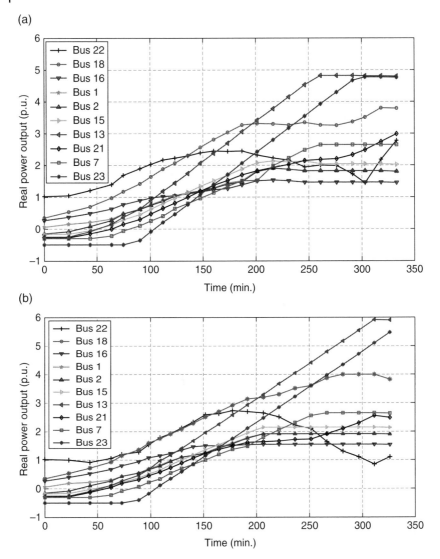

Figure 7.14 Real power output of generation buses. (a) With reserve constraints. (b) Without reserve constraints.

constraints become binding from 220 minutes onward (see Figure 7.15a). The power system can survive the largest loss of energy contingency even relying merely on governor response.

Network operations are conducted to establish a path to load buses, to close loop, and for reactive control, as necessary when each stage begins, as

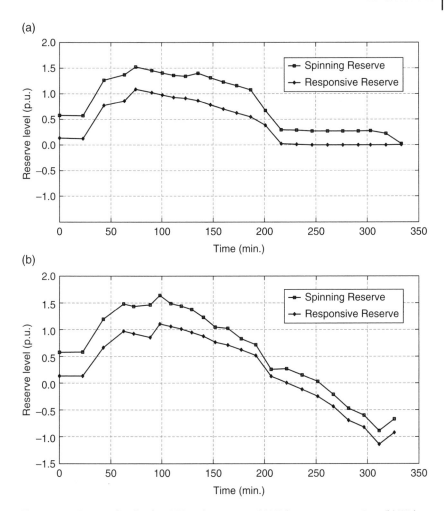

Figure 7.15 Reserve levels of an RTS 24-bus system. (a) With reserve constraints. (b) Without reserve constraints.

listed in Table 7.8. The condenser on bus 10 is dispatched as a reactive source of the TO in following MINLR models after it is put back on. The secure dispatch of network operations can be found in Martins et al. and Viana et al. [16, 17].

7.3.3 Optimal Load Restoration Strategies for IEEE 118-Bus System

The second example is the IEEE 118-bus system with 91 load buses. Assume that the load in each load bus is fed equally by 10 feeders. This case assumes

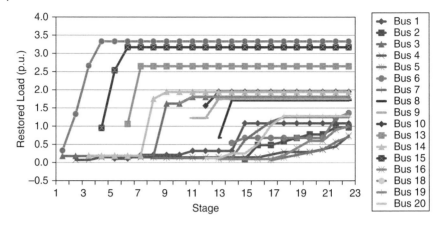

Figure 7.16 Restored loads in each load bus with reserve constraints.

Table 7.8 Network operation at the beginning of stages.

Stage	Operations
8	Energize path 9-11
9	Energize path 1-2
10	Energize path 16-19-20-23
12	Energize path 1-5-10, put the condenser back on
14	Energize path 2-6-10, put the reactor back on
15	Energize path 9-12-13
21	Energize path 10-12-23
22	Energize path 15-21-22

that 637 load increments (70%) are interrupted. The biggest load-increment branching method succeeds in establishing a restoration plan considering the cold load effect, while the other two described in Section 7.2 fail to do so, even with the aforementioned cutting planes.

The IEEE 118-bus system has a total load demand of 3668 MW under normal conditions. The data of this system are taken from [29, 30]. 22 units and 14 transmission paths are offline in the initial state, as shown in Figure 7.17. The offline units, including on buses 4, 6, 8, 15, 19, 24, 27, 31, 34, 40, 42, 70, 72, 73, 74, 85, 90, 91, 103, 107, 110, and 116, will not be cranked during the load restoration in this case. Part of the offline transmission paths will be restored during the load restoration as necessary.

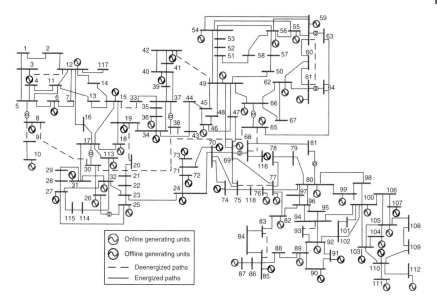

Figure 7.17 Initial state of the IEEE 118-bus system.

The bus voltages are restricted to 0.88–1.08 p.u. Let $I_i = 1$, $\varepsilon = 0.05$, $\beta_i = 4$, $a_i = 0.4$, $b_i = 0.3$, $c_i = 0.3$, and $\rho_i = 5\%$. At each stage, the B&C solver is terminated if the number of nodes in the B&C tree at this stage reaches 1000.

A load restoration plan is successfully established using the biggest-load-increment branching with the cutting planes. This plan consists of 34 stages. The estimated duration of the restoration process is 236.6 minutes. The total number of nodes in B&C trees reaches 1451. The CPU time to figure out this plan is 17 minutes. The reserve levels, the aggregate restored load level, the real power outputs of generating units, and the restored load level on each load bus are shown in Figure 7.18 and Figure 7.19.

Through these case studies, the following observations and discussions are summarized.

1) The restoration strategy provides GOs and TOs with power set points for frequency/voltage control. The load pickup amount will be the baseline for DOs to restore service.
2) The load level that can be restored in the actual outage scenarios depends on the availability of generation and transmission resources. The case studies just demonstrate the efficiency of this methodology during the entire restoration time-horizon.

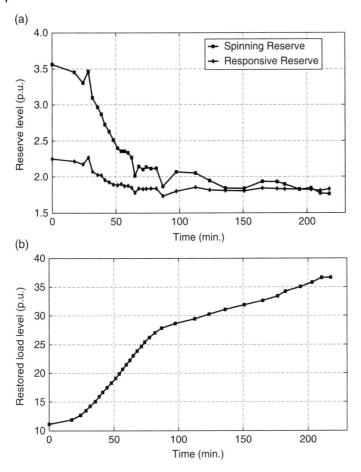

Figure 7.18 Reserve level and aggregate restored load level of IEEE 118-bus system
(a) Reserve level. (b) Aggregate restored load level.

3) The time for communication/interaction and for the execution within power plants/substations are not counted in this chapter in order to emphasize the ramping time of units. It can be considered by adding extra time duration between stages.

4) The merit of BLB results from the fact that it selects branching variables regardless of the value of $\hat{u}_j$. By contrast, general-purpose branching methods fail to identify the important branching variable if some $\hat{u}_j$ tend to 0 (which is the case in the early steps of load restoration) while others tend to 1.

5) The load pickup amounts on load buses are ready for dynamic simulation, which can be considered in future work.

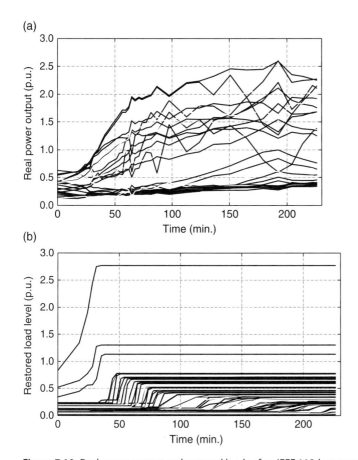

Figure 7.19 Real power output and restored loads of an IEEE 118-bus system. (a) Generator output curves. (b) Restored load levels.

References

1 Adibi, M., Clelland, P., Fink, L. et al. (1987). Power system restoration – a task force report. *IEEE Trans. Power Syst.* 2 (2): 271–277.

2 Liu, S., Hou, Y., Liu, C.-C., and Podmore, R. (2014). The healing touch: tools and challenges for smart grid restoration. *IEEE Power Energ. Mag.* 12 (1): 54–63.

3 Feltes, J. and Grande-Moran, C. (2014). Down, but not out: a brief overview of restoration issues. *IEEE Power Energ. Mag.* 12 (1): 34–43.

4 Adibi, M.M. and Fink, L.H. (2006). Overcoming restoration challenges associated with major power system disturbances – restoration from cascading failures. *IEEE Power Energ. Mag.* 4 (5): 68–77.

5 Fink, L.H., Liou, K.-L., and Liu, C.-C. (1995). From generic restoration actions to specific restoration strategies. *IEEE Trans. Power Syst.* 10 (2): 745–752.

6 Adibi, M.M. (2000). *Power System Restoration: Methodologies & Implementation Strategies.* New York: IEEE Press Series on Power Engineering.

7 Adibi, M.M., Alexander, R.W., and Milanicz, D.P. (1999). Energizing high and extra-high voltage lines during restoration. *IEEE Trans. Power Syst.* 14 (3): 1121–1126.

8 Feltes, J.W. and Grande-Moran, C. (2008). Black start studies for system restoration, *Proceedings of the IEEE Power Engineering Society General Meeting*, Pittsburgh, PA, Jul. 20–24, 2008.

9 Lindenmeyer, D., Dommel, H.W., and Adibi, M.M. (2001). Power system restoration – a bibliographical survey. *Int. J. Elec. Power* 23 (3): 219–227.

10 Hou, Y., Liu, C.-C., Sun, K. et al. (2011). Computation of milestones for decision support during system restoration. *IEEE Trans. Power Syst.* 26 (3): 1399–1409.

11 Kafka, R.J. (2008). Review of PJM restoration practices and NERC restoration standards, *Proceedings of the 2008 IEEE Power and Energy Society General Meeting, 2008.*

12 IESO (2013). Ontario power system restoration plan, Dec. 2013 [Online]. Available at: www.ieso.ca/rules/-/media/24a42a9238b6482687c2ec2203a1 e4e4.ashx.

13 PJM (2014). PJM Manual 36: system restoration, Jun. 2014 [Online]. Available at: http://www.pjm.com/~/media/documents/manuals/m36.ashx.

14 Adibi, M.M. and Martins, N. (2008). Power system restoration dynamics issues, in *Proceedings of the 2008 IEEE Power and Energy Society General Meeting, July 2008.*

15 Sun, W., Liu, C.-C., and Zhang, L. (2011). Optimal generator start-up strategy for bulk power system restoration. *IEEE Trans. Power Syst.* 26 (3): 1357–1366.

16 Martins, N., de Oliveira, E.J., Moreira, W.C. et al. (2008). Redispatch to reduce rotor shaft impacts upon transmission loop closure. *IEEE Trans. Power Syst.* 23 (2): 592–600.

17 Viana, E.M., de Oliveira, E.J., Martins, N. et al. (2013). An optimal power flow function to aid restoration studies of long transmission segments. *IEEE Trans. Power Syst.* 28 (1): 121–129.

18 MISO (2013). MISO real time operation: power system restoration plan, 2013 [Online]. Available at: https://old.misoenergy.org/_layouts/MISO/ECM/ Redirect.aspx?ID=145003.

19 ENTSO-E (2014). Current practices in Europe on emergency and restoration, May 2014 [Online]. Available at: https://www.entsoe.eu/Documents/ Network%20codes%20documents/NC%20ER/140527_NC_ER_Current_ practices_on_Emergency_and_Restoration.pdf.

20 North American Electric Reliability Corporation (NERC) (2013). EOP-005-2: system restoration from blackstart resources, November 21 2013 [Online]. Available at: http://www.nerc.com/files/EOP-005-2.pdf.

21 Ortega-Vazquez, M.A. and Kirschen, D.S. (2007). Optimizing the spinning reserve requirements using a cost/benefit analysis. *IEEE Trans. Power Syst.* 22 (1): 24–33.

22 IEEE Power System Relaying Committee, Cold load pickup issues, May 2008 [Online]. Available at: http://resourcecenter.ieee-pes.org/pes/product/technical-publications/PESTR0038.

23 Adibi, M.M. and Milanicz, D.P. (1999). Estimating restoration duration. *IEEE Trans. Power Syst.* 14 (4): 1493–1498.

24 Mota, A.A., Mota, L.T.M., and Morelato, A. (2004) Dynamic evaluation of reenergization times during power systems restoration, in *Proceedings of the 2004 IEEE/PES Transmission and Distribution Conference and Exposition: Latin America, November 2004*.

25 Wolsey, L.A. (1998). *Integer Programming*. New York: Wiley Interscience.

26 Mahajan, A. (2010). Presolving mixed-integer linear programs, Jul. 2010 [online]. Available: http://www.mcs.anl.gov/papers/P1752.pdf.

27 Achterberg, T., Koch, T., and Martin, A. (2005). Branching rules revisited. *Oper. Res. Lett.* 33 (1): 42–54.

28 IEEE Probability Methods Subcommittee (1979). IEEE reliability test system. *IEEE Trans. Power App. Syst.* PAS-98 (6): 2047–2054.

29 IEEE IEEE 118-bus system data, Available at: http://icseg.iti.illinois.edu/ieee-118-bus-system/.

30 IEEE, IEEE 118-bus system generation data, Available at: http://motor.ece.iit.edu/data/JEAS_IEEE118.doc.

8

Renewable and Energy Storage in System Restoration

8.1 Planning of Renewable Generators in System Restoration

8.1.1 Renewables for System Restoration

As the US transitions to the smart grid era, the use of renewable energy is growing rapidly. It is predicted that 50% of electricity will be generated from wind and solar power by 2050 [1]. The increased reliance on renewable energy could adversely impact power grid resilience and system restoration. While renewables bring us clean and inexpensive energy, their inherent variability and uncertainty present challenges for power grids.

Although great effort has been devoted to improving the tractability of renewables in normal operating conditions [2, 3], research on the role of renewables in emergencies (i.e., power system restoration) is limited to date. Most independent system operators (ISOs) are conservative in employing renewable energy for system restoration. For instance, the independent electricity system operator excludes renewable energy resources from the restoration process and keeps them out of service until the latter stages of restoration [4–7]. This conservative attitude is due to a lack of knowledge about handling the variability and uncertainty of renewables in system restoration. However, with the increasing penetration levels, exclusion of wind or solar power will prolong recovery time and leave the vast majority of loads unserved. Thus, a new restoration paradigm with contributions from wind or solar power is urgently needed.

Research on utilizing large-scale renewable energy sources for power system restoration is limited to date. Zhu and Liu [8] discussed several aspects of power system restoration considering wind farm participation. They proposed a supplementary control on doubly fed induction generators (DFIGs) to reduce

Power System Control Under Cascading Failures: Understanding, Mitigation, and System Restoration, First Edition. Kai Sun, Yunhe Hou, Wei Sun and Junjian Qi.
© 2019 John Wiley & Sons Ltd. Published 2019 by John Wiley & Sons Ltd.
Companion website: www.wiley.com/go/sun/cascade

the negative effect of wind power on system restoration. Aktarujjaman et al. [9] discussed DFIG wind generators participating as black-start resources. The integration of energy storage in the DC link of DFIGs is proposed to ensure a smooth load pickup process. In Seca et al. [10], a power system restoration strategy using utility-scale wind parks with HVDC connection is presented. Wind parks can participate in primary frequency control at the initial restoration phase to improve system stability. An optimal restoration time of renewable energy sources is discussed in El-Zonkoly [11]. The results show that inclusion of renewable energy sources can help reduce the unserved energy.

Most of the aforementioned studies focused on the control aspects of wind generators, and the control strategies were implemented based on a predefined restoration plan. Thus, the most challenging problem of wind variability and uncertainty during the restoration period remained unsolved and will be addressed in this chapter. In the next section, we will introduce an offline restoration planning tool that can be utilized by transmission system operators (TSOs) in the planning phase of restoration to effectively and securely harness the wind energy. Also, this tool can be adopted for training of system operators at control centers, which enables them to carry out a study on the impacts of wind farm location, penetration level, inertial response, and fluctuations. We will use wind power as an example, and solar power can also be integrated into the tool.

8.1.2 The Offline Restoration Tool Using Renewable Energy Resources

After a blackout, power system operators work diligently to bring the system back to its normal state. System restoration consists of the following tasks: preparation and planning, black-start unit (BSU) start-up, transmission line energization, non-black-start unit (NBSUs) start-up, and load pickup [12].

Wind generator can be considered as a third type of generating unit. Like BSUs, wind generators can supply cranking power to other generators. Like NBSUs, wind generators can provide reactive power to energize transmission lines and restore critical loads. Wind generators can also contribute to system frequency control. On the other hand, wind generators differ from conventional BSUs and NBSUs in the following aspects: (i) they are intermittent and weather dependent in nature, (ii) their power can fluctuate highly and cause large ramping events, and (iii) they may reduce system inertia (without supplementary control). Additionally, they have a high ramping rate and can be started quickly. With these characteristics, harnessing wind power in system restoration needs a thorough investigation and formulation.

Following a widespread blackout, the affected TSOs control and coordinate the operation of generating sources, transmission facilities, and loads within their footprints [13]. For instance, TSOs coordinate with generator operators (GOs) for reconnection of tripped generators and with distribution system operators (DSOs) for reconnection of loads. After obtaining the required data from the GOs and DSOs, TSOs provide the restoration plans based on the current status of the system. Once TSOs submit its restoration plan, other contributors should comply with the instructions.

In this restoration tool, depicted in Figure 8.1, initially, the TSO identifies the amount of loads and generators affected or area of blackout. Having accomplished the initial assessment, the TSO obtains the maximum generation capability and cranking power of available generation units from the GOs. Assuming that the wind generators are operated like an NBSU, the forecasted wind power is obtained from wind turbine operators (WTOs). With these data at hand, a number of constraints, including generator start-up and line/bus energization constraints, will be constructed. In addition, the TSO requests GOs and DSOs to identify the generators' ramping capabilities and the amount and locations of loads. Once these data are acquired, the load pickup, dynamic reserve, and wind generator constraints are incorporated into the optimization tool, which is solved by assuming the perfect wind power forecasting data. Then, this deterministic formulation is extended to a stochastic optimization with scenarios generated from forecasted wind power data representing real-world uncertainties.

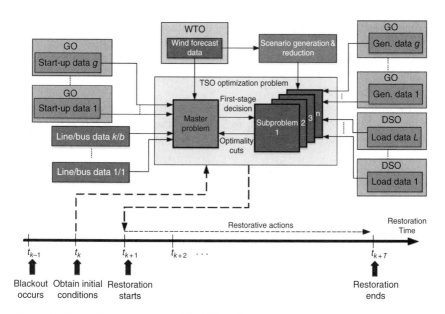

Figure 8.1 The offline restoration tool for TSOs to harness renewable energy.

With the complex stochastic mixed-integer linear programming (MILP) formulation, a solution methodology is proposed with a two-stage decomposition. The first-stage decisions are obtained from the master problem prior to the uncertainty unfolding. The second-stage decisions are scenario dependent and obtained from the subproblems. Then, an optimality cut is generated and fed back to the first stage problem to re-solve the updated master problem. This iterative process continues until the convergence criteria are met.

DSOs are responsible for carrying out the load pickup actions whose values and locations are determined after solving the TSO's optimization problem. Likewise, GOs should be notified concerning the start-up and synchronization times of the generators, and WTOs should obtain the connection times as well as the scheduled power of wind energy sources from the TSO's optimization problem.

8.1.3 System Restoration with Renewables' Participation

8.1.3.1 Problem Formulation

The objective is to maximize the expected value of total restored load during the restoration period in all scenarios, as shown in the following:

$$\text{Max} \ \sum_{s=1}^{N_s} \pi_s \sum_{t=1}^{T} \sum_{l=1}^{N_L} \alpha_l P_{load_l}^{t,s}, \tag{8.1}$$

where π_s denotes the probability of scenario s, P_{load} is the total active load at bus l with priority as α_l, Ns represents the total number of representative scenarios, N_L is the total number of load, and T is the given total restoration time.

The constraints are shown in the following:

8.1.3.1.1 Power Balance

Active and reactive power balance constraints at each time in each scenario are shown in the following equations. To find the voltage and angle of system bus, AC load flow equations have been linearized as stated in Trodden et al. [14]:

$$\sum_{i=1}^{N_G} P_{gen_i}^{s,t} + \sum_{i=1}^{N_n} P_{w_n}^{s,t} - \sum_{l=1}^{N_L} P_{load_l}^{s,t} = \sum_{m=1}^{N_M} P_{f_m}^{s,t}, \forall s,t \tag{8.2}$$

$$\sum_{i=1}^{N_G} Q_{gen_i}^{s,t} + \sum_{i=1}^{N_n} Q_{w_n}^{s,t} - \sum_{l=1}^{N_L} Q_{load_l}^{s,t} = \sum_{m=1}^{N_M} Q_{f_m}^{s,t}, \forall s,t, \tag{8.3}$$

where, P_{gen} and Q_{gen} are real and reactive power output of conventional generators, P_w and Q_w are real and reactive power output of wind generators, Q_{load} is reactive load, P_f and Q_f are real and reactive power flow, and N_g, N_M, and N_n represent total number of traditional generators, lines, and wind generators, respectively.

8.1.3.1.2   Generation and Transmission Restoration

Generation restoration is finding the optimal generator start-up sequence, constrained by the starting time, cranking time to ramp up and parallel with system, ramping rate, and generation capacity. The detailed constraints can be found in Sun et al. [15]. Transmission restoration is finding the shortest transmission path for BSUs to send cranking power to start NBSUs. The detailed formulation of constraints can be found in Sun et al. [16].

8.1.3.1.3   Generator Ramping Limit

Considering wind variability, ramping up and down limits of generators are modeled as

$$P_{gen_i}^{s,t+1} - P_{gen_i}^{s,t} \le RU(i), \forall i,s,t \tag{8.4}$$

$$P_{gen_i}^{s,t} - P_{gen_i}^{s,t-1} \le RD(i), \forall i,s,t, \tag{8.5}$$

where RU and RD are the ramping up and down limits.

8.1.3.1.4   Load Shedding Limit

The following constraints are added to avoid any active or reactive load shedding during restoration.

$$P_{load_l}^{s,t+1} \ge P_{load_l}^{s,t}, \forall l,s,t \tag{8.6}$$

$$Q_{load_l}^{s,t+1} \ge Q_{load_l}^{s,t}, \forall l,s,t \tag{8.7}$$

8.1.3.1.5   Load Pickup Limit

In Adibi et al. [17], frequency response rates (FRRs) are defined as the maximum load each generator can pick up with an acceptable frequency dip due to the sudden change of load. This limit is represented as β^t, and the constraint in Eq. (8.8) is added. Reactive load pickup limit is presented in Eq. (8.9), where pf denotes the power factor of each load.

$$P_{load_l}^{s,t+1} - P_{load_l}^{s,t} \le \beta^{s,t} \tag{8.8}$$

$$Q_{load_l}^{s,t} \le P_{load_l}^{s,t} \tan\left(\arccos\left(pf_l\right)\right), \forall l,s,t \tag{8.9}$$

8.1.3.1.6   Line Flow Limit

Each line flow should be within its thermal limits P_f^{min} and P_f^{max}, as given by

$$P_{f_m}^{min} \le P_{f_m}^{s,t} \le P_{f_m}^{max}, \forall m,s,t. \tag{8.10}$$

8.1.3.1.7 Generator Output Limit

Generator real and reactive power output should be within the limit as given by

$$0 \le P_{gen_i}^{s,t} \le P_{gen_i}^{\max}, \forall i, s, t \tag{8.11}$$

$$Q_{gen_i}^{\min} \le Q_{gen_i}^{s,t} \le Q_{gen_i}^{\max}, \forall i, s, t. \tag{8.12}$$

8.1.3.2 Solution Algorithm

The wind power available to participate in restoration is highly dependent on forecasting accuracy. In the aforementioned formulation, the maximum available wind power is forecasted, which can never be perfect. To handle the uncertainty, the deterministic formulation is now extended to stochastic optimization with scenarios.

The forecasting errors are modeled with scenarios generated using Latin hypercube sampling. A large number of scenarios are generated to completely describe the stochastic nature of the wind power. These scenarios are generated using the forecasted data, mean, and standard deviation of the forecast error at every restoration time step. Wind power forecast error is described using normal distribution function with a mean of zero and a standard deviation of $\alpha\%$, where parameter α can be adjusted by system operators. Latin hypercube sampling can accurately recreate the input distribution through sampling in fewer iterations compared with the Monte Carlo method. In this way, the probability density function curve is divided into N nonoverlapping equiprobable intervals within which random sampling is performed. It guarantees that there will be precisely one sample in each interval such that the entire probability density function space, including the tails, will be covered.

Having a large number of scenarios renders a large-scale mathematical problem. To mitigate the computational burden resulting from the problem size, an efficient scenario reduction algorithm is applied to produce a set of representative scenarios combined with their associated probabilities. A fast-forward reduction technique based on Kantorovich distance has been adopted [18]. The aim is to select a subset S from the generated scenario set in a way that the representative scenarios have the shortest distance to the remaining scenarios.

With the generated scenarios, the stochastic optimization model can be developed. In each scenario, the simulated wind power takes the place of the forecasted wind power, introducing different load pickup sequences and values. The set of decision variables that are contingent upon the scenarios and times include real and reactive power generation output, power flow, reserve amount, and load pickup amount. Even with a limited number of scenarios, the resulting problem requires the use of decomposition techniques so as to attain computational tractability.

In Goshani et al. [19], a two-stage decomposition is employed to solve this large-scale combinatorial problem. The first-stage problem determines the online times of all generators, as well as line and bus energization times. These decisions are common to any wind scenarios. The master problem contains a function representing the objective function of subproblems. Subproblems incorporate all decisions related to the load pickup values and locations, real and reactive power of all generators, and dynamic reserves. The subproblem decisions must be taken relative to the realization of uncertain wind power output.

The integer L-shaped algorithm is applied to solve the two-stage stochastic restoration problems [19]. This algorithm was proposed by Laporte and Louveaux and applied to solve the stochastic integer program [20]. In this method, the master problem is relaxed and solved using branch-and-cut algorithm. For any feasible integer solution found, the subproblem is solved for each scenario and a set of optimality cuts are generated. Then in the second iteration, the master problem is re-solved with optimality cuts generated in the previous iteration to obtain a new feasible solution. This process iteratively continues until the convergence criterion is met. The details can be found in Golshani et al. [19].

8.1.3.3 Simulation Results
The impact of wind participation in power system restoration is thoroughly tested. Various factors are explored, including wind generator location, penetration, fluctuation, and uncertainty. The stochastic optimization is also compared with the deterministic formulation to show the improvement in the total served energy.

The test case is a modified IEEE 57-bus system with the total restorable load of 1,250 MW. The characteristics of buses, transmission lines, and loads are taken from IEEE [21]. G1 is the BSU and the rest of units are NBSUs. A wind farm is connected at bus 38 with total installed capacity of 200 MW. In all cases the base power is assumed to be 100 MW and each restoration time step is 10 minutes (1 p.u.), which is required for preparation and frequency stabilization. Also, the wind farm is operated at unity power factor. The integer L-shape algorithm is coded in C++ using CALLBACK function of IBM ILOG CPLEX 12.6. All simulations were executed on a PC with Intel CoreTM i5 CPU @3.30 GHz and 8 GB RAM.

To investigate the wind impact, a base case benchmark is first established where the wind farm is excluded from the restoration process. The BSU G1 comes online at $t = 2$ restoration time. The first load bus is energized at $t = 3$ when the load pickup process can be started. The first NBSU becomes online after $t = 7$ restoration time. Figure 8.2 depicts the total generation and load pickup curves of the base case, where the total restored load is 936 MW and the total energy served is 1.66 GWh. Note that without wind participation, the

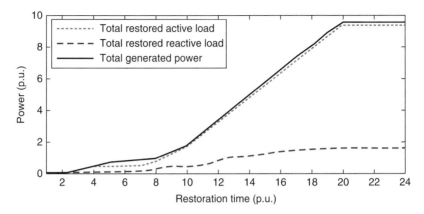

Figure 8.2 Total generated power and real and reactive load pickup in the base case.

system cannot be completely restored. Now assume that the wind farm is allowed to fully participate in the restoration process.

The wind impact is investigated from the following aspects: location, penetration, fluctuation, and inertia capability.

8.1.3.3.1 Impact of Wind Location

Wind location has a direct impact on the entire restoration process. Figure 8.3 compares the total energy served with respect to different wind farm locations. It shows the percentage of increase in the total served energy compared with the base case. For instance, the 200 MW wind farm installed at bus 15 will improve the total energy served by 1.18% and 18.9% versus that installed at bus 38 and the base case, respectively. From Figure 8.4 it is found that the farther it is placed from the BSU, the less contribution will be acquired from the wind farm.

8.1.3.3.2 Impact of Wind Penetration

The wind farm installed at bus 38 has a capacity of 200 MW, representing 16% penetration in the 57-bus test case. We then increase the wind farm capacity to 400 MW, representing 32% penetration. The restoration results are shown in Figures 8.4 and 8.5. Comparing two cases, one can see that high wind penetration leads to better restoration performance. Specifically, in the low-penetration case, the total energy served is 1.99 GWh, while this number is improved to 2.15 GWh in the high-penetration case. Note that in both cases, wind energy spillage exists, that is, wind energy is not fully used up to its forecasted value. This phenomenon could be alleviated using storage devices, as introduced in next section.

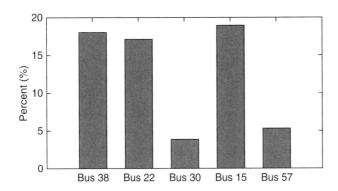

Figure 8.3 Total energy served at different wind farm locations.

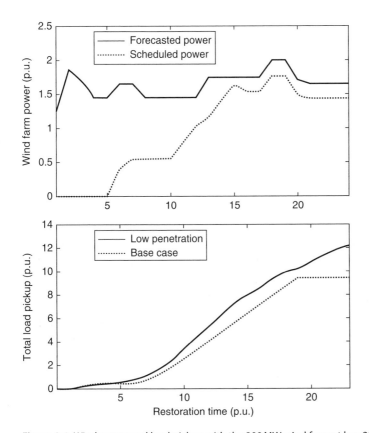

Figure 8.4 Wind power and load pickup with the 200 MW wind farm at bus 38.

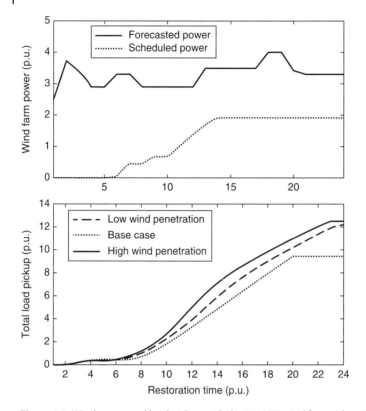

Figure 8.5 Wind power and load pickup with the 400 MW wind farm at bus 38.

8.1.3.3.3 *Impact of Wind Fluctuation*

Wind fluctuation levels can differ from time to time. Highly fluctuated wind power could result in severe ramping events. In Figure 8.6a, one can observe a large fall and rise in the wind farm power where the minimum value reached 50 MW at t = 10. The total energy served is 1.93 GWh, a slight decrease from that in Figure 8.4. If this ramping event occurs at t = 15 as shown in Figure 8.6b, the wind generation curve will be different from that at t = 10. More important, the total restored energy is now decreased to 1.86 GWh, and the load pickup curve gets close to the base case curve at t = 16 and t = 17. The results indicate that not only the fluctuation level but also the fluctuation times play a critical role in deploying wind power for restoration.

For the IEEE 57-bus system, the total number of integer variables in the first stage is 6,050, and the total number of variables in the second stage is 28,525 per scenario. The convergence time for one scenario without adopting the proposed decomposition approach is 1,094.4 seconds. However, after adopting the decomposition method, the solution time decreased to 86.8 seconds. In case of

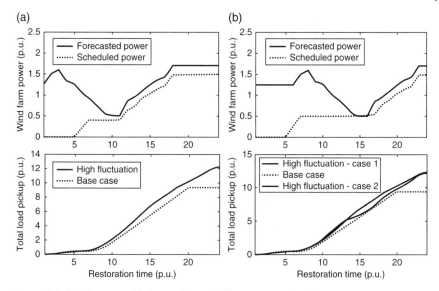

Figure 8.6 Wind power with fluctuations at different restoration times.

10 scenarios, we enabled the multiple thread feature of CPLEX, and the convergence time with 0.5% optimality gap reached 774.8 seconds.

8.2 Operation and Control of Renewable Generators in System Restoration

Increasing reinforcement and expansion of electrical grids to extensive blackouts are caused by the utilization of long-distance high-power exchanges and heavier loading of transmission facilities in deregulated power systems. If a system disturbance provokes a total or partial outage, black-starting and load pickup services are required to promote a smooth and safe reconstruction of AC grids. In general, the alternative strategic choice for system restoration is either a "build-down" strategy or a "build-up" strategy. The build-down strategy is simpler but involves more tasks when a critical time of restoration is required. Conversely, the build-up strategy can appreciably reduce the outage duration but is limited by availability of the resources.

Typically, system restoration under the build-up strategy includes two vital stages, namely, black-starting and stand-alone loaded stages before the external grid is rescued. The advent of voltage-source converter (VSC) technology makes the build-up strategy more preferable for system restoration. That is because the extent of sectionalization over build-up restoration can be guaranteed by exploiting the power generation and control capabilities of VSC systems to provide fast black-starting further afield, as well as to provide a loaded service in the stand-alone mode. Moreira et al. and Jiang-Hafner et al. [22, 23] have described the feasibility and operational concept of using VSC-controlled equipment (like VSC-HVDC and some microgrid units) for black-start and stand-alone loaded services. Analogously, Type 3 wind turbines (WTs), embedded with back-to-back VSCs, can also be expected to function as powerful standby facilities for system restoration.

Nevertheless, the methods to allow Type 3 WTs for system restoration have barely been discussed before. An intuitive thought readily emerges from the studies of the operational concept of synchronous generators (SGs) once independently paired with unpredictable local loads [24–26]. SGs possess the inherent relationships between active power and frequency (P-f) as well as reactive power and voltage (Q-V) [24, 25]. It reveals that the implementation of the build-up restoration strategy to form an islanded AC system requires an independent power source that is equivalent to a voltage oscillator with oscillation frequency and amplitude spontaneously controlled by its exported active/reactive power, both in the black-start and stand-alone loaded stages [22]. Therefore, the converter control available for Type 3 WTs in system restoration needs to be modified and thereby ensure their operation as a power-controlled voltage supply in black-start and stand-alone loaded stages.

Inspired by this, this chapter first studies the black-start capacity of Type 3 WTs without any external grid information [27]. The key requirements for boosting a Type 3 WT to feature black-starting include (i) at least one energy-providing element (like a WT) and at least one energy-dissipating element (internal loads) for maintaining the real-time balance between the power generation and dissipation, (ii) a modified control system for fabricating a steady voltage-source interface that provides the electricity to external loads, and (iii) an available sequence of independent startup for ensuring a safe and smooth restoration. In this way, the P-f and Q-V relationships of Type 3 WTs can be maintained but still masked by phase-locked loop (PLL)-based vector control.

Subsequently, this chapter addresses the frequency control and performance analysis when Type 3 WTs step into the stand-alone loaded stage. From the perspective of P-f and Q-V relationships, the key points for energizing passive loads in stand-alone mode are as follows: (i) presenting an insight into Type 3 WTs' P-f and Q-V relationships, which are similar to those of SGs, but their dynamic processes are in two fundamentally different ways, and (ii) modifying the control system to develop an autonomous mechanism between voltage/frequency and output power [28–31]. On this basis, the impacts of Type 3 WTs on system frequency performance can be mathematically evaluated. Furthermore, the coordinated frequency control of a multiple-WT-based system in the practical application is also discussed.

8.2.1 Prerequisites of Type 3 WTs for System Restoration

The identification of Type 3 WT characteristics is crucial for the build-up strategy. This requires a deep knowledge of islanded AC systems. The reason is that the operational criteria of islanded AC systems not only impose the prerequisites to local black start and stand-alone loaded services but also guide the converter control design for VSC systems. In this regard, this section introduces the *P-f* and *Q-V* relationships to reveal the insight into the operational concept of islanded AC systems, as well as to present the prerequisites for Type 3 WTs in system restoration application.

An islanded AC system utilizes its black-start resources to energize apparatus while maintaining the instantaneous balance between power generation and consumption. Meanwhile, the system needs to accommodate a stable frequency and voltage in the presence of arbitrarily varying loads. Therefore, the *P-f* and *Q-V* relationships are the foundation to monitor and stabilize islanded AC systems and needed to be further analyzed.

8.2.1.1 *P-f* Relationship in Islanded AC Systems

The frequency of an islanded AC system is an indicator of the balance between the active power generation and consumption. The inherent relationship between active power and frequency (*P-f*) is the foundation for the operation and frequency control deployment of an AC system.

A conventional AC system (shown in Figure 8.7) consists of some SGs, local loads, transformers, and so on. Herein, an autonomous regulation of the system frequency is maintained through an internal negative-feedback loop. It is provided by the governor response (slow-action *P-f* relationship) and the inertial response (fast-action *P-f* relationship) of SGs. In case of a power perturbation provoking a significant frequency excursion, SGs will spontaneously redress their exported power, using the kinetic energy stored in rotors and the power reserve extracted from prime movers. It should be noted that local loads can also contribute to the frequency support and typically provide a positive *P-f* association, which differs from that of SGs. Therefore, system frequency

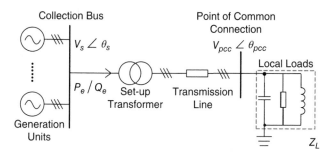

Figure 8.7 General configuration of an AC system.

behaviors are dominated by the dynamic process of power exchange between the online generation equipment and the loads. In order to retain the AC system frequency inside its allowable operating range under different local-load conditions, the main criteria are listed as follows: (i) generation units are essential for power supply and to naturally sustain the P-f relationship, (ii) the maximum power production must be ample to accommodate the total local loads, (iii) the system frequency is able to be self-maintained via negative P-f association (like SGs' governor response).

Two additional criteria allow the AC system to survive permanently: (i) the restoration of system frequency requires the integral control of frequency bias, deployed in a longer timescale (like SGs' automatic generation control), and (ii) an appropriate load shedding process needs to be developed with a target to guard the system against the frequency collapse and to satisfy a sustainable power transmission. However, they are out of the scope and will not be discussed in this chapter.

8.2.1.2   Q-V Relationship in Islanded AC Systems

The voltage level variation reveals the reactive power flow dynamics. It implies that the production or absorption of reactive power contributes to an increase or decrease of voltage level, respectively. However, this action is dispersed since the AC voltage is local and the reactive power dissipates over long-distance transmission lines. For this reason, the Q-V interaction is also dispersed and it is essential to locally deploy some proper equipment for maintaining reactive power balance and voltage level. Conventionally, SGs accommodate the basic means of voltage regulation and the terminal voltage control, for example, automatic voltage regulator to maintain the required voltage level. Other reactive power compensation equipment and processes (like on-load voltage regulation) are additionally introduced to achieve voltage regulation throughout the system.

8.2.1.3   Prerequisites for Type 3 WT-Based System Restoration

As discussed previously, the P-f and Q-V relationships are significant to restore islanded AC systems. The on-line resources and apparatus provide the inherent P-f and Q-V relationships. They ensure the equilibrium conditions of voltage and frequency stability being automatically attained via holding sufficient reactive and active power support. Therefore, Type 3 WTs need to voluntarily self-maintain their frequency and voltage level by executing inherent P-f and Q-V relationships as qualified system restoration resources. Therein, appropriate control modification can be exploited.

8.2.2   Problem Setup of Type 3 WTs for System Restoration

Despite the aforementioned P-f and Q-V relationships, Type 3 WTs fail to attain system restoration without appropriate designed control strategies.

This section continues to describe the *P-f* and *Q-V* relationships to reveal the inherent problems of typical Type 3 WTs to serve as power-controlled voltage sources without any control updating. Importantly, the following discussion provides a guide to design the black-starting and stand-alone control.

8.2.2.1 Problem Sets of Type 3 WTs Applied in System Restoration

Back-to-back VSC can mimic two forms of active/reactive power and voltage magnitude/frequency relationships through the corresponding control strategies of *Q-f* and *P-V* control or *P-f* and *Q-V* control. The former is proposed in Blasco-Gimenez et al. [32] and tailored for the combined operation between offshore WTs and diode-based HVDC. The latter is prevalent in practice and valued by manufacturers [28, 33]. In this sense, only the Type 3 WTs based on *P-f* and *Q-V* control are discussed in this chapter.

With reference to [28], the basic control of a Type 3 WT involves a multitimescale, as illustrated in Figure 8.8. The stator voltage orientation and dual-loop control are adopted. Therein, the active and reactive power are independently controlled by the outer speed control and voltage control, respectively, combined with the inner current control.

Actually, the actions of the multitimescale controllers are sequential [34]. Therefore, the electromechanical timescale dynamic characteristics of Type 3 WTs are mainly dependent on the behaviors of speed control and the machine itself, so that the impacts arising from the responses of the rotor side converter (RSC) current control and grid side converter (GSC) control are readily discarded. It is interesting to note that the typical time constant of PLL is about 40 ms [28, 34], which would be prolonged for inertia support afterwards.

Nevertheless, the general control configuration of Type 3 WTs only supports the normal operation in grid-connected mode [29]. To use Type 3 WTs as back-up resources for system restoration, there are two vital problems to be considered. (i) For the black-start stage, no power can be fed from the external systems to support the converter DC voltage and build a unit stator voltage. (ii) For the stand-alone loaded stage, PLL-based vector control not only masks Type 3 WTs' *P-f* and *Q-V* relationships but also introduces a positive frequency feedback through constant impedance loads.

These two problems require a more intuitive description to facilitate the control design of Type 3 WTs for system restoration. Therefore, subsection 8.2.2.2 gives an elaborated physical insight into the *P-f* and *Q-V* relationships.

8.2.2.2 The Elaboration of Problem Sets

The complex nature of Type 3 WTs makes it highly essential to develop a generalized approach to characterize their inherent properties and defects. From the perspective of internal voltage, the indirect and masked *P-f* and *Q-V* relationships are elaborated in this subsection.

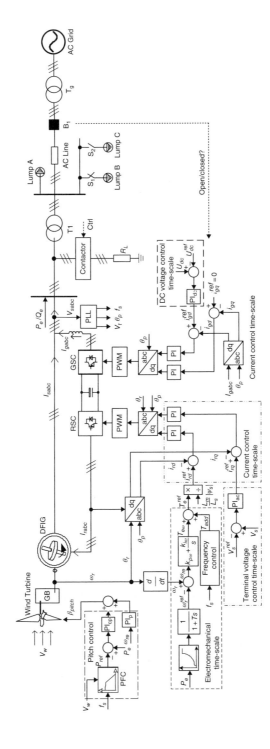

Figure 8.8 Typical control scheme for a Type 3 WT with stator voltage oriented.

According to the derivation in [31], Type 3 WTs can be treated as a simplified mathematical model that is similar to salient pole SGs. Also, their equivalent one-line circuit model in the stand-alone mode is derived, as shown in Figure 8.9. Z_s represents the output impedance of Type 3 WTs. For convenience, the impedances of transformer and transmission line are summed up into RLC (resistor (R), inductor (L), capacitor (C)) loads, represented by Z_L.

In the dq-frame, the steady-state voltage equation can be written as (neglecting the stator resistor R_s)

$$V_s = -j\omega_1 L_s I_s + j\omega_1 L_m I_r, \tag{8.13}$$

where L_s and L_m are stator inductance and mutual inductance, respectively. Referring to [31], the internal voltage of Type 3 WTs is defined based on the rotor current, where

$$E_d = -\omega_1 L_m i_{rq}, \; E_q = \omega_1 L_m i_{rd}. \tag{8.14}$$

Considering the stator voltage orientation (i.e., $V_{sq} = 0$), the power exported from the stator can be expressed as

$$P_s = V_{sd} i_{sd}, \; Q_s = -V_{sd} i_{sq}. \tag{8.15}$$

Since only the slip active power is exported from GSC, then the total output power can be written as

$$P_e = V_{sd} i_d = (1 - \omega_{slip}) P_s, \; Q_e = -V_{sd} i_q = Q_s. \tag{8.16}$$

Equations (8.15) and (8.16) yield

$$i_{sd} = i_d / (1 - \omega_{slip}), \; i_{sq} = i_q. \tag{8.17}$$

Substituting Eqs. (8.14) and (8.17) into the dq form of Eq. (8.13), we get

$$V_{sd} = \omega_1 L_s i_q + E_d, \; V_{sq} = (\omega_1 L_s i_d)/(1 - \omega_{slip}) + E_q. \tag{8.18}$$

The slip term in Eq. (8.18) is generated due to the action of GSC. Normally, the output angle of PLL relative to the terminal voltage angle is sufficiently small. Therefore, Eqs. (8.14) and (8.18) are also applicable for the PLL frame.

In Figure 8.9, V_s is the feedback to the PLL and can be written as

$$V_s e^{j\theta_s} = K_z E_m e^{j(\theta_E + \varphi_z)} \tag{8.19}$$

$$\theta_s = \theta_E + \varphi_z = \theta_{in} + \theta_p + \varphi_z \tag{8.20}$$

$$K_z = \left| Z_L / (Z_L + Z_s) \right|, \varphi_z = phase \left[Z_L / (Z_L + Z_s) \right], \tag{8.21}$$

where φ_z is the phase-shift angle and K_z is the voltage-divider coefficient.

V_{sq} is used to actuate the tracking of PLL and takes responsibility for retaining the frequency at an appropriate level, that is, around 50 Hz. Equations (8.19) and (8.20) yield the signal V_{sq} (given in Eq. (8.22)) that is the error input of the PI compensator in PLL. Since the values of K_z and E_m are nonzero, V_{sq} equals zero if and only if the value of $(\theta_{in} + \varphi_z)$ is zero. It reveals that a positive value of $(\theta_{in} + \varphi_z)$ will lead to the acceleration of ω_p, otherwise ω_p will decelerate. Only when $\theta_{in} + \varphi_z = 0$, ω_p becomes a constant, so that PLL is stable. In short, the frequency stability of PLL is dependent on whether the value of V_{sq} or $(\theta_{in} + \varphi_z)$ is equal to zero or not.

$$V_{sq} = K_z E_m \sin\left(\theta_E + \varphi_z - \theta_p\right) = K_z E_m \sin\left(\theta_{in} + \varphi_z\right) \tag{8.22}$$

The *P-f* relationship of Type 3 WTs is generally masked by the fast action of PLL [31]. As such, it must be reinforced no matter what kind of control for system restoration is adopted.

It is known that the converter control is devised to control the exported active power and the reactive power based on the terminal voltage V_s while updating the internal voltage E. Figure 8.9 demonstrates that the phase displacement θ_E of E is synthesized by θ_{in} and θ_p. θ_{in} represents the location of E in the PLL reference frame and depends on the output of power/excitation control in RSC, while θ_p is the output of PLL and tracks θ_s as far and accurate as possible. That is, the proximal real-time tracking of PLL shields the Type 3 WTs' output power from the frequency perturbations; hence, only a shielded *P-f* relationship can be observed.

In addition, unlike the SGs' *P-f* relationship that is direct and inherently bridged through the rotor, the *P-f* relationship of Type 3 WTs is indirect and softly linked. This is because throughout the *P-f* loop, there is no one physical medium generated by a mechanical component such as the rotor in SGs. The lack of straightforward interpretation of Type 3 WTs' *P-f* relationship becomes an obstacle to the system restoration application. To throw light on Type 3 WTs' *P-f* relationship, it is critical to distinguish the medium here.

According to He et al., Wang et al., and Hu et al. [31, 35, 36], a model based on *P-f* relationship can be built to facilitate the analysis. As an example, a tackled model (see Figure 8.10) is derived from the following equations.

$$\Delta P_e = K_\delta \left(\Delta \theta_E - \Delta \theta_s\right) \tag{8.23}$$

$$\Delta \theta_E = \Delta \theta_{in} + \Delta \theta_p \tag{8.24}$$

$$\Delta \theta_{in} = \left[(Ds+1)\big/Ms^2\right]\left(\Delta P_m - \Delta P_e\right) \tag{8.25}$$

$$\Delta \theta_p = \left[\left(k_p s + k_i\right)\big/\left(s^2 + k_p s + k_i\right)\right]\Delta \theta_s \tag{8.26}$$

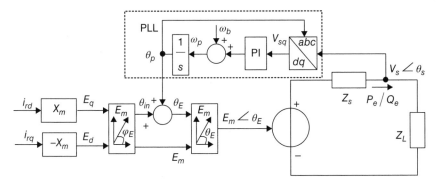

Figure 8.9 One-line circuit model of a Type 3 WT–based islanded system.

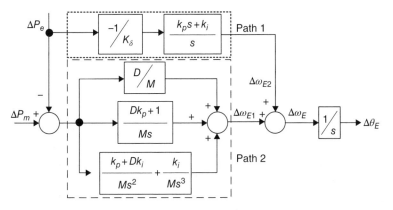

Figure 8.10 Equivalent form seen from a *P-f* relationship.

Combining Eqs. (8.23)–(8.26) and eliminating θ_{in}, θ_p, and θ_s, we get

$$\Delta\theta_E = \left[1+\left(k_p s+k_i\right)/s^2\right]\left[(Ds+1)/Ms^2\right]\left(\Delta P_m - \Delta P_e\right) - \left(1/K_\delta\right)\left[\left(k_p s + k_i\right)/s^2\right]\Delta P_e \tag{8.27}$$

$$M = \left(2H\omega_{r0}\right)/k_{i\omega}K_{eq}X_m \ \text{and}\ D = k_{p\omega}/k_{i\omega}, \tag{8.28}$$

where the values of M and D are given in Eq. (8.28) and other key parameter settings are the same as given in Zhao et al. [31]. The equivalent model suggests that Type 3 WTs share strong similarities with SGs in the foundational mechanism, mathematical model, and power transmitting process. Therein, the *P-f* relationship is constructed through two portions of the internal feedback loop: (i) the DFIG and its decoupled power control in RSC (Path 1) and (ii) the stator PLL (Path 2). The *Q-V* relationship is built by the terminal voltage control. Indeed, the model is different from [31] in two aspects: (i) The input of Path 1 is the power perturbation ΔP_e, instead of the imbalance power ($\Delta P_m - \Delta P_e$); (ii) the inertia laying in Path 2 is rearranged based on the order and meaning of each component.

Therefore, ΔP_e drives the motion of $\Delta \theta_E$ through Path 1 directly and Path 2 indirectly. Path 1 presents the direct effects of PLL, while Path 2 contains the inherent rotor mass, that is, a real energy source, which attains the absorption and release of power imbalance. In other words, Path 2 dominates the process of power transmitting from Type 3 WTs and plays a similar role to rotors in the SGs. It also implies that Path 2 acts as a controlled medium, linking the power imbalance (i.e., the difference between ΔP_m and ΔP_e) and internal voltage frequency $\Delta \omega_{E1}$. Besides, the synchronization mechanism between PLL-based vector controlled devices can be explained via regarding electromagnetic power transmission path as the communicating medium [37].

Unlike the SGs' rotor, the controlled medium (i.e., Path 2) of Type 3 WTs is controllable and multitimescaled due to the inclusion of converter control, such as rotor speed and PLL control. In addition, the controlled medium provides the possibility for modifying the frequency response of Type 3 WTs through complementary feedback control and suitable parameter settings.

There are two general ways to procure the Q-V relationship in Type 3 WTs: reactive power control with a predefined power factor and AC terminal voltage control. In the islanded system, Type 3 WTs are assigned the task of maintaining the load voltage. Additionally, a dual-loop AC terminal voltage control is readily adopted in RSC while unity power factor control is adopted in GSC. It implies that Type 3 WTs feed the reactive power completely through the stator and the Q-V relationship just depends on DFIG itself. Therefore, the Q-V relationship of Type 3 WTs is identical to that of SGs and will not be discussed further.

8.2.3 Black-Starting Control and Sequence of Type 3 WTs

Actually, the traditional control configuration in Type 3 WTs makes them incompetent to serve as restoration sources. This section proposes a modified control strategy for them, so that they can act as restoration sources in the black-start stage (namely the first stage of system restoration). As already discussed, the black-start capacity of Type 3 WTs can be developed by modifying the typical converter control to provide a constant voltage and frequency interface. Due to the absence of the external gird, the AC voltage reference would be created with a modified PLL and RSC control. Furthermore, RSC is also used to generate the desired excitation current and GSC is controlled to transmit the slip power. Thereafter, the sequence for black-starting action can be designed.

8.2.3.1 Proposed Control Strategy

The task of black-starting Type 3 WTs is to form the inherently indirect *P-f* and *Q-V* relationships, so that an interface is fabricated to accommodate a self-built and constant voltage. Furthermore, the power extracted from mechanical elements should be kept equal to the total power dissipation over the start-up duration. Hence, in this study, Type 3 WTs are black-started with a resistive load supply (for power dissipation), a specific orientation algorithm of PLL and a modified deployment of RSC, GSC. It is stated in detail next.

During the start-up procedure, PLL is modified to fabricate a synchronous reference frame that provides the foundation for the converter control. This is executed by clamping the output of PLL at the integral of the normative frequency, such as 60 Hz.

In addition, the magnitude of terminal voltage is maintained at its unit value by RSC. For minimizing the iron rotor loss, the start-up speed of Type 3 WTs is set near the synchronous speed [38]. Therefore, $\omega_{slip} = 0$ in Eq. (8.18). In addition, if we assume that $R_s = 0$, $R_L = Z_L$ and the desirable terminal voltage $V_s e^{j\theta_s} = (1 + j0)$ p.u. in Figure 8.9, we can obtain

$$E_d = 1, E_q = \omega_1 L_s / R_L. \tag{8.29}$$

Hence, the rotor current references are derived as

$$i_{rd}^{rerf} = L_s / (L_m R_L), i_{rq}^{ref} = -1/\omega_1 L_m. \tag{8.30}$$

Equation (8.30) indicates that the expected values of rotor current depend just on the parameters of DFIG for a specific loaded case. Thereafter, the speed control acting as a compensation of the *d*-axis rotor current is used to maintain rotor speed, while the AC voltage control is superposed to the *q*-axis rotor current to render an expected voltage.

GSC is used to transfer the power extracted from the rotor to the external system. The alternative control strategy is constant active power transmission control or DC-voltage control, depending on whether a DC battery is embedded or not. Specifically, the existence of a DC battery ensures a constant DC voltage, and the active current command i_{gd}^{ref} is hence obtained from Eq. (8.31). Otherwise, i_{gd}^{ref} is derived from the output of DC-voltage control. i_{gq}^{ref} is permanently equal to zero.

$$i_{gd}^{ref} = (\omega_{slip} P_e) / [(1 - \omega_{slip}) V_s] \tag{8.31}$$

8.2.3.2 Sequence of Black-Starting Actions

Once a system collapses, an island waits to be established (given in Figure 8.8). Type 3 WTs are required to undertake a task for black-starting and picking up proximate loads. Meanwhile, fault assessment and verification, as well as the achievement of goals and objectives for restoration, require the implementation of the following actions.

1) Detect the state of breaker B1. Is it open or closed?
2) If it is open, Type 3 WTs and local loads are entirely blocked.
3) Subsequently, the energy storage element is deblocked to power the converter and WT control systems, protection devices, and metering components (e.g., tachometer and voltage/current sensors), as well as to build the DC link voltage (i.e., charging DC capacitor).
4) The control system is deblocked. It has to be noted that a modified PLL is engaged in fabricating voltage-orientation while maintaining the voltage and frequency reference at the unit and the base value, respectively. In addition, RSC and GSC should come with the proposed control strategies according to the Eqs. (8.29) and (8.30). Therein, it is beneficial to build the stator voltage gradually by setting its reference gradually ramping up to the unit value.
5) A control signal (Ctrl) stemming from the control system is used to attain the contactor closure. In this way, the resistor R_L along with other dissipative elements (like the mechanical components and hardware devices) takes responsibility for power dissipation. It is noteworthy that the power absorption by dissipative elements and the storage element is sufficient to dissipate the power generation from the generator.
6) Mechanical components (including WT, blade, yaw, and DFIG) are then activated with the help of control systems. In this step, the generator is driven with the proximally synchronous speed to minimize power loss [38].
7) DC storage is cut off and GSC is transferred to DC-voltage control simultaneously. In this segment, the DC storage needs to be blocked to avoid the increased burden on RSC while slip rising [39]. Therefore, GSC is responsible for maintaining DC voltage once DFIG is fully activated.
8) AC terminal voltage control is engaged to maintain the output voltage, while the normal PLL is unblocked to achieve a realistic stator voltage-orientation. The output signal of PLL, including the magnitude, phase, and frequency of stator voltage, is then directed to the converter control (RSC and GSC).

This operation is well applicable at the intervals of the mild and sustainable wind speed at which Type 3 WTs hold a sufficient power margin in a stand-by mode for black-starting [39]. The flowchart in Figure 8.11 demonstrates the sequence for black-starting Type 3 WTs. In this way, they get fully started and here RSC should adhere to necessary capacities to provide self-maintained frequency regulation for the subsequent stand-alone loaded service.

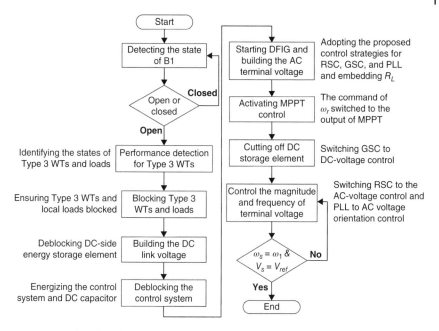

Figure 8.11 Flowchart for Type 3 WTs' black-starting action sequence.

8.2.4 Autonomous Frequency Mechanism of a Type 3 WT-Based Stand-Alone System

This section proposes a step-by-step approach to feed the local loads and concomitantly cut off the additional resistor in the stand-alone loaded stage, after the completion of Type 3 WTs' start-up. At first, the inner frequency adjustment mechanism of Type 3 WTs is presented. To achieve the operation criteria in AC systems, the inherent *P-f* relationship of Type 3 WTs needs to be highlighted to guarantee the stability of the stand-alone system. Therein, a complementary frequency feedback control is adopted to ensure the self-built islanded local system during the grid collapse. In this regard, this section then presents the detailed implementation of frequency control and the evaluation of dynamic frequency performance. Finally, to facilitate the practical application, the frequency regulation mechanism of multiple WTs is also discussed.

8.2.4.1 Frequency Adjustment Mechanism

For a drop in θ_s ($\Delta\theta_s < 0$), the exported power from Type 3 WTs tends to increase and allows $\Delta\omega_E$ to decelerate through both Path 1 and Path 2 (see Figure 8.8). Therein, the condition $\Delta\theta_E < 0$ is created. If the external grid is positive, $\Delta\theta_s$ can be simultaneously maintained, and $\Delta\theta_E$ can be ultimately driven to synchronize with $\Delta\theta_s$ due to the presence of PLL. In this case, both ΔP_e and $\Delta\omega_E$ would vanish while maintaining frequency stability. In contrast, the basic criterion $(\theta_{in} + \varphi_z = 0)$ of PLL stability in stand-alone mode is fragile in front of system disturbances. The reason is that a negative $\Delta\varphi_z$ is generated over the duration of $\Delta\omega_E$ decreasing. According to Eq. (8.20), $\Delta\theta_s$ will further keep on decreasing following the decline of $\Delta\theta_E$ and $\Delta\varphi_z$, such that the active power variation would reach a constant value ($\Delta P_e = -K_\delta\Delta\varphi_z$) rather than zero. Thus, the output frequency of Type 3 WTs would keep decreasing. From the perspective of frequency stability, Type 3 WTs' output frequency forms an open or positive feedback loop (if the load impedance angle has no or positive association with frequency) through passive loads, rather than a negative feedback loop via the active power in the grid [40]. This is analogous to the instability mechanism of SGs' rotor in the case of employing SGs to pick up local loads alone without extra frequency feedback control. Therefore, a negative-feedback frequency control, analogizing SGs' governor control, is naturally adopted in Type 3 WTs' control system.

8.2.4.2 Self-Maintained Frequency Service

According to Kundur and Machowski et al. [24, 25], the inherent P-f relationship in an SG includes the faster inertial response and the slower primary frequency response. Actually, the inertial response restrains the frequency decaying speed while the later promotes an ability for autonomous frequency regulation.

The artificially enhanced P-f relationship of Type 3 WTs imitates the inherent and direct P-f relationship of SGs. In particular, an enhanced inertia can be achieved via inertia control [28–31], while droop control [34] and deloaded control [30] or primary frequency control [28] are used to facilitate the frequency stability and power reserves. Advantageously, the presence of the self-maintained frequency services supports power balance to cope with the wind fluctuations and load variations throughout islanded operation.

As shown in Figure 8.8, the frequency services of Type 3 WTs are implemented in this chapter through the control surrounded by the dashed line. Therein, the parameter adjustments of PLL are deployed to attain a larger equivalent inertia and to avoid a sharp frequency fluctuation. The droop control is simultaneously implemented and allows Type 3 WTs to retain its inner frequency voluntarily. Furthermore, power reserves are held with the help of deloaded control to cope with unpredicted load transitions. These three controls have been separately explored in many studies, and they are all emphasized here as the basic services for Type 3 WTs operating in grid-disconnected mode.

8.2.4.3 Mathematical Performance Analysis

Performance analysis provides a way to the parameter settings and optimization of the original and additional control. From internal voltage and power

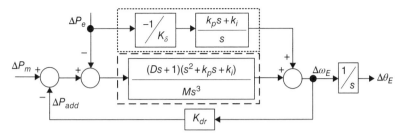

Figure 8.12 Diagram of the equivalent form of Type 3 WTs with self-maintained frequency services.

transmission process, a Type 3 WT can be represented by a generalized model that is similar to an SG, illustrated in Figure 8.12. Droop control is included as well. It is noteworthy that the impacts from the deloaded action are not evaluated here, since the deloaded control is mainly executed by pitching and its response is relatively slow due to the constraints of mechanical components.

Just like previous studies, we are only interested in a perturbation of the external system, represented by ΔP_e. Therefore, characterized by a sudden load upset, we deduce the P-f relationship of this system as given in Eq. (8.32)

$$\Delta\omega_E = -\frac{M\left(k_p s + k_i\right) + K_\delta\left(Ds+1\right)\left[1+\left(k_p s + k_i\right)/s^2\right]}{K_\delta\left\{Ms + K_{dr}\left(Ds+1\right)\left[1+\left(k_p s + k_i\right)/s^2\right]\right\}}\Delta P_e. \tag{8.32}$$

Thanks to $D \gg 1$ [28], Eq. (8.32) can be written as

$$\Delta\omega_E = -\left[\frac{Mk_i s}{K_\delta\left(M + K_{dr}D\right)}\frac{T_R s + 1}{s^2 + 2\zeta\omega_n s + \omega_n^2} + \frac{1}{K_{dr}}\right]\Delta P_e \tag{8.33}$$

and

$$T_R = \left(k_p - K_d/K_{dr}\right)\!\big/k_i, \omega_n = \sqrt{k_i\big/\left[1+M/\left(DK_{dr}\right)\right]},$$
$$\zeta = \left(k_p/2\sqrt{k_i}\right)\sqrt{1/\left[1+M/\left(DK_{dr}\right)\right]} \text{ and } \omega_d = \omega_n\sqrt{1-\zeta^2} \tag{8.34}$$

For a step function of ΔP_e that we usually fasten, we have $\Delta P_e = P_{step}/s$. From Eq. (8.34), we get

$$\Delta\omega_E\left(t\right) = -\frac{P_{step}}{K_{dr}} - K\sqrt{T_R^2 + \left(\frac{1-T_R\zeta\omega_n}{\omega_d}\right)^2}\, e^{-\zeta\omega_n t}\sin\left(\omega_d t + \varphi\right), \tag{8.35}$$

Table 8.1 Impacts of control parameters on system frequency performances.

	Rotor Inertia	Speed Control		PLL Parameters		Droop Gain
	H	$k_{p\omega}$	$k_{i\omega}$	k_p	k_i	K_{dr}
ζ	↓	↑	–	↑	↓	↑
ω_n	↓	–	–	–	↑	↑
T_R	–	–	–	↑	↓	↑

where

$$K = \frac{Mk_i P_{step}}{K_\delta \left(M + K_{dr} D \right)} \text{ and } \varphi = \arctan\left(\frac{T_R \omega_d}{1 - T_R \zeta \omega_n} \right).$$

Equations (8.34) and (8.35) provide a way to predict the impacts of control parameters upon dynamic frequency performance. According to Eq. (8.34), the impacts of the key parameters on the damping ratio (ζ), oscillation frequency (ω_n), and time constant (T_R) are tabulated in Table 8.1. Let ↑, ↓, and – denote positive, negative, and poor associations, respectively. The results suggest that the achievement of the desirable ζ, ω_n, and T_R values is attributed to the parameter settings of the control pertaining to the electromechanical timescale. Therein, k_i is inversely proportional to T_R, and the increment of k_i would weaken the damping effect but result in a larger oscillation frequency. In contrast, k_p is directly proportional to ζ and has a positive association with T_R. Additionally, a larger droop gain K_{dr} represents an enhanced damping effect, in addition to the augmented ω_n and T_R. It is interesting to note that if the prerequisite $D>>1$ is false, the consistent results can also be derived, except for a more complicated quantification.

8.2.5 Simulation Study

For the simulations, the system topology in Figure 8.8 is conducted in Matlab/Simulink. B1 is open once the AC grid experiences a blackout, and then Type 3 WTs can be deployed to black start the local system and restore unserved loads. Type 3 WT base is aggregated by 1,000 Type 3 WTs with rated power at 1,500 MW (1,000 × 1.5 MW). Thereafter, two major simulations are performed:

1) Black-start validation: demonstrate the black-starting actions of Type 3 WTs by means of the proposed control strategies and sequence.
2) Stand-alone loaded validation: evaluate the performance of the fabricated *P-f* relationship and verify the results by analyzing frequency dynamics for different parameter settings in the stand-alone loaded condition.

8.2.5.1 Black-Starting Validation

In this segment, the wind speed is assumed to be constant at $10\,\text{m}\,\text{s}^{-1}$ and the additional dissipated resistor R_L is 0.5 pu. The black-starting actions are illustrated in Figure 8.13 with the adoption of the proposed control strategy and start-up sequence. In order to clearly describe Type 3 WTs' behaviors, operational time in each step is prolonged.

The output power, rotor speed, and current diagrams demonstrate that a smooth start-up of Type 3 WTs is procured. The start-up duration ends at $t = 120\,\text{s}$ and then follows the islanded operation. The satisfactory performances of the proposed RSC and GSC strategies are also verified in Figure 8.13, since the levels of the terminal AC voltage and DC voltage are not considerably affected throughout the start-up period.

8.2.5.2 Stand-Alone Loaded Validation

After the restoration of the Type 3 WTs, the frequency self-maintained control is triggered. R_L is shed gradually while local loads are picked up successively. The local loads comprise two parts, that is, the permanent Lump A ($670\,\text{MW} - j100\,\text{Mvar}$), and the interruptible Lump B ($50\,\text{MW}$) and Lump C ($50\,\text{Mvar}$).

Figure 8.14a–c illustrates the response of a Type 3 WT in an islanded system when the steady wind speed is $10\,\text{m}\,\text{s}^{-1}$. In this condition, the primary frequency control is enabled and wind turbines are operated in deloading mode. The 7.5% local active power load is shed at $t = 5$ seconds (breaking S_1) and reconnected at $t = 85$ seconds (closing S_1), respectively. Then 7.5% inductance reactive power load is added at $t = 165$ seconds (closing S_2). The wind speed

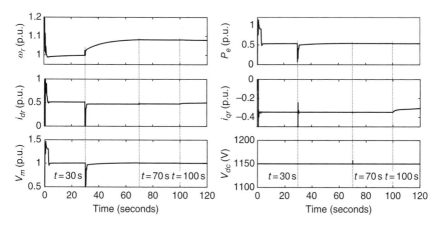

Figure 8.13 Black-starting of Type 3 WTs with sequential actions: (i) $t < 30$ s, starting up with a synchronous speed command; (ii) $t = 30$ s, controlled from MPPT [28]; (iii) $t = 70$ s, cutting off the DC battery; and (iv) $t = 100$ s, embedding AC voltage control and normal PLL.

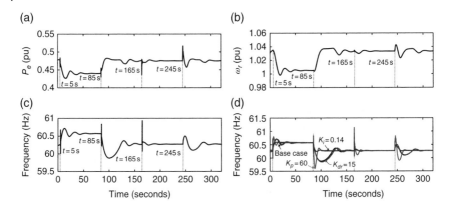

Figure 8.14 Type 3 WTs' response in stand-alone operation: (a) the exported power (p.u.); (b) the rotor speed (p.u.); (c) the system frequency (Hz); and (d) the system frequency response under different parameter settings (Hz).

rises from 10 to $11\,\mathrm{m\,s^{-1}}$ at $t = 245$ seconds. The simulation results indicate that the system frequency can be self-maintained via the autonomous frequency control.

The parameter adjustments of PLL (k_i is usually smaller to attain a noticeable inertia) and droop control impose distinct impacts on the islanded system frequency dynamics, as shown in Figure 8.14d. It reveals that, under a frequency decay event, the decrease of k_i attenuates the frequency overshoot but exaggerates its oscillation frequency, whereas a larger k_p setting provokes an extensive frequency overshoot caused by the action of a larger T_R. In addition, droop control secures the fundamental closed-loop stability for grid-connected operation, and a larger K_{dr} promotes a larger effective damping and T_R when confronted with power perturbations. Therefore, the sections of k_p, k_i, and K_{dr} should be deliberate to ensure the islanded system has a desirable performance.

This chapter proves that Type 3 WTs can be utilized as the alternative resources for system restoration both in the black-starting and stand-alone loaded stages. From the view of P-f and Q-V relationships, this chapter reveals that Type 3 WTs must feature a power-controlled voltage source throughout the build-up system restoration process, and furthermore a positive-feedback effect caused by passive loads is the key challenge for Type 3 WTs operating in stand-alone mode. In this regard, the main conclusions are summarized as follows.

1) The proposed black-starting control and sequence promote Type 3 WTs well applying in system restoration and smoothly self-building its terminal voltage. In addition, the proposed black-starting procedure is applicable for other PLL-based VSC generations (like photovoltaics) to sectionalize an islanded system with passive local loads.

2) Type 3 WTs are able to self-maintain the system frequency in the standalone loaded stage, by means of the supplemented frequency control. This is because the inherent *P-f* relationship of Type 3 WTs is linked via a controlled medium, which can be artificially devised by reconstructing the control system.

3) The impacts of control parameters on the frequency response of Type 3 WTs are mathematically evaluated via damping ratio, oscillation frequency, and zero-point value. The results contribute to control design and provide physical insight into the relationship between control parameters and the output frequency characteristics of Type 3 WTs.

8.3 Energy Storage in System Restoration

8.3.1 Pumped-Storage Hydro Units in Restoration

Energy storages are viable solutions to accommodate wind power uncertainty and variability due to their flexible characteristics. Among different storage technologies, pumped storage hydro (PSH) is the most mature and economical option for large-scale applications. PSH units can switch between pumping and generation modes to provide fast-response energy and reserve. Coordinating with PSH units, wind energy resources will not impede but rather expedite the restoration process. An optimal coordination will minimize wind power curtailment (or spillage) during the restoration process, and hence a faster recovery can be achieved.

Research on employing wind power in system restoration is mainly focused on the operation and control of wind turbines [8–11, 41–42]. In Zhu and Liu. [8], different aspects of power system restoration together with constraints of wind farm restoration are discussed. In Aktarujjaman [9], connecting a battery storage to the DC link capacitor of a doubly fed induction generator wind turbine can result in black-start functionality. An offshore wind park connected to the HVDC link provided additional voltage control capability [10]. The Firefly algorithm was used to find the optimal restoration plan with the aid of renewable sources [11]. The coordination of wind and PSH units has also been studied, but mostly under normal operation conditions, such as economic dispatch and unit commitment problems [41, 42]. The study on coordinating wind energy and PSH in emergency conditions is limited to date.

Wind generators differ from conventional generators in two aspects: (i) they are intermittent and weather dependent in nature, and (ii) their power can highly fluctuate and cause large ramping events. PSH units can be employed to compensate the wind power variability and uncertainty. The electricity absorbed or generated by PSH units needs to be coordinated to follow the wind power profile. At times of high wind power generation, instead of curtailing, wind power can be harnessed to pump water from a lower reservoir and store it in the upper reservoir. This usually occurs at the initial phase of restoration when available online generators and transmission paths are limited. As the restoration progresses, the energy stored in the upper reservoir can be unleashed so that PSH units either participate in the load pickup process or compensate wind ramping events.

The wind-PSH coordinated power system restoration problem is formulated as a two-stage adaptive robust optimization problem in Golshani et al. [43]. The structure of wind-PSH-assisted restoration is depicted in Figure 8.15. After a major contingency or blackout, the initial conditions of power grid are obtained to determine the extent of power outage. An optimal restoration plan is then devised with the aid of wind and PSH units. Wind uncertainty and variability as well as system reliability requirements are taken into account. It is assumed that wind turbines are operated similar to an NBSU. That is, wind turbines can participate in the restoration process after the establishment of transmission paths. However, unlike conventional NBSUs, they have the capability of fast starting without demanding cranking power, and having fast ramping rate capability.

A two-stage adaptive robust optimization technique is incorporated. The first-stage objective is to maximize total generation capacity, while the second-stage objective is to minimize the total unserved load. In the first stage,

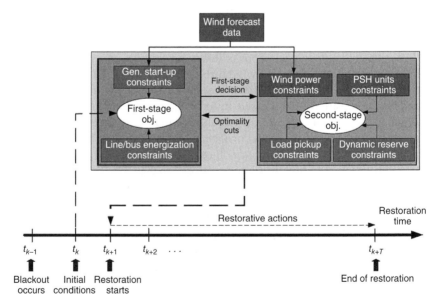

Figure 8.15 Structure of wind-PSH-assisted restoration process.

start-up sequences of BSU, NBSUs, and wind generators as well as energization times of transmission paths are determined as integer variables, given generation and transmission constraints. In the second stage, load pickup sequences, wind power dispatch levels, and operating modes of PSH units are determined, given wind power and PSH unit constraints as well as load pickup and dynamic reserve constraints. Note that the first-stage decisions are fixed when solving the second-stage problem. Once the worst-case realization of the uncertainty is found, the optimality cut is generated and fed back to adjust the first-stage decisions. The detailed mathematical formulation can be found in Golshani et al. [43], and the main problem structure is summarized hereafter.

8.3.1.1 Problem Formulation

The objective in the first-stage problem is to maximize the total generation capability minus $f(x)$, which is the total unserved load reflected in the second-stage problem. The adaptive second-stage objective is to minimize the total unserved load under the worst-case realization of uncertain parameters, which are related to wind energy.

First-stage constraints include the initial conditions, generator start-up function, and energization constraints. Second-stage constraints include power balance constraints, load pickup and dynamic reserve constraints, wind farm constraints, uncertainty set of wind power, and PSH constraints. For a robust optimization model, a polyhedral uncertainty set is defined containing wind power forecast along with the maximum forecast error [43]. As the uncertainty bound increases, the size of the uncertainty set enlarges and the resulting solutions become more conservative. Then the budget of uncertainty is defined to control the total deviation of wind farm power output from the forecasted value. Constraints related with PSH units are presented in detail as follows.

Generally, power output of a hydro unit is a nonlinear, nonconvex function of the turbine discharge rate and the net head [44]. Variations on the net water head are neglected, and it is assumed that each PHS unit has one water-to-power curve for the generation mode and one power-to-water curve for the pumping mode [42]. The difference between the two curves is defined as PSH efficiency. Their piecewise-linear approximations are obtained by using auxiliary binary variables similar to the method presented in Conejo et al. [44]. In Eq. (8.36), binary variables $I_{h,t}^g, I_{h,t}^p, I_{h,t}^l$ indicate that hydro unit h is in generation, pumping, or idling mode, and these modes are mutually exclusive in each restoration time t. Reservoir volume limits are shown in Eq. (8.37). Reservoir volume relationship with PSH discharge rate is shown in Eq. (8.38). In (8.39), the net power output of PSH unit h is computed. Eqs. (8.40)–(8.43) are applied to both pumping (p) and generation (g) modes, where $P_{h,t}$ shows the power output and $q_{h,t}$ is the water discharge rate of PSH unit h. The generation (g) and pumping (p) mode capacity limits are satisfied in Eqs. (8.40)–(8.43).

$$I_{h,t}^g + I_{h,t}^p + I_{h,t}^l \leq 1 \tag{8.36}$$

$$Vol^{\min} \leq Vol_t \leq Vol^{\max} \tag{8.37}$$

$$Vol_{t+1} = Vol_t - q_{h,t}\Delta T \tag{8.38}$$

$$P_{h,t} = P_{h,t}^g - P_{h,t}^p \tag{8.39}$$

$$q_h^{\min} u_{b_h,t} \leq q_{h,t}^{p(g)} \leq q_h^{\max} u_{b_h,t} \tag{8.40}$$

$$q_h^{\min,p(g)} u_{b_h,t} \leq q_{h,t}^{p(g)} \leq q_h^{\max,p(g)} u_{b_h,t} \tag{8.41}$$

$$q_h^{\min} I_{h,t}^{p(g)} \leq q_{h,t}^{p(g)} \leq q_{h,t}^{\max} I_{h,t}^{p(g)} \tag{8.42}$$

$$q_h^{\min.p(g)} I_{h,t}^{p(g)} \leq q_{h,t}^{p(g)} \leq q_{h,t}^{\max.p(g)} I_{h,t}^{p(g)} \tag{8.43}$$

8.3.1.2 Solution Algorithm

In the robust optimization, the probability function of uncertain parameters is unknown, and only an uncertainty set is defined based on historical data or system operators' experience. The robust optimization solution is immune against all realizations of uncertain parameters, including worst-case conditions. This approach was first introduced by Soyster [45] to handle parameter uncertainties in linear programming. As it provides the protection for all possible outcomes of the uncertain parameters, the solution could be overly conservative. To this end, several approaches to balance the robustness and total cost (in minimization problem) have been proposed [46]. Recently, more research works have focused on the adaptive robust optimization and their applications in management, business, and power systems [41, 47]. Adaptive robust optimization models are usually transformed to multilevel optimization problems, which brings great challenges to the solution algorithms. In addition to approximation algorithms, decomposition-based algorithms have been proposed, especially for two-stage adaptive robust optimization, for example, dual cutting planes [48] and column-and-constraint generation [49, 50].

In Golshani et al. [43], a decomposition algorithm, column-and-constraint generation (C&CG), is employed to solve this two-stage adaptive robust optimization problem with mixed-integer recourse. The nested C&CG is applied to decompose the original problem to a master problem with an outer-level C&CG algorithm and a MILP subproblem with an inner-level C&CG algorithm. The inner-level C&CG algorithm can solve subproblems and identify the worst-case realization of wind uncertainty, and the outer-level C&CG algorithm can solve the master problem so as to determine the optimal solution in the first stage, including the on time of generation units and energization time of transmission lines and buses.

8.3.1.3 Simulation Results

A modified IEEE 39-bus system is used for testing. The system contains 10 generators, a 600 MW wind farm located at bus 17, and two PSH units. The characteristics of PSH units are given in Table 8.2. Two cases, low and high wind power fluctuations, are considered to examine the effectiveness of wind-PSH coordination in the restoration process. In both cases, the base power is 100 MW and each restoration time step is 10 minutes (1 p.u.) for preparation and stabilization. The characteristics of buses, transmission lines, and loads are taken from IEEE [51]. The confidence interval of wind farm power output is set as 10% of the forecasted value. The proposed C&CG algorithm is implemented in C++ using ILOG CPLEX 12.6 and the Concert Library. All simulations were executed on a PC with Intel CoreTM i5 CPU @3.30 GHz and 8 GB RAM.

An illustrative example with perfect wind forecast data is first presented to investigate the wind spillage phenomenon. Energy storage is not considered in the example. Neglecting wind uncertainty, the optimization problem turns to a deterministic MILP problem. Table 8.3 summarizes generator online times. G10 is the BSU with a self-starting capability that comes online after

Table 8.2 PSH units' characteristics.

Parameter	Value	Parameter	Value
q_h^{min}	0.05 ($Hm^3 h^{-1}$)	$P_h^{min,p}$	20 (MW)
q_h^{max}	0.75 ($Hm^3 h^{-1}$)	$P_h^{max,p}$	250 (MW)
$P_h^{min,g}$	16 (MW)	Vol^{max}	10 (Hm^3)
$P_h^{max,g}$	180 (MW)	Vol^{min}	3 (Hm^3)

Table 8.3 Generators' optimal on-time for all cases.

Gen. No.	Bus No.	Time (p.u.)	Type Unit	Gen. No.	Bus No.	Time (p.u.)	Type unit
G1	31	12	CT	G6	36	14	CT
G2	32	13	ST	G7	37	9	CT
G3	33	13	ST	G8	38	15	ST
G4	34	14	ST	G9	39	14	ST
G5	35	14	ST	G10	30	2	CT

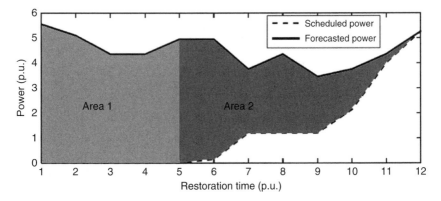

Figure 8.16 Forecasted and scheduled wind farm power output without PSH contribution and wind uncertainty.

restoration time $t = 2$. The forecasted and scheduled wind power is depicted for the early stages of restoration in Figure 8.16. Note that before restoration time $t = 5$ the wind farm generates zero power. Subsequently, wind power increases slightly, with a large portion curtailed. Finally, at $t = 12$ all wind power is utilized in the restoration process. The total energy served is 11.70 GWh and total wind energy spillage is 631.75 MWh.

The shaded area between two curves can be divided into two regions: Region 1 represents the amount of wind energy that is spilled due to the unavailability of transmission path, and Region 2 represents the amount of wind energy that is curtained to maintain power system reliability. Wind energy spillage in area 2 can be reduced using energy storage.

8.3.1.3.1 *Restoration with/without PSH Contribution*

Now let us consider wind-PSH coordination in restoration. Assume that the two PSH units are installed at bus 17 where the wind farm is located. Figure 8.17 compares the scheduled wind power with/without PSH units' contributions. Note that wind uncertainty is not yet included, that is, $\Delta_w = 0$. It can be seen that at restoration time $t = 6$ the scheduled wind power reaches its forecasts. The total wind energy spillage reduces to 358.75 MWh, and the total energy severed increases to 12.33 GWh. Among the increased wind power, a small portion is used to serve loads and provide cranking power to other NBSUs, while the larger portion is used for pumping water to fill the upper reservoir. This helps to boost the total dynamic reserve level when a limited number of units are on. As the restoration process moves forward, more wind power can be accommodated into the system so that more capacities on NBSUs are released, which improves the total load pickup capability of the system at later stages. Figure 8.18 compares the load restoration curves with/without PSH unit contribution.

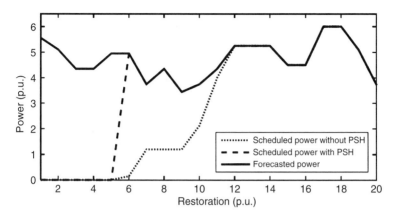

Figure 8.17 Scheduled and forecasted wind farm power output with/without PSH units' contributions.

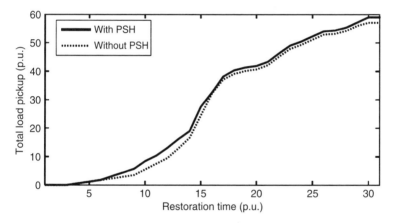

Figure 8.18 Total load pickup curves with and without PSH units' contributions.

8.3.1.3.2 *PSH Unit Location*

The impact of PSH location is studied here. It is assumed that the PSH units are installed at bus 8 (far from the wind farm location). Note that in this scenario, although the wind farm comes online after restoration time $t = 6$, the PSH units cannot store wind farm energy until $t = 9$. This is due to the fact that bus 8 is only energized after $t = 9$, thus imposing limitation on the operation of PSH units. Figure 8.19 compares the reservoir volume for different PSH locations. It can be seen that the pumping mode operation is postponed due to the unavailability of transmission path from $t = 6$ to $t = 9$. In this case, the total served energy is 12.163 GWh and total wind energy spillage is 542.74 MWh.

8.3.1.3.3 *Wind Power Uncertainty*

To model wind randomness in power system restoration, it is assumed that uncertain wind farm power lies within the uncertainty set. The total restoration horizon in this study is set to 30 p.u.; thus, the budget of uncertainty can vary within the interval (0, 30). From Figure 8.20, one can observe that when the budget of uncertainty increases, total energy served will decrease in both scenarios. It also shows that with wind-PSH coordination, the total served energy is always greater than that without PSH units. In particular, utilizing PSH units under the worst-case conditions ($\Delta_w = 30$) outperforms the best case condition without PSH units, where $\Delta_w = 0$. Therefore, PSH units not only compensate the wind uncertainty, they also facilitate load pickup.

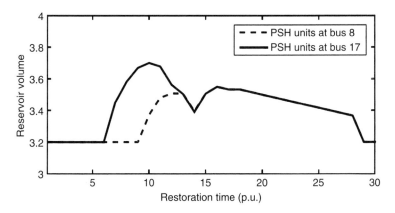

Figure 8.19 Reservoir volume (Hm^3) for different PSH locations.

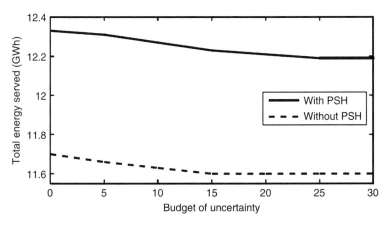

Figure 8.20 Impact of budget of uncertainty on total energy served with and without PSH units' contribution.

8.3.1.3.4 Wind Power Fluctuation

Now assume that wind power presents a large ramping behavior during the early stages of restoration between $t = 5$ and $t = 15$. Figures 8.21 and 8.22 show that the PSH units are capable of handling large wind power fluctuations. Without PSH contribution, the wind power is curtailed to avoid load shedding and meet dynamic reserve requirement. However, coordinating with the PSH units, all wind energy can be harnessed despite the large fluctuations. The uncertainty study in Figure 8.22 also demonstrates the

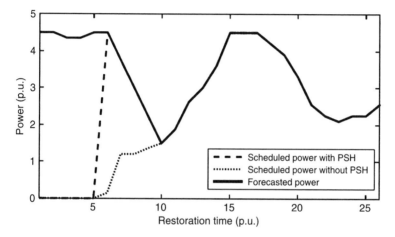

Figure 8.21 Scheduled and forecasted wind farm power with and without PSH units' contribution in large fluctuation case.

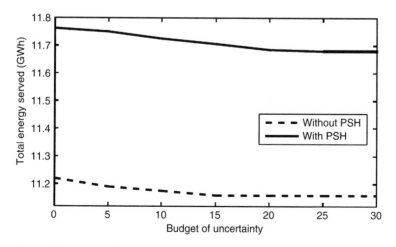

Figure 8.22 Impact of budget of uncertainty on total energy served.

effectiveness of PSH units in the case of high wind fluctuations. The PSH coordination consistently outperforms that without PSH units. The total served energy without PSH contribution is 11.22 GWh, whereas with PSH contribution it increases to 11.77 GWh; and total wind energy spillage reduces from 544.65 to 332.50 MWh. It is worth noting that under wind power fluctuations and by utilizing PSH units, the total energy served becomes greater than the case where the wind power presents low fluctuations without utilizing PSH.

8.3.2 Batteries for System Restoration

Frequency deviation caused by load pickup in each restoration stage is a major concern for system operators. Therefore, loads need to be restored in small blocks to avoid the violation of prime movers' constraints [52]. Generally, hydro (HY), combustion turbine (CT), and steam turbine (ST) units can only pick up the load with 15%, 10%, and 5% of each unit's capacity, respectively [17]. These numbers represent the maximum load each generator can pick up with acceptable frequency dip due to the sudden change of load, defined as frequency response rates (FRRs). PJM uses these FRRs to maintain the frequency within 60 ± 0.5 Hz [6].

There are several efforts to address the frequency issue in system restoration. An analytical model was developed in Adibi et al. [17] to calculate the FRR of different prime movers to the sudden change of load. In Yari et al. [53], authors used FRRs to determine the best sequence of load pickup in an interconnected system of hydro and steam units. It was recommended to pick up load first by HY units then ST units to maintain the maximum frequency dip within limits. A wide-area measurement system was applied in Yari et al. [54] to determine a suitable amount of load to pick up or generation to increase by predicting the progression of frequency stability. Considering different issues related to load pickup, an average system frequency model was developed in Hanbing and Yutian [55] to determine the amount of load that can be safely picked up at a substation.

To overcome the deficiency or surplus of generation during load pickup, a battery storage system (BSS) can be used to compensate the imbalance and expedite the load pickup process. Currently, batteries are used to establish a communication system or serve critical load in the early stage of system restoration [56]. FERC order 784 [57] is designed to promote the utilization of batteries as fast-response storage units for frequency regulation. It also opens the opportunity to use high-capacity BSSs to accelerate the restoration process. The largest tested battery system is a 34 MW NaS battery system installed in Rokkasho village, Aomori, Japan. AES has installed a 12 MW Li-ion battery system for frequency regulation [58]. Therefore, the utility-scale BSS can serve as either load or generation in system restoration process.

In Kadel et al. [59], a battery-dispatching strategy is developed for faster and more reliable load pickup in power system restoration. During different stages of system restoration, batteries can be used as either load or generation. Initially, batteries can be operated in the discharging mode as BS units to crank NBS units or pick up critical load. When generators begin to ramp up, loads need to be picked up to maintain system frequency. If there is more generation than the maximum amount of load that can be restored, batteries can be operated in charging mode. If there is not enough generation to restore load, rather than waiting for the generation ramping up, batteries can be operated in the discharging mode as a power source. This process will continue until the whole system is restored back to normal operating conditions. By considering the FRR of different types of generators and state of charge (SoC) constraints of batteries, the charging/discharging sequence and time has been determined for BSS. The detailed mathematical formulation can be found in Kadel et al. [59], and the key structure is summarized in the following sections.

8.3.2.1 Problem Formulation

The objective is to maximize the total restored energy during the restoration period. Constraints include power balance, generator output function, generator frequency response rate, and state of charge of batteries. The first two groups of constraints are similar to models developed in previous sections, and the last two groups are new constraints and will be presented hereafter.

8.3.2.1.1 *Generator Frequency Response Rate*

In [17], Adibi et al. have developed FRRs for typical prime movers (CT, ST, and HY) that are used in the initial phase of power system restoration. It has been simulated for a percentage range of generators' loads L and for percentage range of sudden load pickups ΔL. It has been shown that the prime movers' frequency dip ΔF is (i) independent of the generators' loading L but depends on their types and sizes (capacities) and (ii) proportional to the size of the sudden load pickup ΔL. These two attributes develop the FRRs $\Delta F/\Delta L$ that is a constant number (index) for a given type of prime movers to be used in determining the allowable sudden load pickup and to determine the effective reserve distribution. In this model, for 0.5 Hz maximum frequency dip, FRRs of CT, ST, and HY units are 12%, 5%, and 13%, respectively. Then at each time of picking up load, the total amount of restored load should be smaller or equal to the summation of maximum allowable load pickup for all generators, as given by

$$\sum_{j\in\Omega_L} u_{jt_{kstart}}\left(1-u_{jt_{kstart}-1}\right)P_{Lj} \leq \sum_{i\in\Omega_G}\left(\text{FRR}_i \times P_{Capi}\right), \quad \forall k\in\Omega_L, \tag{8.44}$$

where, P_{Capi} is the generation capacity of generator i. In Eq. (8.44), only load to be picked up at time t_{kstart} will be included.

8.3.2.1.2 *State of Charge of Batteries*

In order to operate batteries at high efficiency and also maintain the cycle life of batteries, battery SoC should be within certain limits, as given by

$$SoC_{min} \leq SoC_i(t) \leq SoC_{max}, \quad \forall i \in \Omega_B,$$ (8.45)

where, SoC_{min} and SoC_{max} are the lower and upper limits of SoC. SoCs of all units are maintained between 20% and 80% [60]. The relationship between functions of SoC and the battery output can be achieved as follows.

$$SoC_i(t) = \begin{cases} SoC_i(t-1) + \eta_c \times P_{Bi}(t-1)\Delta t, & \text{charging} \\ SoC_i(t-1) - (1/\eta_d) \times P_{Bi}(t-1)\Delta t, & \text{discharging} \end{cases},$$ (8.46)

where, η_c and η_d are the charging and discharging efficiency.

8.3.2.2 Solution Algorithm

A rule-based search algorithm is proposed in Kadel et al. [59] to solve this battery-assisted restoration problem. There are two stages of battery application in system restoration: crank NBS units and pickup load. If there is surplus generation, the battery will be charged to store the energy; otherwise, the battery will be discharged to provide the cranking power to start NBS units. If there is not enough generation, the battery will be discharged to pick up load; otherwise, the battery will be charged to store the energy. The SoC constraint of batteries will always be checked. The process will continue until all loads are restored.

8.3.2.3 Simulation Results

Three case studies have been performed for illustration of the proposed model and solution methodology. Two BS units and one NBS unit are used to pick up loads. The generator characteristics are shown in Table 8.4 [61]. Different load profiles are used in three cases. Batteries from 100 kWh to 6.78 MWh are used to compare the contribution of different sizes of batteries in load pickup. Load

Table 8.4 Generator characteristics.

Unit	Type	MW Cap. (MW)	Ramp Rate (MW/h)	Start-Up Req. (MW)
Chester_4–6	CT	39	120	N/A
Conowingo_1–11	Hydro	385	384	N/A
Schuykill_1	Steam	135	135	2.7

pickup actions are performed every 10 minutes (1 p.u. time), and batteries have 100% efficiency in a fully charged state in the beginning.

It is assumed that batteries can respond to a control signal without any delay. Practically, batteries are connected through the control and management system to the grid. It is important to consider the dynamic performance of BSS. In the context of batteries' application in load pickup, the FRR of BSS will greatly impact the maximum load that can be picked up without violating the frequency limit. In the absence of battery dynamics associated with converters, inverters, and transformer equipment, several reasonable $\Delta F/\Delta L$ have been assumed in this discussion.

The test system has three types of generating units, CT, ST, and HY. Generator parameters can be found in Adibi et al. [17]. The FRRs of three types of prime movers are calculated in the same study [17]. Different FRRs are assumed for BSS. In Case 1, there is no battery, as the base case. In Case 2, one 27 MW battery with 40% FRR is added to the system. In Case 3, the FRR of the battery is assumed to be 100%, which is approaching the isochronal operation. In Case 4, the battery is assumed to pick as much load as its capacity. A load increase of 50 MW happens to the system and there is no adjustment in the governor's speed change position. The characteristics of generators and batteries are shown in Table 8.5.

The calculation of different FRRs of three batteries is shown as follows.

$$\text{Battery I}: \Delta F/\Delta L = -2.50\,\text{Hz/p.u.} \quad \Delta L/\Delta F = -40\%/\text{Hz} \left(-40\right)$$

$$\text{Battery II}: \Delta F/\Delta L = -1.00\,\text{Hz/p.u.} \quad \Delta L/\Delta F = -100\%/\text{Hz} \left(-100\right)$$

$$\text{Battery III}: \Delta F/\Delta L = -0.00\,\text{Hz/p.u.} \quad \Delta L/\Delta F = \text{large}\%/\text{Hz} \left(\text{V large}\right)$$

The comparison of FRRs of ST, CT, HY, and three batteries is shown in Figure 8.23.

Table 8.5 Generator and battery characteristics.

Type		No. of Unit	Unit Rate (MW)	Response Rate (% Hz^{-1})	Response Rate (MW Hz^{-1})
ST		2	135	−12.6	−34.0
CT		3	16	−21.7	−13.9
HY		4	32	−29.4	−28.2
Battery	Case 1	N/A	N/A	N/A	N/A
	Case 2	1	27	−40.0	−10.8
	Case 3	1	27	−100.0	−27.0
	Case 4	1	27	N/A	N/A

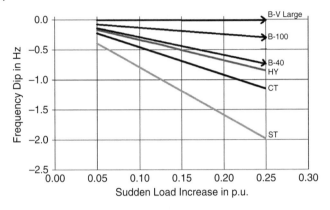

Figure 8.23 FRRs of ST, CT, HY, and three batteries.

Table 8.6 Load pickup and frequency dip.

Case	Load Pickup (MW)				Max Freq. Dip (Hz)
	ST	CT	HY	Battery	
I	22.34	9.13	18.53	0.00	0.66
II	19.55	7.99	16.22	6.21	0.58
III	16.49	7.74	13.68	13.10	0.49
IV	10.20	4.17	8.46	27.00	0.30

Each unit and battery response to the sudden change of 50 MW load based on their FRRs is recorded. The results are shown in Table 8.6. System maximum frequency dip can be calculated in the following equation:

$$Max.\ Freq.\ Dip(Hz) = Total\ load\ (MW) \big/ Total\ FRR(MW/Hz).$$

For example, in Case I, system FRR is $76.1\ MW\,Hz^{-1}$ ($= 34 + 13.9 + 28.2$), and Max. Freq. Dip is 0.66 Hz ($= 50/76.1$).

It can be observed that batteries can help decrease maximum frequency dip. The higher battery FRR, the more benefit to system frequency stability. Therefore, a battery with high FRR can help pick up more loads with the same amount of maximum frequency dip. Also, without considering battery FRR, the system will pick up more loads, which may cause frequency instability. Therefore, the accurate value of battery FRR is critical to maximize the benefit of batteries and maintain system frequency in load pickup.

8.3.2.3.1 *Case 1 – One BS Unit*

One BS unit (CT) is used to serve 11 loads, and each load is 3 MW with equal priority. The ramp rate of CT is 20 MW pu^{-1}, and in each step, the maximum load that CT can pick up is 4.68 MW using 12% FRR. Therefore, the surplus generation can charge the battery. Considering the SoC constraints, the charging/discharging schedule of the battery will be determined. Three scenarios are analyzed: no battery, one 100 kWh battery, and one 450 kWh battery. The comparison of load pickup curves is shown in Figure 8.24.

In the case of no battery, a 3 MW load is picked up at each step, and the load pickup curve is shown as the curve with stars. If using one 100 kWh battery, the load pickup curve is shown as the curve with square. At each unit of time, the surplus power between generation and load is 1.68 MW. The load pickup action is performed every 10 minutes, and the total surplus energy in that time duration is 280 kWh. This is larger than the battery energy capacity of 100 kWh. Therefore, this battery cannot be charged and will not contribute to the load pickup. If using one 450 kWh battery, the battery can contribute to the load pickup, shown as the curve with circle. The restoration time is decreased and total restored energy is increased. The SoC of this 450 kWh battery is shown in Figure 8.25. Using a 450 kWh battery increased the total restored energy (MWp.u.) by 13.13% compared with no battery or a 100 kWh battery. Based on this case study, it was shown that the contribution of a battery in load pickup depends on the size of the battery, generator ramp rate and FRR, load size, and the time duration between restoration actions.

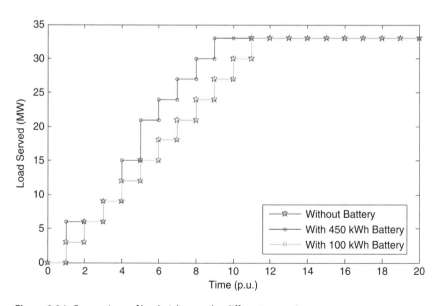

Figure 8.24 Comparison of load pickup under different scenarios.

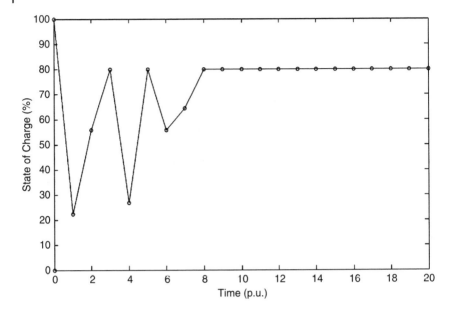

Figure 8.25 SoC of the 450 kWh battery.

8.3.2.3.2 Case 2 – One BS Unit and One NBS Unit

One BS unit (CT), one NBS unit (ST), and one 2.7 MWh battery are used to serve part of the load profile in [57]. After starting up the BS unit, CT provides 2.7 MW to crank the NBS units. Considering 12% FRR of CT and 5% FRR of ST, the maximum load that two units together can pick up is 11.43 MW. Two scenarios are analyzed: no battery and one 2.7 MWh battery. The comparison of load pickup curves is shown in Figure 8.26.

In the case of no battery, the load is picked up according to their priority level, shown as the curve with square. When using the 2.7 MWh battery, the battery can contribute to decrease total restoration time and increase total restored energy, shown as the curve with circle. The SoC of the 2.7 MWh battery is shown in Figure 8.27. It can be observed that during the first five time intervals, the battery is operated in the discharging mode to pick up more load than the case without a battery, and the SoC is reduced to the lower limit of 20%.

8.3.2.3.3 Case 3 – Two BS Units and One NBS Unit

Two BS units (CT and HY), one NBS unit (ST), and two different batteries of 3.33 and 6.78 MWh were tested to serve the load. The load profile in Perez-Guerrero et al. [62] was doubled to match the total generation capability. After starting up the BS unit, CT and hydro units provided 2.7 MW to crank the NBS units. Considering 12% FRR of CT, 13% FRR of HY, and 5% FRR of ST, the maximum load that three units together could serve was 61.48 MW. Three scenarios

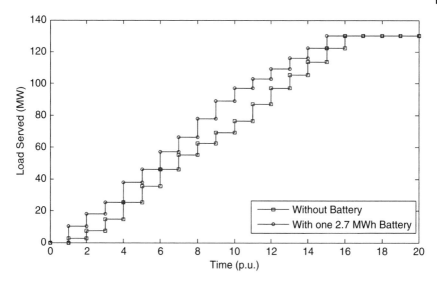

Figure 8.26 Comparison of load pickup without battery and with one 2.7 MWh battery.

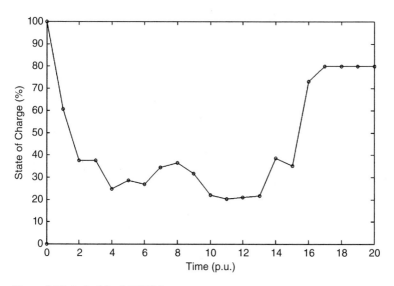

Figure 8.27 SoC of the 2.7 MWh battery.

were analyzed: no battery, one 3.33 MWh battery, and one 6.78 MWh battery. The comparison of load pickup curves is shown in Figure 8.28. In the case of no battery, the loads were served according to their priority level, shown as the curve with square. Different from the previous two cases, some loads could not be restored due to the deficiency of available generation to pick up the smallest size of

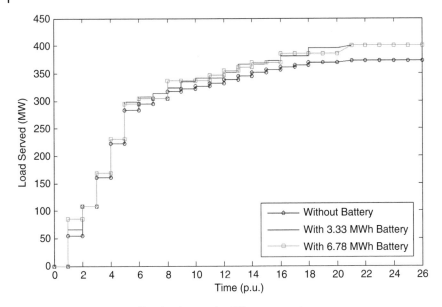

Figure 8.28 Comparison of load pickup under different scenarios.

remaining load, during the time period t_{22}–t_{25}. The total restored load was 375.6 MW. If a 3.33 or 6.78 MWh battery was used, a total of 400.6 MW load was served, as shown by the black line and squares, respectively. Comparing these two scenarios, more loads were picked up in the early stage using the 6.78 MWh battery, which brings more total restored energy. The restored energy (MW p.u.) increases by almost 5% for scenarios with batteries compared to the scenario without battery contribution. The load restoration time was reduced to 19 p.u. with the battery contribution. It can be observed that more energy can be restored and then used to pick up more load using a 6.78 MWh battery. The SOC of batteries in two scenarios are shown in Figure 8.29. The state of charge of the batteries used is limited to 20–80% at any time interval.

The summary of different scenarios in three cases is shown in Table 8.7. It can be observed that batteries can support expedition of the load restoration process. In Case 1, a smaller size battery does not contribute to load restoration. In Case 2, a battery can help reduce restoration time and increase total restored energy. In Case 3, a larger battery can contribute to restoring more energy, but the restoration time does not change.

8.3.3 Electric Vehicles in System Restoration

A plug-in hybrid electric vehicle (PHEV), also known as a plug-in hybrid, is a hybrid electric vehicle with rechargeable batteries that can be fully charged by

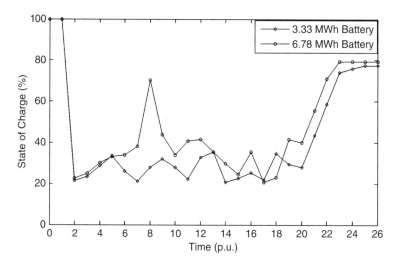

Figure 8.29 SOC of the 3.33 and 6.8 MWh battery.

Table 8.7 Comparison of load restoration with and without batteries.

Cases		Total Restoration Time (p.u.)	Total Restored Energy (MW p.u.)	% Increase in Restored Energy
Case 1	Without battery	11	297	0.00
	With 100 kWh battery	11	297	0.00
	With 450 kWh battery	9	336	13.13
Case 2	Without battery	16	1,441.9	0.00
	With 2.7 MWh battery	15	1,627.4	12.86
Case 3	Without battery	21	5,955.6	0.00
	With 3.33 MWh battery	19	6,248.9	4.92
	With 6.78 MWh battery	19	6,264.2	5.18

connecting to an external electric power source. Technically, PHEVs are defined as any vehicle containing a battery storage system of 4 kWh or greater capacity, with the intent of using the stored energy for the operation of a vehicle for a minimum distance of 10 miles in all electric models [63]. They share the characteristics of a conventional electric vehicle and a combustion engine vehicle. The PHEV uses a gasoline engine and an electric motor for propulsion with a larger battery pack that can be recharged, allowing operation in the all-electric mode until the battery is depleted. PHEVs have lower operating and maintenance costs compared to vehicles with internal combustion engines while they are operating in all-electric mode. PHEVs eliminate the problem of range anxiety for customers because the internal combustion engine starts to work after the battery state of charge is reduced to a certain level.

Numerous works have been developed to illustrate the significance of PHEV aggregation on power grids to maintain system reliability and capacity. Kempton et al. in 2005 [64] analyzed business models and strategies to utilize electric vehicles for grid stabilization and other purposes. Pang et al. in 2012 [65] discussed the potential benefits of PHEVs as a dynamically configurable energy source in power systems. They presented the PHEVs as the mobile energy storage capable of behaving as either loads or generation sources depending on the power system requirement. PHEVs were presented as a solution to reduce the demand side management and outage management during peak load and outage conditions. Zigkiri et al. in 2013 [66] utilized PHEVs in the distribution system to support cold load pickup during PSR. The high current requirement during the restoration of cold load was compensated through the stored energy in PHEVs. This study only demonstrated the application of PHEVs during discharging of the storage. Yuchao et al. in 2013 [67] analyzed the integration of an electric vehicle storage system with a power system model. They found that the integration made a minor contribution in the distribution system losses and voltage regulation. Yilmaz et al. in 2013 presented a review of the current status and the impact of PHEV implementation in distribution systems [68]. The authors concluded that if a proper charging/discharging scheme is applied with adequate power electronics like bidirectional chargers and inverters to interact between the grid and vehicles, PHEVs could be utilized as the storage source for the grid with high benefits. Chunhua et al. in 2013 investigated the opportunities and challenges associated with the integration of PHEVs into the existing grid. They discussed three frameworks for electric vehicles' integration into the grid: vehicle-to-grid, vehicle-to-home, and vehicle-to-vehicle. A genetic algorithm was used to determine the optimal size of the aggregation to maintain the power quality [69].

According to the International Energy Agency, the penetration of electric vehicles was 0.02% of total vehicle fleet in 2012, and it is predicted that the penetration level will be 2% by 2020 [70]. The increasing number of PHEVs also brings the capability of flexible generation and distributed energy storage to distribution systems. Most PHEVs are parked at home or in parking lots/ garages 95% of the time [71]. The parking lot or garage can act as an aggregator to collect the SoC and other parameters of PHEV batteries, communicate with upper-level operators, and decide the charging or discharging mode of each PHEV. A control system or PHEV aggregator acts as an interface between PHEVs, the electricity market, and the grid. It has the most significant task of generating proper control signals to maintain grid stability through a proper communication channel. The PHEV aggregator collects data from each individual vehicle in the aggregation. State of charge, driven distance, health of battery, and current charging/discharging limits are the important data fed into the aggregator. The aggregator decides the mode of operation of the battery energy storage within each vehicle.

In a distribution network, PHEVs offer a promising solution for bottom-up restoration. PHEVs can be utilized as fast-response energy resources to buffer energy for grid stabilization and speed up distribution system recovery. During system restoration, PHEVs can be utilized as either a generation source or load. PHEVs parked in a garage or parking lot can be used to supply the feeder containing prioritized loads immediately after an outage or blackout. In the case of distributed generation connected to different feeders, PHEVs can also serve as BS sources to crank NBS generators. Considering generation availability and different feeder characteristics, the charging/discharging sequence of battery storage system in PHEVs can be determined. If the available generation exceeds the maximum amount of load that can be restored, PHEVs can be operated in the charging mode. Otherwise, if there is not enough generation to restore the feeder load, instead of waiting for extra generation from online units, PHEVs can be operated in the discharging mode. Therefore, the aggregated PHEVs can be regarded as one generation or load in a distribution system. PHEVs' contribution to distribution system restoration is investigated in [72], and key models and observations are summarized in the following discussion.

As shown in Figure 8.30, the example system consists of several feeders with loads and aggregated PHEVs connected to each feeder. To overcome the deficiency or surplus of generation during load pickup, the energy available in

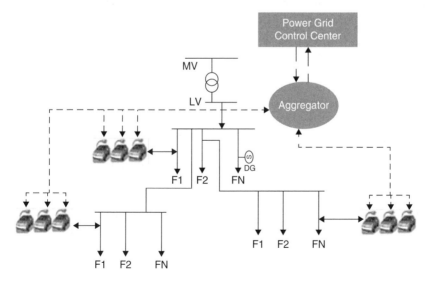

Figure 8.30 Illustration of connections between vehicles, aggregator, and upper-level operators.

PHEVs can be used to compensate for the imbalance between generation and load, which expedites the load pickup process. Initially, PHEVs were assumed to be fully charged, and operated in the discharging mode to serve the load with the highest priority in distribution network. Then, considering generation availability and different feeder characteristics, the charging/discharging sequence of battery storage system in PHEVs can be determined. If the available generation exceeds the maximum amount of load that can be restored, PHEVs can be operated in the charging mode. Otherwise, if there is not enough generation to restore the feeder load, instead of waiting for extra generation from online units, PHEVs can be operated in the discharging mode.

8.3.3.1 Problem Formulation

PHEVs' mobility data can be derived from the National Transportation Survey, which has been conducted in various countries. Having determined PHEVs' mobility, the required probability data to build a Markov chain transition matrix can be calculated. The Markov chain model is employed to generate a synthesized driving cycle. A Monte Carlo-type simulation method is used to estimate the expected mean value of PHEV parameters at each specific time of day. The result of the Monte Carlo simulation is then used to predict the impact of PHEVs on the distribution system restoration.

First, the total capacity of the battery in MWh is the sum of the individual battery capacity. Vehicle usage and driving patterns determine the state of charge of the battery available in the PHEV. This work does not consider the impact of driver behavior and assumes all the participating PHEVs are at an 80% charged state in the beginning of the restoration process. Based on the information from individual PHEVs, the aggregator is assumed to make the decisions for the whole aggregation of the PHEV vehicle fleet. To determine the number of PHEVs, the analysis is given in the following:

8.3.3.1.1 *Determining the Number of Vehicles*

Based on data from the US Energy Information Administration, the average consumption of a US residential utility customer was 903 kWh per month [66]. Then, the daily demand of each household is approximately 30 kWh. It is assumed that each household has two vehicles. Given the total demand of each test system, the total number of vehicles can be obtained by

$$\text{Total \# of vehicles} = \left(\text{Total Daily Demand}\right)\Big/\left(30\,\text{kWh}\right) * 2. \quad (8.47)$$

Using PHEV existence level 0.02% in 2012 and 2% in 2020, the total number of PHEVs is

$$\text{Total \# of PHEVs} = \left(\text{Total \# of vehicles}\right) * \left(\text{PHEV penetration\%}\right). \quad (8.48)$$

8.3.3.1.2 *Determining the Capacity of Aggregated PHEVs*

In each PHEV, the capacity of battery is 15 kWh. Considering different driving scenarios, it is assumed that each PHEV has 10 kWh energy capacity of battery available during system restoration, with full efficiency and an 80% charged state in the beginning.

$$\text{Capacity of PHEVs} = (\text{Total \# of PHEVs}) * (10\text{kWh}) \tag{8.49}$$

Second, a Markov chain model is an effective way to generate representative driving pattern in a statistical way and has been applied in studies [73, 74]. The Markov property says that when time proceeds from t to $t+1$, the next state of the process depends only on the present state and not on the previous states. A discrete-time Markov chain is a sequence of random variables X_1, X_2, X_3, ... with the Markov property that can be expressed as

$$\mathbb{P}(X_{t+1} = k | X_0 = x_0, \dots, X_t = x_t) = \mathbb{P}(X_{t+1} = k | X_t = x_t)$$
$$\forall t \geq 0, \{k, x_0, \dots, x_t\} \in S. \tag{8.50}$$

It is assumed that each PHEV can be in three different parking states: parked in an industrial area (P_I), parked in a commercial area (P_C), or parked in a residential area (P_R). We also defined three driving states D_A, D_B, and D_C representing different types of trips performed by PHEVs between the different parking locations. Therefore, our model comprises three parking and driving states, as shown in Figure 8.31. Let us assume that vehicle m occupied parking

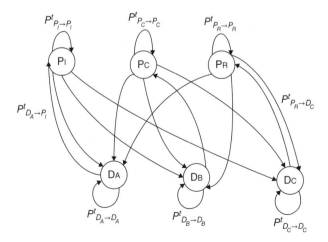

Figure 8.31 Discrete-time Markov chain states representation.

state P_I at time step t. The probability of vehicle m remaining at the same state at time $t+1$ is $P^t_{P_I \to P_I}$. Whereas the probability of transition from parking state I to driving states D_A is $P^t_{P_I \to D_A}$. Here it is assumed that a vehicle in driving state D_A can only park in parking state P_I, due to the type of trip. We made the same assumption for vehicles in driving states D_b and D_c: they can only park in parking states P_C and P_R, respectively. Since the transition probabilities depend on t, the process is known as an inhomogeneous Markov chain. For a given time t, this leads to the formation of a transition probability matrix T^t in the following form:

$$
T^t = \begin{bmatrix}
P^t_{P_I \to P_I} & 0 & 0 & P^t_{P_I \to D_A} & P^t_{P_I \to D_B} & P^t_{P_I \to D_C} \\
0 & P^t_{P_c \to P_c} & 0 & P^t_{P_c \to D_A} & P^t_{P_c \to D_B} & P^t_{P_c \to D_C} \\
0 & 0 & P^t_{P_R \to P_R} & P^t_{P_R \to D_A} & P^t_{P_R \to D_B} & P^t_{P_R \to D_C} \\
P^t_{D_A \to P_I} & 0 & 0 & P^t_{D_A \to D_A} & 0 & 0 \\
0 & P^t_{D_B \to P_c} & 0 & 0 & P^t_{D_B \to D_B} & 0 \\
0 & 0 & P^t_{D_C \to P_R} & 0 & 0 & P^t_{D_C \to D_C}
\end{bmatrix}. \tag{8.51}
$$

Note that changing from one driving state to another requires that the PHEV first occupy a parking state. Thus, several elements of the transition matrix are zero. The sum of the probabilities in each row of the transition matrix (8.51) equals one. The Markov chain transition matrix is cyclostationary and can be calculated for a period of one week. Having defined the initial state probabilities and state transition matrix, the Markov chain model can be fully characterized and will be periodically repeated for weekly cycles.

The detailed mathematical formulation can be referred to in [72]. The objective is to maximize the total restored energy in the loads and PHEV batteries during the restoration period. Constraints include real and reactive power balance, SoC of PHEV batteries, and feeder load status.

8.3.3.2 Simulation Results

8.3.3.2.1 Impact of PHEV Penetration Levels on System Restoration

Aggregation of PHEVs was modeled as one single energy storage source for simulation purposes. Four scenarios of projected PHEV penetration, no PHEV, 0.02% in 2012, 2% in 2020, and 0.2% in between were considered to examine the impact of different PHEV penetration levels on distribution restoration. The predicted number of PHEVs for each scenario for the test cases is summarized in Table 8.8.

To compare the contribution in the restoration process, different numbers of PHEVs were chosen in each system. In the base scenario, there is no PHEV. Three scenarios of projected PHEV percentages, 0.02% in 2012, 2% in 2020, and 0.2% in between, are considered to examine the impact of different PHEV

Table 8.8 Number of total load, houses, vehicles, and PHEVs.

System	Total Load (MW)	Total House #	Total Vehicle #	PHEV # in 2012	PHEV # in 2020
4-feeder	22.8	18 240	36 480	7.3	730
100-feeder	679	543 200	1 086 400	217.3	21 728

Table 8.9 Number of PHEVs in different scenarios.

System	Base Scenario	0.02% Penetration	0.2% Penetration	2% Penetration
4- feeder	0	7	73	730
100-feeder	0	217	2173	21 730

penetration levels on distribution restoration. These five scenarios are summarized in Table 8.9.

The generation and load profile in each test system can be found in [75]. Load pickup actions were performed every 10 minutes (1 p.u. time). It was assumed a total blackout happened to the system. Two case studies are considered, 4-feeder and 100-feeder systems.

PHEV owners' daily driving patterns need to be determined by using the Markov chain method. Driving is assumed to be related to three different velocities sampled from normal distributions with 20% standard deviation of mean 50 (km h^{-1}). Maximum and minimum battery size are assumed to be uniformly distributed in the intervals (16, 20) and (4, 6) kWh, respectively. Fast, medium, and slow charging power for three stations P_I, P_C, and P_R are set to 5.5, 3.7, and 2.3 kW, respectively. Initial SoC is assumed to be 50% of maximum battery size. The lower bound of SoC is 20%, and the upper bound of SoC is 90%. Travel behavior data were extracted from [76], by which the state transition matrix (2) can be calculated.

Figure 8.32 shows the transition states for three vehicles starting their trips from parking states P_I, P_C, and P_R. One can observe that a sample vehicle occupying parking state P_I starts driving state D_A at 6:00 a.m. before returning to its initial state (P_I) at 6:30 a.m. Whereas another vehicle occupying initial state P_C will leave its current state at 8:00 a.m. by starting type of trip D_A. One thousand random samples were generated to model the driving behavior of vehicles in different states by using state transition matrix (2). The Monte Carlo simulation method is employed to calculate the expected mean value of PHEV parameters at time t. Figure 8.33 shows the expected mean value of SoC at different times of day for two vehicles in parking state P_R and driving state D_C. In Figure 8.34,

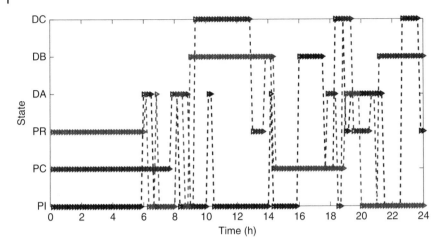

Figure 8.32 Synthetic driving patterns for a sample vehicle in various parking states.

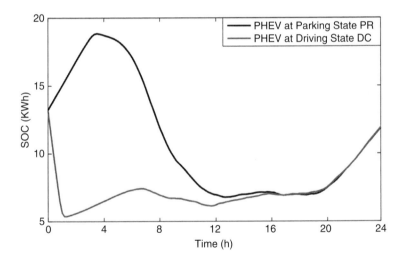

Figure 8.33 Expected mean value of SoC in parking state P_R and driving state D_C.

we observe the convergence of the expected mean value of SoC for a PHEV in parking state P_R at 6:00 a.m. and 6:00 p.m. using 1000 samples, to ensure that the number of samples is sufficient.

Three cases are studied: in Case 1, PHEVs will not participate in the restoration process, whereas PHEVs will feed power into the grid in Cases 2 and 3. It is assumed that once an outage occurs, PHEVs occupying one of the parking states will receive an incentive signal to stay at their current parking lots up to the end of restoration time to aid the restoration process. The time of outage is assumed to be at 6:00 a.m. in Case 2 and 6:00 p.m. in Case 3. The restoration

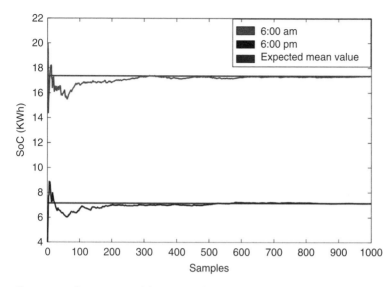

Figure 8.34 Convergence of the expected mean value of SoC for a PHEV in parking state P_R at 6:00 a.m. and 6:00 p.m.

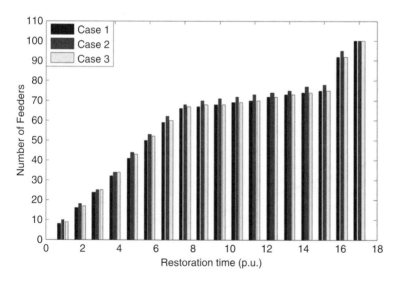

Figure 8.35 Number of energized feeders in different case studies.

optimization problem (3) is run for a 100-feeder test system considering PHEVs' driving patterns. Figure 8.35 shows the number of energized feeders at each restoration time. It can be observed that in Case 2 and at the initial phase of restoration, 10 feeders have been energized, whereas in Cases 1 and 3, this value decreased to 8 and 9 feeders, respectively. As the restoration process proceeds, Case 2 takes the priority over the other cases in terms of the number of energized feeders. However, at the final stage (after 18 p.u. restoration time) all feeders are energized and their corresponding loads are picked up. This arises from the fact that, in Case 2, the number of PHEVs and the expected SoC level are significantly greater than in Case 3. With this in mind, PHEVs in Case 2 can help to improve the load pickup capability of the system by discharging the energy stored in their batteries. This is particularly true at the initial phase of restoration, when most conventional NBS units have not been started. Table 8.10 compares the total served energy throughout the restoration period in different cases. Case 2 is the best case and Case 1 is the worst case in terms of total served energy.

Figure 8.36 shows the charging/discharging sequence of PHEV batteries. As stated earlier, the charging process of PHEVs may occur when the available generation exceeds the maximum amount of load that can be restored. Note that in this study the lowest priority has been assigned to the PHEV loads. Therefore, in Case 2 after 17 p.u. restoration time, the charging process of PHEV batteries will be started with the target of charging all PHEVs to their Socmax level. However, in Case 3 we see the first recharging cycle started at $t = 8$. This

Table 8.10 Total energy served in different cases.

Case	Case 1	Case 2	Case 3
Served energy (GWh)	1.174	1.219	1.192

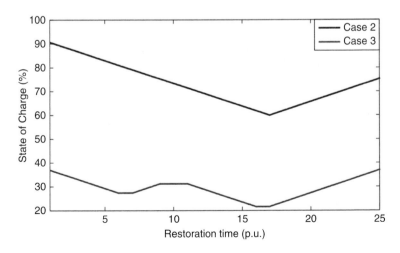

Figure 8.36 State of charge of PHEV batteries in different case studies.

is due to the fact that in Case 2 and after 6 p.u. restoration time, the SoC level of PHEVs are not enough to pick up a block of load. Instead, generation-load mismatch can charge PHEVs at $t = 7$. Similar to Case 2, the final charging cycle commenced after full restoration of all loads at $t = 17$.

References

1 National Renewable Energy Laboratory (2012). (ed. M.M. Hand et al.) *Renewable Electricity Futures Study*, Vol. 4. [Online]. Available at: https://www.nrel.gov/docs/fy12osti/52409-1.pdf) Golden, CO: National Renewable Energy Laboratory.

2 Vittal, E., O'Malley, M., and Keane, A. (2010). A steady-state voltage stability analysis of power systems with high penetrations of wind. *IEEE Trans. Power Syst.* 25 (1): 433–442.

3 Hughes, F.M., Anaya-Lara, O., Jenkins, N., and Strbac, G. (2005). Control of DFIG-based wind generation for power network support. *IEEE Trans. Power Syst.* 20 (4): 1958–1966.

4 Ontario Power System Restoration Plan (2016). [Online]. Available at: http://www.ieso.ca/documents/systemOps/soOntPowerSysRestorePlan.pdf.

5 Aktarujjaman, M., Kashem, M.A., Negnevitsky, M., and Ledwich, G. (2006). Black start with DFIG based distributed generation after major emergencies, in *Proceedings of the Power Electronics, Drives and Energy Systems, PEDES '06 International Conference*, 12–15 December 2006, pp. 1–6.

6 PJM Interconnection. PJM Manual 36: System Restoration. [Online]. Available at: http://www.pjm.com/~/media/documents/manuals/m36.ashx. June 20, 2013.

7 MISO (2012). MISO Real Time Operations – Power System Restoration Plan, Version 4.2. [Online]. Available at: https://www.misoenergy.org/_layouts/miso/ecm/redirect.aspx?id=145003. December 1, 2012.

8 Zhu, H. and Liu, Y. (2012). Aspects of power system restoration considering wind farms, in *Proceedings of the International Conference on Sustainable Power Generation and Supply, September 2012*, pp. 1–5.

9 Aktarujjaman, M. Kashem, M.A. Negnevitsky, M. and Ledwich, G. (2006). Black start with DFIG based distributed generation after major emergencies, in *Proceedings of the International Conference on Power Electronics, Drives and Energy Systems, December 2006*, pp. 1–6.

10 Seca, L., Costa, H., Moreira, C.L., and Lopes, J.A. (2013). An innovative strategy for power system restoration using utility scale wind parks, in *Proceedings of the Bulk Power System Dynamics and Control-IX Optimization, Security and Control of the Emerging Power Grid, August 2013*, pp. 1–8.

11 El-Zonkoly, A.M. (2015). Renewable energy sources for complete optimal power system black-start restoration. *IET Gener. Transm. Distrib.* 9 (6): 531–539.

12 Adibi, M.M. and Fink, L.H. (1992). Special considerations in power system restoration. *IEEE Trans. Power Syst.* 7 (4): 1419–1427.

13 MISO Power System Restoration Plan (2016). [Online]. Available at: https://www.misoenergy.org/Library/Repository/Procedure/RTO-PSR-001MISO Power System Restoration Plan Manual Volume I Version9.1.pdf.

14 Trodden, P.A., Bukhsh, W.A., Grothey, A., and McKinnon, K.I.M. (2014). Optimization-based islanding of power networks using piecewise linear AC power flow. *IEEE Trans. Power Syst.* 29 (3): 1212–1220.

15 Sun, W., Liu, C.-C., and Zhang, L. (2011). Optimal generator start-up strategy for bulk power system restoration. *IEEE Trans. Power Syst.* 26 (3): 1357–1366.

16 Sun, W. and Liu, C.C. (2013). Optimal transmission path search in power system restoration, in *Bulk Power System Dynamics and Control - IX Optimization, Security and Control of the Emerging Power Grid (IREP), 2013 IREP Symposium*, pp. 1–5, 25–30 August 2013.

17 Adibi, M.M., Borkoski, J.N., Kafa, R.J., and Volkmann, T.L. (1999). Frequency response of prime movers during restoration. *IEEE Trans. Power Syst.* 14 (2): 751–756.

18 Dupacov, J., Grwe-Kuska, N., and Rmisch, W. (2003). Scenario reduction in stochastic programming: an approach using probability metrics. *Math. Program. Ser. A* 3: 493–511.

19 Golshani, A., Sun, W., Zhou, Q., Zheng, Q.P., and Hou, Y. (2017). Incorporating wind energy in power system restoration planning, *IEEE Trans. Smart Grid.* doi: 10.1109/TSG.2017.2729592.

20 Laporte, G. and Louveaux, F.V. (1993). The integer L-shaped method for stochastic integer programs with complete recourse, *Oper. Res. Lett.* 13: 133–142, April.

21 IEEE 57-bus data. [Online]. Available at: http://icseg.iti.illinois.edu/ieee-57-bus-system.

22 Moreira, C.L., Resende, F.O., and Peas Lopes, J.A. (2007). Using low voltage microgrids for service restoration. *IEEE Trans. Power Syst.* 22 (1): 395–403.

23 Jiang-Hafner, Y., Duchen, H., Karlsson, M., Ronstrom, L., and Abrahamsson, B. (2008). HVDC with voltage source converters-a powerful standby black start facility, in *Proceedings of the 2008 IEEE/PES Transmission and Distribution Conference and Exposition, 2008*, pp. 1–9.

24 Kundur, P. (1994). *Power System Stability and Control*. New York: McGraw-Hill.

25 Machowski, J., Bialek, J., and Bumby, J. (2011). *Power System Dynamics: Stability and Control*. Wiley.

26 Demello, F.P. and Concordia, C. (1969). Concepts of synchronous machine stability as affected by excitation control. *IEEE Trans. Power Syst.* 88 (4): 316–329.

27 Sun, L., Peng, C., Hu, J., and Hou, Y. (2018). Application of Type 3 wind turbines in system restoration, *IEEE Trans. Power Syst.* 33 (3): 3040–3051.

28 Clark, K., Miller, N.W., and Sanchez-Gasca, J.J. (2010). Modeling of GE wind turbine generators for grid studies. GE Energy, 4 [Online] Available at: https://

www.researchgate.net/profile/Kara_Clark/publication267218696_Modeling_
of_GE_Wind_Turbine-Generators_for_Grid_Studies_Prepared_by/
links/566ef77308ae4d4dc8f861ef/Modeling-of-GE-Wind-Turbine-Generators-
for-Grid-Studies-Prepared-by.pdf

29 Zhang, Y. and Ooi, B.T. (2013). Stand-alone doubly-fed induction generators
(DFIGs) with autonomous frequency control. *IEEE Trans. Power Delivery*
28 (2): 752–760.

30 Wang, S., Hu, J., and Yuan, X. (2015). Virtual synchronous control for grid-
connected DFIG-based wind turbines. *IEEE J. Emerg. Sel. Top. Power Electron.*
3 (4): 932–944.

31 He, W., Yuan, X., and Hu, J. (2017). Inertia provision and estimation of
PLL-based DFIG wind turbines. *IEEE Trans. Power Syst.* 32 (1): 510–521.

32 Blasco-Gimenez, R., Ano-Villalba, S., Rodríguez-D'Derlée, J. et al. (2011).
Diode-based HVdc link for the connection of large offshore wind farms.
IEEE Trans. Energy Convers. 26 (2): 615–626.

33 Pourbeik, P. (2013). Proposed changes to the WECC WT3 generic model for
type 3 wind turbine generators. Prepared under Subcontract No. NFT-1-11342-01
with NREL, Issued to WECC REMTF and IEC TC88 WG27, 2013.

34 Zhao, M., Yuan, X., Hu, J., and Yan, Y. (2016). Voltage dynamics of current
control time-scale in a VSC-connected weak grid. *IEEE Trans. Power System*
31 (4): 2925–2937.

35 Wang, S., Hu, J., Yuan, X., and Sun, L. (2015). On inertial dynamics of virtual-
synchronous-controlled DFIG-based wind turbines. *IEEE Trans. Energy
Convers.* 30 (4): 1691–1702.

36 Hu, J., Sun, L., Yuan, X., and Wang, S. (2017). Modeling of type 3 wind turbine
with df/dt inertia control for system frequency response study. *IEEE Trans.
Power Syst.* 32: 2799–2809.

37 Zhang, L., Harnefors, L., and Nee, H.P. (2011). Interconnection of two very
weak ac systems by VSC-HVDC links using power-synchronization control.
IEEE Trans. Power Syst. 26 (1): 344–355.

38 Karakasis, N., Mademlis, C., and Kioskeridis, I. (2014). Improved start-up
procedure of a stand-alone wind system with doubly-fed induction generator,
in *7th IET International Conference on Power Electronics, Machines and
Drives (PEMD 2014), IET, 2014.*

39 REN21 Secretariat (2016). Renewables 2016 Global Status Report, Paris,
France. [Online]. Available at: http://www.ren21.net/wp-content/
uploads/2016/06/GSR_2016_Full_Report_REN21.pdf.

40 Rudraraju, V.R.R., Nagamani, C., and Ganesan, S.I. (2016). A control strategy
for reliable power output from a standalone WRIG with battery supported DC
link. *IEEE Trans. Power Electron.* 32 (6): 4334–4343.

41 Jiang, R., Wang, J., and Guan, Y. (2012). Robust unit commitment with
wind power and pumped storage hydro. *IEEE Trans. Power Syst.* 27 (2):
800–810.

42 Khodayar, M.E., Shahidehpour, M., and Wu, L. (2013). Enhancing the
dispatchability of variable wind generation by coordination with

pumped-storage hydro units in stochastic power systems. *IEEE Trans. Power Syst.* 28 (3): 2808–2818.

43 Golshani, A., Sun, W., Zhou, Q. et al. (2018). Coordination of wind and pumped-storage hydro units in power system restoration, *IEEE Trans. Sustain. Energy.* doi: 10.1109/TSTE.2018.2819133.

44 Conejo, A.J., Arroyo, J.M., Contreras, J., and Villamor, F.A. (2002). Self-scheduling of a hydro producer in a pool-based electricity market. *IEEE Trans. Power Syst.* 17 (4): 1265–1272.

45 Soyster, A.L. (1973). Convex programming with set-inclusive constraints and applications to inexact linear programming. *Oper. Res.* 11: 1154–1157.

46 Ben-Tal, A. and Nemirovski, A. (1998). Robust convex optimization. *Math. Oper. Res.* 23: 769–805.

47 Zhao, C., Wang, J., Watson, J.P., and Guan, Y. (2013). Multi-stage robust unit commitment considering wind and demand response uncertainties, *IEEE Trans. Power Syst.*, 28 (3): 2708–2717, August.

48 Bertsimas, D., Litvinov, E., Sun, X. et al. (2013). Adaptive robust optimization for the security constrained unit commitment problem. *IEEE Trans. Power Syst.* 28 (1): 52–63.

49 Hu, B. and Wu, L. (2016). Robust SCUC considering continuous/discrete uncertainties and quick-start units: a two-stage robust optimization with mixed-integer recourse. *IEEE Trans. Power Syst.* 31 (2): 1407–1419.

50 Zeng, B. and Zhao, L. (2013). Solving two-stage robust optimization using a column-and-constraint generation method. *IEEE Oper. Res. Lett.* 41 (5): 457–461.

51 IEEE 39 Bus System, Illinois Center for a Smarter Electric Grid (ICSEG). [Online]. Available at: http://publish.illinois.edu/smartergrid/ieee.

52 Adibi, M.M., Borkoski, J.N., and Kafka, R.J. (1987). Power system restoration - the second task force report. *IEEE Trans. Power Syst.* 2 (4): 927–932.

53 Yari, V., Nourizadeh, S., and Ranjbar, A.M. (2010). Determining the best sequence of load pickup during power system restoration, in *Environment and Electrical Engineering (EEEIC), 2010 9th International Conference, 16–19 May 2010*, pp. 1–4.

54 Yari, V., Nourizadeh, S., and Ranjbar, A.M. (2010). Wide-area frequency control during power system restoration, in *Electric Power and Energy Conference (EPEC), 2010 IEEE, 25-27 August 2010*, pp. 1–4.

55 Hanbing, Q. and Yutian, L. (2012). General model for determining maximum restorable load, in *Power and Energy Society General Meeting, 2012 IEEE, 22-26 July 2012*, pp. 1–6.

56 Edstrom, F. and Soder, L. (2011). A circuit breaker reliability model for restoration planning considering risk of communication outage, in *PowerTech, 2011 IEEE Trondheim, 19-23 June 2011*, pp. 1–6.

57 FERC (2013). Order 784: third-party provision of ancillary services; accounting and financial reporting for new electric storage technologies. FERC, USA, July 18, 2013.

58 Hsieh, E. and Johnson, R. (2012). Frequency response from autonomous battery energy storage, in *CIGRE US National Committee, 2012 Grid of the Future Symposium*.

59 Kadel, N., Sun, W., and Zhou, Q. (2014). On battery storage system for load pickup in power system restoration, in *Proceedings of the IEEE Power & Energy Society General Meeting, National Harbor, MD, 27–31 July 2014*.

60 Nguyen, M.Y., Nguyen, D.H., and Yoon, Y.T. (2012). A new battery energy storage charging/discharging scheme for wind power producers in real-time markets. *Energies* 5 (12): 5439–5452.

61 Liu, C.C., Liou, K.L., Chu, R.F. et al. (1993). Generation capability dispatch for bulk power system restoration: a knowledge-based approach. *IEEE Trans. Power Syst.* 8 (1): 316–325.

62 Perez-Guerrero, R., Heydt, G.T., Jack, N.J. et al. (2008). Optimal restoration of distribution systems using dynamic programming, *IEEE Trans. Power Delivery*, 23 (3):1589–1596, July 2008.

63 Khaligh, A. and Zhihao, L. (2010). Battery, ultracapacitor, fuel cell, and hybrid energy storage systems for electric, hybrid electric, fuel cell, and plug-in hybrid electric vehicles: state of the art. *IEEE Trans. Veh. Technol.* 59 (6): 2806–2814.

64 Kempton, W. and Tomić, J. (2005). Vehicle-to-grid power implementation: from stabilizing the grid to supporting large-scale renewable energy. *J. Power Sources* 144 (1): 280–294.

65 Pang, C., Dutta, P., and Kezunovic, M. (2012). BEVs/PHEVs as dispersed energy storage for V2B uses in the smart grid. *IEEE Trans. Smart Grid* 3 (1): 473–482.

66 Zigkiri, A. (2013). The role of plug-in electric vehicles in system restoration after black-out. In: *EEH – Power Systems Laboratory*. Zurich: ETH.

67 Ma, T., Houghton, T., Cruden, A., and Infield, D. (2012). Modeling the benefits of vehicle-to-grid technology to a power system. *IEEE Trans. Power Syst.* 27 (2): 1012–1020.

68 Yilmaz, M. and Krein, P.T. (2013). Review of the impact of vehicle-to-grid technologies on distribution systems and utility interfaces. *IEEE Trans. Power Electron.* 28 (12): 5673–5689.

69 Chunhua, L., Chau, K.T., Wu, D. et al. (2013). Opportunities and challenges of vehicle-to-home, vehicle-to-vehicle, and vehicle-to-grid technologies. *Proc. IEEE* 101 (11): 2409–2427.

70 Global EV Outlook, Understanding the Electric Vehicle Landscape to 2020. (2013). International Energy Agency.

71 Tomić, J. and Kempton, W. (2007). Using fleets of electric-drive vehicles for grid support. *J. Power Sources* 168 (2): 459–468.

72 Golshani, A., Sun, W., and Zhou, Q. (2016). PHEVs contribution to self-healing process of distribution systems, in *Proceedings of the IEEE Power & Energy Society General Meeting, Boston, MA, July 2016*.

73 Gong, Q., Tulpule, P., Marano, V., Midlam-Mohler, S., and Rizzoni, G. (2011). The role of ITS in PHEV performance improvement, in *Proceedings of the American Control Conference, San Francisco, 2011*, pp. 2119–2124.

74 Lee, T.K., Adornato, B., and Filipi, Z.S. (2011). Synthesis of real-world driving cycles and their use for estimating PHEV energy consumption and charging opportunities: case study for Midwest/U.S. *IEEE Trans. Veh. Technol.* 60 (9): 4153–4163.

75 Perez-Guerrero, R.E. and Heydt, G.T. (2008). Distribution system restoration via subgradient-based Lagrangian relaxation. *IEEE Trans. Power Syst.* 23 (3): 1162–1169.

76 Mauricio, J. M., Marano, A., Gómez-Expósito, A., and Ramos, J. L. M. (2015). Frequency regulation contribution through variable-speed wind energy conversion systems, *IEEE Trans. Power Syst.* 24 (1): 173–180, October.

9

Emerging Technologies in System Restoration

9.1 Applications of FACTS and HVDC

With the development of power electronics, high-voltage direct current (HVDC) and flexible AC transmission systems (FACTS) are widely used in power systems today. Alternative HVDC transmission has special advantages over AC transmission, while FACTS provide auxiliary service for enhancing AC transmission system security and flexibility. In this chapter, two kinds of the most widely used HVDC technologies, that is LCC-HVDC and VSC-HVDC (line-commutated converter HVDC and voltage-source converter HVDC), are introduced. After a brief introduction to the structures and control strategies of different HVDC systems, the black-start capability and black-start sequence with HVDC are discussed. After that, FACTS technology's potential benefits for system restoration are also explored.

9.1.1 LCC-HVDC Technology for System Restoration

9.1.1.1 Characteristics of LCC-HVDC

The classical LCC-HVDC based on thyristor technology, in use for decades, has proven to be superior to AC transmission in terms of operating cost and reliability [1]. Figure 9.1 depicts an LCC-HVDC using thyristors in a current source converter topology. Two AC systems are connected by the HVDC link. This system includes: the AC/DC and DC/AC converters, filters, converter transformer, DC line, and the smoothing reactors.

The LCC-HVDC systems have special characteristics. Since thyristors can only be turned off at the moment when the current reaches zero, the external voltage source is required for thyristor commutation [2]. Furthermore, the delayed firing of thyristors results in current lagging voltage, LCC-HVDC link absorbs a large amount of reactive power. Current source converters have the ability to withstand a short circuit, since DC inductors can limit the currents

Power System Control Under Cascading Failures: Understanding, Mitigation, and System Restoration, First Edition. Kai Sun, Yunhe Hou, Wei Sun and Junjian Qi.
© 2019 John Wiley & Sons Ltd. Published 2019 by John Wiley & Sons Ltd.
Companion website: www.wiley.com/go/sun/cascade

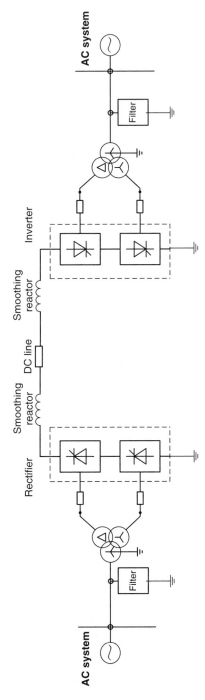

Figure 9.1 LCC-HVDC configuration.

during a fault. However, the power direction could only be changed by reversing the DC voltage instead of current [3].

The low losses and high power level enable LCC-HVDC to be the most economical connection for bulk power transmission over a long distance. The application of HVDC at 800 kV has been efficient, environmentally friendly, and economically attractive for point-to-point power transmission with 6400 MW over more than 1000 km [4].

9.1.1.2 Control of LCC-HVDC

The fast regulation and multiple control modes can be implemented by controlling the trigger phase of the converter. This control strategy provides high flexibility for LCC-HVDC operation [5]. The major control modes employed by HVDC are listed as follows:

- DC current control mode: maintaining constant DC current;
- DC voltage control mode: keeping either the sending terminal voltage or receiving terminal voltage constant or within a given range;
- Firing angle control in the rectifier mode: keeping the firing angle small during normal operation to reduce reactive power consumption as well as hold a regulation margin;
- Extinction angle control in the inverter mode: maintaining extinction angle above a given turn-off margin angle so as to lower the commutation failure risk.

The LCC-HVDC control is achieved by regulating the firing angle on both rectifier and inverter sides. With different control strategies, an HVDC system may display various transmission modes. The following will give a brief review of the control strategies for the rectifier and inverter, respectively.

9.1.1.2.1 *Controller for Rectifier in LCC-HVDC*

- DC current control

The objective of a constant DC current control is to maintain the DC current I_d at or close to the setting level. By comparing the actual current value and the setting, the error is fed back to generate the firing angle in order to eliminate the error, as shown in Figure 9.2. This control strategy performs well under different operating conditions, even after the occurrence of a fault. When a fault occurs in the system, the controller can limit the transient overcurrent to protect the devices.

- Minimum firing angle α_{min} control

The conduction conditions for thyristors are listed as follows: one is the forward voltage across the positive and negative pole of the thyristor; the other is the necessary trigger pulse. If the forward voltage applied across the thyristor

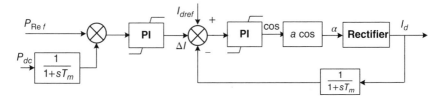

Figure 9.2 DC power controller.

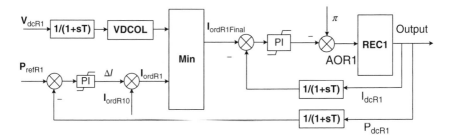

Figure 9.3 Controller for the rectifier.

is too low, the simultaneous conduction of thyristors cannot be achieved. The minimum firing angle control is assigned to solve this problem. In the most practical projects, the angle is set to be 5°. To achieve the α_{min} control on the rectifier side, the additional current margin of 0.1 p.u. can be imposed on the current order. This strategy leads to saturation of the current controller and maintains the output at 5° for the rectifier.

- Voltage-dependent current-order limit (VDCOL)

A disturbance may result in the decrease of AC voltage at the rectifier or inverter accompanied by a DC voltage drop. Under this condition, the system will deteriorate significantly if the converter still strives to maintain the full load current. VDCOL control is introduced in the rectifier to reduce the DC current order in case the DC voltage drops to a certain level and to prevent continuous commutation failures. Using this control strategy, only after the DC voltage recovers to a sufficient level does the DC current gradually return to its original current level. The upper limit of voltage value corresponds to the initial current order, the lower voltage value to the minimum current. In Figure 9.3, DC voltage is sent to VDCOL and the corresponding current value is obtained and compared with the current order from the power controller; the minimum value is chosen for the inner current controller.

9.1.1.2.2   Controller for Inverter in LCC-HVDC

- Constant extinction angle (γ) control

A small extinction angle of an inverter may cause a commutation failure. Under this condition, the thyristor will conduct again under the forward voltage applied since the capability of positive blocking of the thyristor does not recover fully. The continuous commutation failure will disturb the HVDC and should be avoided. As a result, the extinction angle should be kept at an acceptable value.

On the other hand, the increase in extinction angle, accompanied by a declining inverter power factor, can lead to high reactive power consumption. It is suggested that the extinction angle should be kept at a reasonable value. Usually, it is chosen to be 15~18°. The constant extinction angle control is illustrated in Figure 9.4.

- DC current control

According to the current margin method, the inverter should be equipped with a DC current controller. While the current setting of the inverter's controller is smaller than that of the rectifier, the current controller on the inverter side will not be activated when the extinction angle control is involved. When the DC voltage on either the rectifier side or inverter side fluctuates sharply, the control mode will be shifted. That is, the minimum firing angle control of the rectifier takes effect to control the system voltage and the current controller controls the current.

9.1.1.3   LCC-HVDC Black-Start Capability Exploration

Normally, LCC-HVDC does not have black-start capability and cannot supply a passive load because the switching of the thyristors depends on the existence of AC grid voltage. However, some countermeasures have been proposed to overcome this limitation recently. Two possible solution strategies are presented as follows.

Figure 9.4 Controller for inverter.

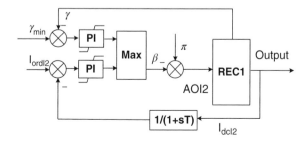

9.1.1.3.1 Control Modification

The recent development performed by Siemens showed that LCC-HVDC might be able to supply a passive load with a novel control strategy [6]. To implement this method, the control software needs some modifications to enable the black-start capability. During the implementation of this method, when the inverter is connected to a passive load, the firing pulses for inverters are synchronized to an independent sinusoidal signal such as 60 or 50 Hz. This is the major difference compared with conventional control strategies of HVDC. Besides, the equivalent circuit on the inverter side has a capacitive feature at the fundamental frequency to ensure normal commutation of the converter. The AC voltage at the inverter stays constant with variable loads by means of controlling DC currents of the converter.

The black-start sequence of LCC-HVDC is

- the inverter filters are connected first,
- the inverter and rectifier are deblocked in sequence.

During this process, the AC voltage of the inverter is raised slowly. The DC current must be large enough to enable the inverter to absorb the reactive power generated by the filters. The DC voltage is kept at a low level to ensure that the rectifiers and inverters operate at a firing angle close to 90°. As a consequence, the converter shall be designed in a range of 90° operation. This strategy can be achieved with moderate ratings of thyristor valves and harmonic filters.

9.1.1.3.2 Auxiliary Device Support

Another potential solution is to resort to auxiliary devices, such as STATCOM or a diesel-driven generator, to build grid voltage. A case study for LCC-HVDC integrating offshore wind power is carried out as illustrated in Figure 9.5. In this system, the corresponding auxiliary service is needed to enable the

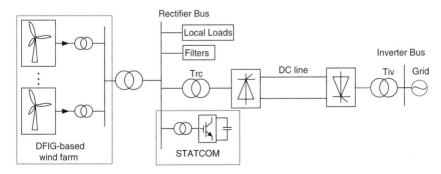

Figure 9.5 LCC-HVDC with STATCOM for wind power integration (DFIG = doubly fed induction generator).

operation of LCC-HVDC during periods with little wind. The STATCOM is installed to provide commutation voltage for the converter and fast reactive power support. These additional components make it feasible for application of LCC-HVDC in wind farm integration [7].

After a partial or complete outage occurs in a system with a DFIG-based wind farm and HVDC system, this system can be restored with the support of the STATCOM. Usually, STATCOM has about 30~40% of the system capacity. For offshore wind farms, due to the difficulty of maintenance, self-recovery capability is of importance [8]. The black start for the system may be described as follows:

- The DC link voltage of STATCOM is built up to the rated value by the backup generator.
- STATCOM is placed into operation and the AC voltage of the rectifier rises to the reference level.
- Wind farm output active power increases gradually to compensate for the losses on the cable and transformer.
- The backup generator is tripped out and the rectifier and inverter are unblocked.
- DC current climbs up and DC voltage reaches the rated value.
- The active power of the wind farm increases quickly and the DC current also increases to the rated value. The black-start process is complete.

9.1.2 VSC-HVDC Technology for System Restoration

9.1.2.1 Characteristics of VSC-HVDC
With the development of power electronic technology, a novel HVDC technology is available. By using insulated gate bipolar transistor components, the VSC technology is developed.

Compared with LCCs, VSCs have the following advantages [9]:

- Self-commutating and no external voltage source is required.
- Independent control of the reactive and active power consumed or generated by the converter.
- The reactive power flow can be independently controlled in each AC network.
- Possibility of connection to a weak AC network or even to the system without AC voltage sources.
- Reduction of the filter and converter size due to the limited injection of low-order harmonic currents and the absence of inverter commutation failures.

9.1.2.2 Regular Control of VSC-HVDC
The general configuration of VSC-HVDC is illustrated in Figure 9.6, where the VSC station is integrated to the point of common coupling through interfacing

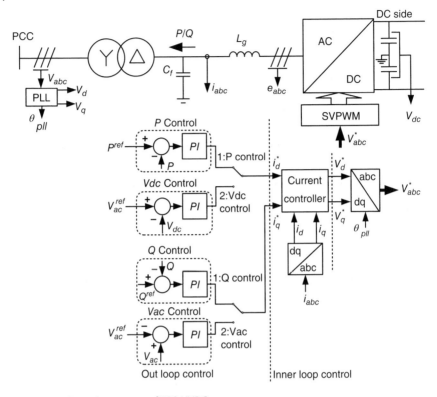

Figure 9.6 Control structures of VSC-HVDC.

a reactor, an AC-side filter, and a coupling transformer. The VSC technology provides two degrees of control freedom by permitting independent modulation of active power and reactive power.

The active power control is based on a direct power set-point or an indirect DC-voltage regulation. In contrast, the reactive power can be controlled by a predefined reactive power value or a mediate AC voltage control. Also worth noting, the power control employs a dual-loop structure and phase-locked loop (PLL)-based vector control. The inner loop adjusts the AC current and the outer loop controls the transferred power. Generally, the outer loops must be slower-acting than the inner loops to guard stability [10].

There are three regular control modes of VSC-HVDC dependent on its outer loop control, including constant DC voltage, constant AC current or active power, and constant AC voltage or reactive power, (see Figure 9.6). However, the combination of these control modes is necessitated and at least one station must be responsible for maintaining DC voltage.

The selection of control modes resorts to the nature of a specific application. For the situation where both stations of VSC-HVDC are connected to strong grids, the constant active power and reactive power control mode is suitable for rectifier while the inverter is controlled in the constant DC voltage and reactive power control mode. In contrast, when VSC-HVDC is used to power passive systems, the task of maintaining DC voltage would be assigned to the sending-end station (connected to the active grid) while the receiving-end station is responsible for controlling the passive system frequency and voltage. In this way, the VSC-HVDC system can be regarded as an ideal synchronous generator, especially since it is able to respond quickly to power perturbations.

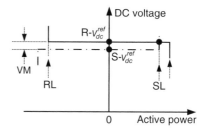

Figure 9.7 Active power–DC voltage relationship.

Sometimes, in order to facilitate active power control and reserve transmission, one of the VSC-HVDC stations can deploy DC-voltage deviation control, as proposed in [11]. Supposing that the sending-end station controls DC voltage and the receiving-end station adopts active power control withholding a DC voltage margin, the relationship between active power transfer and DC voltage can be seen in Figure 9.7. This control mode can readily achieve a forward and reverse active power transfer by devising the receiving-end low-limit value in DC voltage reference (R-$V_{dc}ref$) and varying the voltage margin value, respectively. Furthermore, this control mode is able to ensure a steady start-up while controlling reactive power or AC voltage, even when the other station is locked and out of service. However, the parameter settings of DC-voltage deviation control are more sophisticated and additionally impose serious impacts on control performance. Therefore, due to space restriction, DC-voltage deviation control would be regarded as an alternative control mode for VSC-HVDC in system restoration application, and it is further discussed next.

It is well known that VSCs are self-commutated and can operate indefinitely at zero-power or very low-power transfers [12]. To procure the participation of VSC-HVDC in system restoration, the receiving-end station has the task of forming an outlet with an autonomous capacity of frequency and voltage regulation in order to power a passive system. In this sense, the control mode where the active power and AC voltage control is deployed is preferable in the receiving-end station, while the sending-end station adopts the DC voltage control and reactive power. In addition, there are some auxiliary controls (like frequency support control) added to the outer control referring to the requirements of the overall system.

(a)

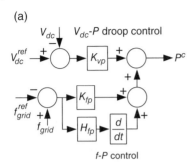

(b)

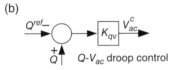

Figure 9.8 Auxiliary control in receiving-end station. (a) V_{dc}- P droop control and f-P control; (b) Q- V_{ac} droop control.

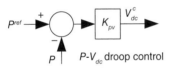

Figure 9.9 Auxiliary P-V_{dc} droop control in sending-end station.

9.1.2.3 Auxiliary Control of VSC-HVDC for Restoration

When AC systems experience a blackout, the connected VSC-HVDC station that is required for system restoration needs to function as a voltage source and provide power for the external systems. In this sense, there are three issues to be identified: (i) holding a capacity of autonomous power sharing in the face of DC-side transients, (ii) playing an active role in supporting the load-side system frequency and voltage, and (iii) ensuring the stability of PLL, especially without the grid information [13]. To cope with these issues, DC voltage-active power (V_{dc}-P) droop control [14], frequency-active power (f-P) droop control [14, 15], and reactive power-AC voltage (Q-V_{ac}) droop control [14] are hence implemented in the receiving-end station control system, as shown in Figure 9.8. In addition, the sending station should simultaneously operate in the active power-DC voltage (P-V_{dc}) droop control mode so as to allow proper power exchange in case of power variations (refer to Figure 9.9).

- V_{dc}-P and f-P droop control

In order to fabricate an ideal voltage-source interface for passive systems, the V_{dc}-P and f-P droop control are requested to be deployed in the receiving-end station. Then the compensation on the active power reference can be addressed as

$$P^c = K_{vp}\left(V_{dc}^{ref} - V_{dc}\right) + \left(K_{fp} + H_{fp}s\right)\left(f_{grid}^{ref} - f_{grid}\right). \tag{9.1}$$

This is on the basis that, in a DC system, the variation of DC-bus voltage level demonstrates whether a fixed active power exchange is retained between the sending-end station and the receiving-end station. By contrast, the frequency of an AC system is an indicator exhibiting whether its active power generation and consumption are balanced. Also worth noting, the derivative control, paralleled with f-P droop control is intuitively added to augment the inertia support ability of VSC-HVDC.

- Q-V_{ac} droop control

It is imperative to introduce Q-V_{ac} droop control in the receiving-end station, such that the AC voltage magnitude can be automatically restored under arbitrary load situations. In this way, the dynamic AC voltage compensation is determined by the exported reactive power, as given by

$$V_{ac}^c = K_{qv}\left(Q-Q^{ref}\right).$$ (9.2)

- P-V_{dc} droop control

For system restoration application and fast-action power transfer, the sending-end station should determine the DC voltage reference as a function of the injected power in the AC system (or the fed DC current). Referring to the flowing control rule, a target $V_{dc}{}^{ref}$ is defined based on the P-V_{dc} droop coefficient:

$$V_{dc}^c = K_{pv}\left(P^{ref} - P\right).$$ (9.3)

- Specific control for starting up a dead grid

During normal power transmission operation, PLL effectively tracks the observed AC voltage in load-side systems and provides the AC voltage reference for power control of the receiving-end inverter. However, when a blackout occurs, the AC voltage reference can be only generated according to the predetermined magnitude and frequency. Therefore, during the start-up procedure, PLL is blocked, and the AC voltage reference is set as its unit value and rotates with a normative frequency.

9.1.2.4 Black Start in a VSC-HVDC Case Study

Figure 9.10 presents the major apparatus of the studied system, which is a mutation of a two-area system in Kundur [16]. Provided that the sending-station rectifier of VSC-HVDC is connected to an active grid (as opposed to a dead/passive grid), the rectifier can be used to energize DC capacitors and hold the DC link voltage to a reference value by operating in DC-voltage control mode. The receiving-end station of VSC-HVDC is used for the services of local black start and load pickup when the system in Area II experiences an outage.

Supposing that a blackout in Area II occurs, the backup battery would retain the control and protection system alive. A few seconds later, the control and protection systems identify that the system in Area II is definitely staying in black, the inverter of the receiving-end station will be blocked, the AC breaker (B1, B2, and B3, see in Figure 9.10) will be opened and the inverter control will be automatically transferred to black-start control. Thereafter, the AC voltage reference is achieved through clamping the magnitude as its unit value and the phase as the integral of a nominated frequency.

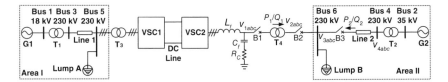

Figure 9.10 Configuration of the black-start study system.

As soon as B1 is open, the inverter will be commanded to deblock and the AC voltage will be gradually ramped up to its reference value. Once a healthy AC voltage of the receiving-end station is built up, the energization of apparatus in station and loads can proceed stepwise, bearing in mind that the loads picked up at any step cannot exceed the available generation capability [17].

The proposed system restoration sequence is the following:

- First, black-start control is activated at $t = 3$ seconds, establishing the AC voltage of the receiving-end station.
- Next, transformer T_4 is energized at $t = 5$ seconds with circuit breaker B1 closed (B2 is kept open).
- At $t = 7$ seconds, the shunt reactor bank (aggregated as Lump B) is energized with B2 closed (B3 is kept open).
- Last, the loads in Area II are rescued by closing circuit breaker B3 at $t = 8$ seconds.

Details about simulation tests are presented as follows. Figure 9.11 shows the reference inverter voltage is ramped up to steady-state value with a time period of 0.8 s. Figures 9.12–9.14 show the voltage response at the bus of transformer, lumped load B, and loads in Area II, respectively. Results indicate that these elements can be restored both quickly and smoothly with VSC-HVDC as the black-start power source.

VSC-HVDC systems provide alternative methods for system restoration [18]. Compared with traditional system restoration, the benefits of VSC-HVDC systems can be summarized as follows:

- Since VSC has black-start capability, it can provide initial energy resource for system restoration. In the traditional power system restoration, only black-start units can be employed for this purpose.
- AC voltage in a VSC-HVDC system is controllable. It provides a flexible resource for voltage and reactive power regulation during system restoration. In traditional power system restoration, only the excitation system of generating units can perform this task.
- A VSC-HVDC system can control the system frequency directly. This characteristic can benefit system restoration during load pickup.
- Due to the controllability during energization of VSC-HVDC, the overvoltage problem as a result of electromagnetic transients can be avoided during the system restoration process.

Figure 9.11 Inverter output reference voltage.

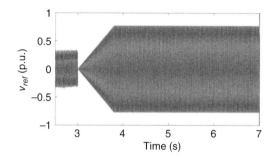

Figure 9.12 Transformer voltage response.

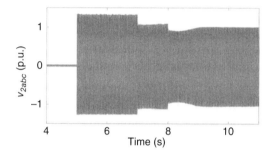

Figure 9.13 Voltage response of lump load B.

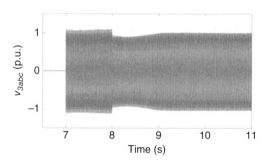

Figure 9.14 Bus 4 voltage response.

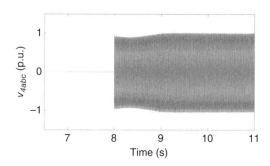

9.1.3 FACTS Technology for System Restoration

9.1.3.1 State-of-the-Art Technologies of FACTS

According to the definition of IEEE DC and FACTS subcommittee, FACTS are "alternating current transmission systems incorporating power electronic-based and other static controllers to enhance controllability and increase power transfer capability." A FACTS controller is defined as "a power electronic-based system and other static equipment that provide control of one or more AC transmission system parameters." Recently, FACTS technologies have been developed to promote the use of power electronic switching devices to replace traditional thyristors.

The primary benefits from the use of FACTS devices are

- Improved capability of a transmission system
- Improved stability and availability of a power system
- Improved power quality to customers
- Increased renewable energy adopted to minimize environmental impacts
- Diminished transmission losses

Today, FACTS technology provides solutions for a more secure, flexible, and cleaner operation of power systems. Development of modern technologies of power electronics also improves the quality and capacity of FACTS devices at a lower cost. Therefore, with more and more FACTS devices in the transmission system, various functions, including system restoration, need to consider the potential applications of FACTS devices.

The main commercial available applications of FACTS controllers include SVC, SC, TCSC, STATCOM, and UPFC. They are briefly introduced as follows.

- SVC (Static var compensator)

SVC is a shunt-connected static var generator or absorber whose output is adjusted to exchange capacitive or inductive current so as to maintain or control specific parameters of the electric power system (typical bus voltage). Many SVCs have been installed on the system around the world for voltage control and stability enhancement (Figure 9.15).

- SC (Fixed series capacitor)

SC is a series capacitor bank that has a reactance or reactances that are defined by the discrete reactances of the capacitors and are not variable. SCs installed on a long-distance transmission line in series can help reduce both angular derivation and voltage drop, resulting in an improved capability and stability of the lines.

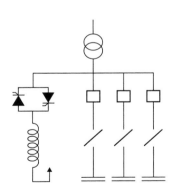

Figure 9.15 Static var compensator.

- TCSC (Thyristor-controlled series capacitor)

TCSC is a capacitive reactance compensator, which consists of a series capacitor bank shunted by a thyristor controlled reactor in order to provide a variable series capacitive reactance. There have been several installations of TCSC in the US. Such installations of

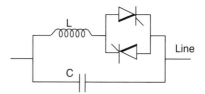

Figure 9.16 Thyristor-controlled series capacitor.

TCSC can benefit the system by increasing power transfer within the line's thermal limit and also by helping to control power flow, line impedance, damp electromechanical power oscillations, and reduce subsynchronous resonance (Figure 9.16).

- STATCOM (Static synchronous compensator)

STATCOM is a static synchronous generator operating as a shunt-connected static var compensator, whose capacitive or inductive output current can be controlled independently of the AC system voltage. The installation of STACOM in Tennessee provides voltage control to the transformer and voltage support to the bus at a high voltage line during various load conditions (Figure 9.17).

- UPFC (Unified power flow controller)

UPFC is a combination of a static synchronous compensator (STATCOM) and a static synchronous series compensator (SSSC). They are coupled via a common DC link to allow bidirectional flow of real power between the series output terminal of SSSC and the shunt output terminal of the STATCOM. It is controlled to provide concurrent real and reactive series line compensation without an external electric energy source. UPFC is able to control, concurrently or selectively, the transmission line voltage, impedance, angle, and the real and reactive power flow on the line (Figure 9.18).

Figure 9.17 Static synchronous compensator.

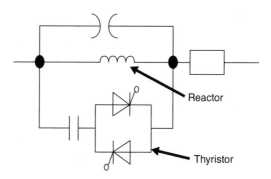

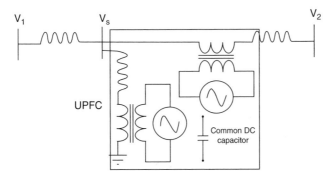

Figure 9.18 Unified power flow controller static.

9.1.3.2 Potential Applications of FACTS in Power System Restoration

During system restoration, while only a few generators are started and some transmission lines are energized, the issues of instability and imbalance of active and reactive voltage control need to be addressed. FACTS is a promising tool to facilitate a faster, smoother, and more reliable restoration process. The major issues that the FACTS controller may be able to address include [19] oscillation stability, voltage stability, transient stability, standing phase angle control, reactive power control, sustained overvoltage, transient overvoltage, reduction in stored energy for circuit breaker, and smooth load pickup. A detailed illustration of these issues for restoration is discussed as follows.

- Oscillation stability

A low-frequency oscillation in voltage magnitude may lead to the collapse of a system. During system restoration, the oscillation may occur due to interconnection of transmission system and poor damping of the system. Therefore, oscillation stability is a major concern in the system restoration procedure.

FACTS controller may help to mitigate power system oscillations during system restoration. For example, STATCOM can provide both capacitive and inductive compensation of reactive power. STATCOM can generate or absorb reactive power. The capability enables STATCOM to damp the voltage oscillations in a system.

SSSC can be used to regulate the difference of both angle and phase between the sending and receiving ends. SSSC can be effectively applied when the power system oscillations occur. UPFC, which consists of a STATCOM and an SSSC, performs well when they are used to damp power system oscillations.

A FACTS stabilizer can perform a similar function of damping power system oscillations. It is more flexible and effective compared with a power system stabilizer (PSS), since it can be installed at a key location of the system but not limited inside the power plants. During the process of system restoration, the generating units with PSS devices may take hours to start. The dispatchers may have better options to connect the FACTS stabilizer by energizing adjunct lines to damp the system oscillations.

● Voltage stability

Voltage stability can be evaluated by indices such as maximum loadability margin, minimal eigenvalue of Jacobi matrix, Kessel–Glavitsch L-indicators, and so on. All such indices can be improved by a reduced line reactance. During system restoration, while some parallel transmission lines are not energized, FACTS controllers can help to adjust the line reactance, resulting in an improvement in voltage stability margin of the system.

The voltage stability index can be improved by bus voltages of sending and receiving ends of the transmission line. FACTS devices, such as STATCOM and UPFC, can perform the function of regulating the bus voltage. Thus, the risks of voltage instability caused by the weak interconnection of transmission lines can be mitigated.

● Transient stability

Transient stability issues may be more serious during the system restoration process. The interconnection of transmission systems is not strong as the system is being reestablished. During energization, the reactance of some transmission lines may be large compared with that of a normal condition of system operation. Also while only part of the load and generators are started, which may be in the forms of islands in the system, sudden change of system load or outage of some generators can be a large disturbance for the system that can cause a transient stability problem. The FACTS controllers have potential applications for the enhancement of transient stability during system restoration.

FACTS can enhance transient stability by adjusting line reactance. FACTS controllers such as SVC and TCSC can perform line compensation to improve transient stability. Some studies show that STATCOM can effectively enhance power system transient stability. This is because during the first swing, STATCOM can supply required current compensation at a lower line voltage as a result of sharply increasing power flow on the transmission line.

● Standing phase angle control

While reenergizing the transmission system or synchronizing two islands during system restoration, the system dispatcher may face an excessive standing phase angle (SPA) at the sending and receiving ends of the transmission line. For a large SPA, closing of the circuit breaker can cause a high impact on the system. It may also lead to cascading events and system outages, as well as damage to equipment.

Practical experience has shown that a large SPA can cause generator outage. Therefore, relays are installed on the transmission lines or transformers to avoid closing on a large SPA. A possible solution to reducing the SPA difference is to adjust the output of available generating units, load tap change of transformers, and so on. Such a procedure of rescheduling may need multiple coordinated actions by generating units and lines. With the availability of FACTS devices, the SPA difference can be rapidly and effectively reduced.

- Reactive power control

An imbalance of reactive power may occur due to increased reactive power caused by line charging currents during energization of high voltage transmission lines. A result of excessive positive reactive power on transmission lines is the sustained overvoltage. A possible solution is to pick up some positive reactive loads. However, such actions will cause a concurrent imbalance of active power, resulting in a sudden drop of the system frequency. An alternative is to use the negative reactive compensators with zero active power, such as synchronous machines, SVCs, and shunt reactors.

A thyristor-based FACTS controller can enhance the capability to absorb reactive power during system restoration. This FACTS device provides an option to mitigate the imbalance of reactive power while energizing the transmission system during system restoration.

- Sustained overvoltage

A sustained overvoltage is caused by line-charging currents of lightly loaded high-voltage transmission lines. The problems caused by sustained overvoltage include overheating of transformers, instability of the power system, overexcited transformers, under- and overexcitation of generators, and harmonic distortions.

Some FACTS devices are available to regulate overvoltage and absorb the large positive charging reactive power of lightly loaded high-voltage transmission lines. Such FACTS controllers include thyristor controlled shunt reactors, thyristor switched shunt capacitors, STATCOM, and SVC. With such FACTS controllers, the high-voltage transmission lines can be reenergized while maintaining the voltage within acceptable limits.

- Transient overvoltage

Transient overvoltage is caused by the closing of power circuit breakers while energizing a large section of a high-voltage transmission line, or by switching capacitive components in the system. Such transient overvoltage may lead to failures of protective relays. For the high-voltage transmission system, where the voltage is above 500 kV, and even 1000 kV for the newly planned extreme high-voltage AC transmission system in China, the operational voltage is close to the arrester operating voltage. Transient overvoltage together with the effect of sustained overvoltage may result in an arrester failure.

A possible solution for transient overvoltage is to utilize FACTS devices. The closure time point can be optimally selected with the instantaneous voltage to avoid switching transients in the system.

- Reduction in stored energy for circuit breaker

Traditional power circuit breakers need stored energy in the batteries or compressed air to control the breaker. The energy for operation of the breaker is depleted during the outage of the equipment or the blackout of the system and needs to be charged by the power system during the restoration procedure. The thyristor-based FACTS device can be an alternative switch with high speed, high power, and minimal required current, to replace the traditional mechanical circuit breakers. Without the consideration of stored energy for the breaker in the substations, power system restoration can be more flexible with a shortened duration and a simplified procedure.

One obstacle to the application of the FACTS device for the power circuit breaker is the considerable loss and limited overload capacity of the FACT device and semiconductor switches. The location and selection of the type of switches in the system need further investigation.

- Smooth load pickup

Frequency stability is a major concern while picking up the load in the initial stage of power system restoration. A frequency dip may occur with the pickup of a large size of load. The system restoration procedure usually restricts the size of incremental load pickup and the frequency drop in Hz in one-minute intervals.

As a summary, FACTS can change the power flow distribution of a system. During system restoration, potential violations or actual violations can be eliminated by FACTS. Potential benefits of FACTS for system restoration include but are not limited to the following:

- Thyristor-controlled phase regulators are used to control the standing phase angles.
- Thyristor-controlled voltage regulators are used for reactive power control of synchronous machines.
- SVCs or STATCOMs can be used to enhance system reliability of energizing long extra high voltage (EHV) transmission lines.
- Switching surges can be minimized by using thyristor-controlled circuit breakers, although this device has not yet been developed due to high losses.
- Thyristor-controlled circuit breakers can be used to reduce the required stored energy at transmission and distribution stations.
- Thyristor-controlled circuit breakers and thyristor-controlled braking resistors can be used to increase transient and dynamic stability.
- Asymmetry in EHV transmission lines can be eliminated by TCSC, shunt SVC, and STATCOM.

9.2 Applications of PMUs

Phasor measurement units (PMUs) can provide accurate and time-stamped power system information under diverse system operating conditions. Power system restoration is performed after a complete or partial outage to restore the system to a normal operating condition. Since system restoration is achieved by a series of operations under abnormal operating conditions, all actions should be implemented efficiently and safely. Understanding the status of the system is a high-priority task. The restoration strategy should be established and implemented according to the available information from the system. PMU, as a state-of-the-art technology, provides an efficient tool for information acquisition. Large-scale installation of PMUs has been motivated by the Northeast US blackout in 2003. Some organizations, such as NERC [20] and North American SynchroPhasor Initiative (NASPI) [21], emphasize the impact of PMUs on system restoration.

9.2.1 Review of PMU

9.2.1.1 Concept of PMU

The objective of PMU is to implement the so-called synchronized phasor concept, that is, the phasor measurements that occur at the same time at different locations. In power systems, enormous sensors have been installed. These sensors monitor information at different locations with considerably high accuracy. However, common time is not available until the invention of PMU. As a result, information at different locations cannot be synchronized by traditional sensors. It significantly challenges the power system operation, which should be balanced instantaneously.

PMU is developed with time-stamped measurements. This concept has been defined by IEEE standard C38.118 [22]. Two important definitions are shown as follows:

- Phasor: A complex equivalent of a simple cosine wave quantity such that the complex modulus is the cosine wave amplitude and the complex angle (in polar form) is the cosine wave phase angle.
- Synchronized phasor: A phasor calculated from data samples using a standard time signal as the reference for the measurement. Synchronized phasors from remote sites have a defined common phase relationship.

According to the definition of IEEE C38.118, currently, both magnitudes and phase angles of the sine waves of voltages and currents are measured at the locations where PMUs are installed. To implement synchronized phasor, PMUs synchronize from the common time source of a global positioning system (GPS) radio clock. The GPS receivers make possible the synchronization of several readings taken at distant points. Based on this technology, PMUs

Figure 9.19 The basic block diagram of PMU.

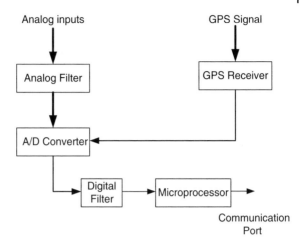

provide truly synchronized voltage and current measurements at diverse locations in a power grid to system operators. Using the accurately time-stamped measurements, it is possible to compare two quantities at remote locations in real time. System status can be assessed by this accurate comparison as well. The basic diagram is illustrated in Figure 9.19.

9.2.1.2 Standards of PMU

To integrate measurement systems into power system environments, standards are critical. With this standard, the data output formats are specified to ensure the measurements produce comparable results. The synchrophasor standard will help ensure maximum benefits from the phasor measurements and allow interchange of data between a wide variety of systems for users of both real-time and off-line phasor measurements.

The need for PMUs' standard as well as the standard for synchrophasors has been recognized by IEEE since 20 years ago. The first standard, IEEE Std 1344-1995 standard for synchrophasors, was completed in 1995 and reaffirmed in 2001. The latest standard, IEEE Std C37.118-2005, was completed in 2005. IEEE Std C37.118-2005 replaced the previous IEEE Std 1344-1995. The standard is not yet comprehensive – it does not attempt to address all factors of PMUs. Some important issues need to be addressed, including the definition of a synchronized phasor, definition of time synchronization, application of time tags, method to verify measurement compliance with the standard, and message formats for communication with a PMU.

Although the uses of PMUs are not limited by this standard, the primary purpose of this standard is to ensure PMUs' interoperability under steady-state conditions, that is during observation, signals of frequency, magnitude, and phase angle are constant. The reason is that in this standard, the time tag is defined as the time of the theoretical phasor represented by the estimated phasor, and then a time near the center of estimation window will be selected normally. Therefore, the straightforward application of PMUs is to provide measurements of voltages and currents under steady-state conditions.

9.2.2 System Restoration with PMU Measurements

Highly efficient restoration strategies' design and implementation are all based on available information. During system restoration, to maintain the safety of a power system, almost all constraints, such as steady-state constraints, dynamic constraints, and even electromagnetic constraints, should be met. At different stages, information requirements are diverse. For instance, at the beginning stage of system restoration, system operators have to identify the system status before establishing a restoration strategy while identifying violations for safety in implementing each restoration action. Furthermore, the system conditions during restoration are significantly different from regular operating conditions. Special considerations associated with different information requirements are needed. As a result, the monitoring system with high precision and communication speed is recognized as a critical element for power system restoration.

9.2.2.1 Observability of Power Systems with PMUs
The problem of PMU placement in power systems to ensure the observability is well recognized [23–25]. In this research, PMUs are assumed to have the capacity to measure voltage phasor at the buses where PMUs are installed and current phasors along the branches that are connected to the bus. Based on this understanding, the optimal PMU placement problem is modeled as a search problem to minimize the numbers of PMUs to cover all of the buses in the network with a depth of one. An illustrative example is shown in Figure 9.20. For instance, if two PMUs are installed at bus 1 and bus 3, respectively, complete observability can be obtained. However, if two PMUs are installed at bus 4 and bus 5, respectively, bus 2 cannot be observed.

In some research works, the optimal PMU placement problem is solved by some heuristic algorithms, such as the tree search algorithm, genetic algorithm, simulated annealing algorithm, and immunity genetic algorithm [23, 25]. As highly efficient heuristic algorithms, nonlinear constraints as well as realistic models of PMU are easy to be integrated. However, convergence property cannot be ensured theoretically at present. Another realistic consideration is the numbers of PMU channels, that is one PMU installed at a bus can only monitor limited phasors of current and voltage.

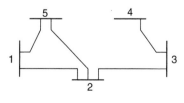

Figure 9.20 An illustrative example.

An algorithm is established to obtain optimal placement of PMU to ensure complete observability of system with limited information channels of each PMU. For a system with N buses, mathematically, the algorithm is formulated as follows:

Algorithm 1

$$\min \quad \mathbf{f}^T \mathbf{X} \tag{9.4}$$

$$\text{s.t. } \mathbf{C}\mathbf{X} > 0 \tag{9.5}$$

where $\mathbf{C}$ is the connection matrix of the power grid, that is,

$$\mathbf{C} = \left[c_{ij} \right] = \begin{cases} 1 & i = j \text{ or } i \text{ and } j \text{ } are \text{ connected directly} \\ 0 & i \text{ and } j \text{ } are \text{ not connected directly} \end{cases} \tag{9.6}$$

$\mathbf{X}$ is the binary decision vector of size N, ith element is 1 if a PMU is installed at bus i and 0 if no PMU is installed at that bus; $\mathbf{f}$ is defined as $\mathbf{Y} = 1 \cdot \mathbf{C}$ if $Y_i \geq M$, $f_i = \inf$, else $f_i = 1$, where M is the limit of channels of a PMU.

Use the network illustrated in Figure 9.20. The connection matrix is

$$\mathbf{C} = \begin{bmatrix} 1 & 1 & 0 & 0 & 1 \\ 1 & 1 & 1 & 0 & 1 \\ 0 & 1 & 1 & 1 & 0 \\ 0 & 0 & 1 & 1 & 0 \\ 1 & 1 & 0 & 0 & 1 \end{bmatrix}.$$

For the ith column, the jth element identifies whether bus j is connected with bus i ($c_{ij} = 1$) or not ($c_{ij} = 0$) by a branch. If two columns, say pth and qth, are added, all nonzero elements indicate that all of the buses connected with bus p and q directly. For example, summation of the second and third columns $(1, 2, 2, 1, 1)^T$ means all of the buses are connected with either bus 2 or bus 3. As a result, if two PMUs are installed at bus 2 and bus 3, respectively, complete observability can be obtained. The optimal placement of PMU is modeled to find minimal numbers of columns of connection matrix with nonzero elements in summarizing vector. Furthermore, to consider the limit of PMU channels, the numbers of branches connected with a bus where the PMU is installed should be limited. In other words, in connection matrix $\mathbf{C}$, the number of nonzero elements of the vector, which describes the candidate bus for PMU installed, should be less than the number of PMU channels.

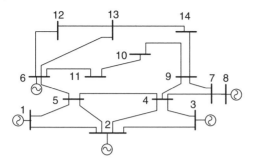

Figure 9.21 Topology of IEEE 14-bus system.

As a binary linear optimization problem, numerous highly efficient algorithms have been developed and can be employed to solve the proposed model with limited computing time. Use the IEEE 14-bus system as an example, in Figure 9.21.

If the limit of channels of PMU is not considered, one of the optimal placements of PMU is on buses 2, 6, 7, and 9. With different numbers of channels of each PMU, the optimal placements of PMUs are listed in Table 9.1.

It should be noted that one channel is used to measure the phasor of the bus voltage. As a result, if the number of channels is M, the branches from the bus are fewer than M.

The optimal placements of the IEEE 30-bus and 57-bus system are also listed in Table 9.2.

Results in Table 9.1 and Table 9.2 confirm the reduction in the number of PMUs using placement based on increasing channels of PMUs.

The case study was also conducted on the IEEE 118-bus system and 300-bus system. When the limit of PMU channels is 10, the optimal placements of PMUs for the IEEE 118-bus system are 3, 5, 9, 12, 15, 17, 20, 23, 28, 30, 36, 40, 44, 46, 51, 54, 57, 62, 63, 68, 71, 75, 77, 80, 85, 86, 90, 94, 101, 105, 110, and 114. For the IEEE 300-bus system the placements are 1, 2, 3, 11, 12, 15, 17, 22, 23, 25, 26, 27, 33, 37, 38, 43, 48, 49, 53, 54, 55, 58, 59, 60, 62, 64, 65, 68, 71, 73, 79, 83, 85, 86, 88, 92, 93, 98, 99, 101, 110, 112, 113, 116, 118, 119, 128, 132, 135, 138, 139, 143, 145, 148, 149, 152, 157, 163, 167, 173, 183, 187, 188, 189, 190, 193, 196, 202, 204, 208, 210, 211, 213, 216, 217, 219, 222, 226, 228, 263, 267, 269, 270, 272, 273, 274, 276, 280, 281, 282, 283, 284, 285, 286, 287, and 294.

Table 9.1 Optimal placements of PMU of IEEE 14-bus system.

Numbers of Channels	Buses with Installed PMUs
6	2, 6, 7, 9
5	1, 3, 7, 10, 13
4	1, 3, 8, 10, 12, 14

Table 9.2 Optimal placements of PMUs of IEEE 30-bus and 57-bus systems.

	Buses with Installed PMUs	
Numbers of Channels	IEEE 30-Bus	IEEE 57-Bus
8	1, 7, 9, 10, 12, 18, 24, 25, 27, 28	1, 4, 6, 13, 20, 22, 25, 27, 29, 32, 36, 39, 41, 45, 47, 51, 54
7	1, 7, 9, 12, 17, 19, 22, 24, 25, 27, 28	1, 2, 6, 10, 19, 22, 26, 29, 30, 32, 36, 39, 41, 44, 46, 49, 54
6	3, 5, 9, 13, 15, 17, 19, 22, 25, 27, 28	1, 4, 7, 10, 20, 23, 27, 30, 32, 36, 39, 41, 45, 46, 49, 52, 54
5	3, 5, 9, 13, 14, 17, 19, 22, 24, 25, 28, 29	3, 5, 8, 14, 16, 17, 19, 22, 26, 29, 30, 32, 36, 39, 42, 43, 45, 48, 51, 54
4	3, 5, 8, 11, 13, 14, 17, 19, 21, 23, 26, 29	2, 6, 12, 19, 21, 23, 27, 30, 33, 35, 39, 40, 42, 43, 44, 45, 46, 47, 50, 52, 54

9.2.2.2 Contributions of PMU for System Restoration

Time-stamped system information from PMUs significantly benefits system restoration in all stages. Generally, system restoration is divided into three stages, that is, the preparation stage, system restoration stage, and load restoration stage. PMUs have different contributions for different stages.

At the preparation stage, evaluation of system status and definition of target system is the major objective. PMUs can help implement the objective of this stage by providing precise system information. With PMUs' information, the remaining system is identified and available components of the system can be detected as well. By the state estimation technologies associated with PMUs, the status of the system can be precisely understood. Particularly, the most essential issue for system restoration – the initial sources – can be detected. This information will help operators initialize the restoration strategy. Furthermore, by detecting available components of the system, the target system can be designed.

At the system restoration stage, reintegration of the bulk network is the major objective. Some loads will be restored as a means of maintaining the stability of the system. Benefits from PMUs include monitoring system status to establish decisions, monitoring system status after each action to ensure security of the system, monitoring standing angles of the branches to ensure stability of the system, monitoring bus voltage magnitudes and phase angles to evaluate system voltage stability and small signal stability, and estimating transient stability before each action.

At load restoration stage, as the last stage of system restoration, PMUs can also be beneficial by providing information to support each restoration action. The benefits at this stage include monitoring steady-state variables of the system, that is, voltage, current, and power flow calculated by voltage and current, to ensure security of the system; monitoring frequency during pickup each load; evaluating voltage stability during each load pickup by the variables provided by PMUs; assessing small signal stability of system after a big load pickup; and evaluating transient stability for load pickup.

To fully implement these benefits of PMUs for system restoration, PMU information should achieve the following requirements:

- To optimize placement of PMUs to achieve complete observability of the grid
- To establish coordination of PMUs information during system restoration
- To design reasonable operation methods to ensure workability of PMUs following an outage

Algorithm 1 presented in this chapter can be used to fully implement the first requirement. For the last two requirements, more sophisticated algorithms are required.

9.2.2.3 Restoration-Oriented PMU Placement

To acquire sufficient and accurate information during system restoration, direct measurements from critical components are required. From the system restoration's viewpoint, the most important components are generating units and critical loads. The PMU placement problem in this context is to minimize numbers of PMUs subject to complete observability and make sure all important components are equipped with PMUs. Based on this idea, Algorithm 1 is modified as follows:

Algorithm 2

$$\min \quad \mathbf{f}^T \mathbf{X} \tag{9.7}$$

$$\text{s.t. } \mathbf{CX} > 0, \tag{9.8}$$

where $\mathbf{C}$ and $\mathbf{X}$ are the same as Algorithm 1; $\mathbf{f}$ is defined as follows: if a generating unit or critical load is connected at bus i, M is a large positive number; else it is based on the rules defined in Algorithm 1.

By setting different factors in vector $\mathbf{f}$, correlative elements of generating units and loads will be sent as a negative number. As a minimization problem, the installation on these buses can be ensured.

For the IEEE 14-bus system illustrated in Figure 9.21, only considering generating units at buses 1, 2, 3, 6, and 8, the solution is 1, 2, 3, 6, 8, and 9. All the generating units are installed. The complete observability is obtained as well. Compared with the result in Section 9.2.2 (1), more PMUs are installed because all generating units are equipped with PMUs.

This algorithm is also tested on the IEEE 30-bus and 57-bus systems. In the IEEE 30-bus system, generating units are installed at buses 1, 2, 13, 23, and 27. One of the solutions for PMU installation is 1, 2, 6, 9, 10, 12, 13, 18, 22, 23, 25,

Table 9.3 Restoration-oriented optimal placements of PMUs in IEEE 118-bus and 300-bus systems.

System	Bus with Generating Units	PMU Placement
IEEE 118-Bus	1, 4, 6, 8, 10, 12, 15, 18, 19, 24, 25, 26, 27, 31, 32, 34, 36, 40, 42, 46, 49, 54, 55, 56, 59, 61, 62, 65, 66, 69, 70, 72, 73, 74, 76, 77, 80, 85, 87, 89, 90, 91, 92, 99, 100, 103, 104, 105, 107, 110, 111, 112, 113, 116	1, 4, 6, 8, 10, 12, 15, 18, 19, 22, 24, 25, 26, 27, 31, 32, 34, 36, 40, 42, 45, 46, 49, 53, 54, 55, 56, 59, 61, 62, 65, 66, 69, 70, 72, 73, 74, 76, 77, 80, 85, 87, 89, 90, 91, 92, 96, 99, 100, 103, 104, 105, 107, 110, 111, 112, 113, 116
IEEE 300-Bus	8, 10, 19, 55, 63, 69, 76, 77, 80, 88, 98, 103, 104, 117, 120, 122, 125, 126, 128, 131, 132, 135, 149, 150, 155, 156, 164, 165, 166, 169, 170, 177, 192, 199, 200, 201, 206, 209, 212, 215, 217, 218, 220, 221, 222, 247, 248, 249, 250, 251, 252, 253, 254, 255, 256, 257, 258, 259, 260, 261, 262, 263, 264, 265, 267, 292, 294, 295, 296	7, 8, 10, 11, 16, 19, 23, 25, 27, 35, 37, 48, 51, 54, 55, 58, 60, 62, 63, 64, 68, 69, 71, 72, 73, 76, 77, 80, 81, 85, 88, 92, 93, 98, 99, 101, 103, 104, 109, 113, 117, 118, 120, 122, 125, 126, 128, 131, 132, 135, 138, 143, 145, 148, 149, 150, 155, 156, 157, 164, 165, 166, 169, 170, 173, 177, 183, 187, 189, 190, 192, 194, 199, 200, 201, 205, 206, 209, 212, 213, 215, 217, 218, 219, 220, 221, 222, 226, 228, 247, 248, 249, 250, 251, 252, 253, 254, 255, 256, 257, 258, 259, 260, 261, 262, 263, 264, 265, 267, 268, 269, 270, 272, 273, 274, 276, 292, 294, 295, 296

and 27. For the IEEE 57-bus system, generating units are installed at buses 1, 2, 3, 6, 8, 9, and 12. One of the solutions for PMU installation is 1, 2, 3, 6, 8, 9, 12, 15, 19, 22, 26, 29, 30, 32, 36, 39, 41, 45, 47, 50, and 53. For these two systems, similar results to the IEEE 14-bus system are obtained. With one more constraint, more PMUs are needed for complete observability. For the IEEE 118-bus system and 300-bus system, the results are listed in Table 9.3.

9.2.2.4   Establish Restoration Strategy with PMU Measurements

As described before, for the purpose of system restoration, with complete observability by PMU, all of the buses with generating units and critical loads are equipped with PMUs. During the restoration process, the establishment of each transmission path should ensure observability. Currently, the restoration decision support systems for transmission path establishment only consider steady-state or dynamic constraints, and information acquisition methods are not involved yet [26, 27]. In this context, usually, the charging current of each path is employed as the weight. As a result, the shortest path means the lowest risk for voltage violation, as proposed in Hou et al. [26, 27]. Here a sophisticated algorithm, which integrates PMU information and charging current of each line, is proposed. This method is modified from Algorithm 2 of Hou et al. [26].

To obtain an objective bus B from the energized block set Ω_E, the following steps are used.

Step 1: Establish the distance matrix, that is

$$\mathbf{DM} = \left[d_{ij} \right] =$$
$$\begin{cases} 0, & \text{if } i \text{ and } j \in \Omega_E \\ \text{charging current of line } i - j, & \text{if } i \text{ or } j \notin \Omega_E \cap i \text{ and } j \\ & \text{are observed with PMU} \\ \text{a large number } \rho, & \text{if } i - j \text{ is a transformer} \\ & \cap i \text{ or } j \notin \Omega_E \cup i \text{ or } j \\ & \text{are not observed with PMU} \end{cases} \quad (9.9)$$

Step 2: $\forall i \in \Omega_E$, find the shortest path from i to B by Dijkstra's algorithm [13] as $\{n_k, k = 1, 2, ..., m, \text{and } n_1 = i\}$, where n_k is a bus through the shortest path and the number of buses is m;

Step 3: Find $n_\lambda \in \Omega_E$ and $n_{\lambda+1} \notin \Omega_E$, where $1 \leq \lambda < m$

In this step, $n_{\lambda+1}$ is the first bus outside the block and all buses within the path after $n_{\lambda+1}$ are outside the block.

Step 4: Output Path = $\{n_k\}$, $k = \lambda + 1, \lambda + 2, ..., m$

The idea of this algorithm is to connect all buses within the block by a zero length line first. Therefore, the shortest path from any bus within this block to the object bus is the shortest path from this block to that bus. For the path with an unobservable bus, a large number is set as the penalty.

The proposed method is tested on the IEEE 14-bus system. As analyzed in Section 9.2.2 (2), PMUs are installed at buses 1, 2, 3, 6, 8, and 9. Assuming that the only generating unit at bus 1 is a black-start unit, according to the algorithm proposed in [26], the sequence of restoration is shown in Table 9.4. At each step, observability is obtained.

To sum up, PMUs and associated communication technology provide state-of-the-art technologies for system status monitoring.

Table 9.4 Sequence for restoration of generating units.

Step	Restoration Action	Path
1	Restart BS at 1	–
2	Crank NBS at 2	1–2
3	Crank NBS at 6	1–5–6
4	Crank NBS at 8	2–4–9–7–8
5	Crank NBS at 3	4–3

9.3 Microgrid in System Restoration

Natural disasters have been causing severe power outages in recent years. For example, in 2012, after Hurricane Sandy struck the East Coast of the US, approximately 8.35 million customers were reported to be without power [28]. Weather-related power outages have introduced tremendous economic loss and significant life risk, highlighting the importance of enhanced power grid resilience [29]. Rapid and effective response for electric service restoration is one of the critical requirements of a resilient power grid, as most recovery activities greatly depend on a reliable power supply [30]. However, a natural disaster can cause widespread and severe damage to power grids, leaving numerous customers without power for days, sometimes for over a week. For faster restoration, resilient response strategies are critically necessary when communities are threatened by natural disasters.

This chapter proposes dispatching microgrids (MGs) in system restoration. MGs can restore critical loads even when operating in islanded mode. The objective is to minimize the outage duration of loads, considering their priorities and demand sizes. Major decision variables include the following:

- Allocating mobile energy resources (MERs): to assign MERs to candidate nodes in the distribution system (DS)
- Forming MGs powered by MERs: to open or close each line to form MGs, and to pick up or not pick up each load

Conditions and constraints to be considered include

- Capacity differences of MERs
- Priorities of critical loads
- DS element damage
- Road network (RN) damage/congestion
- Radial topology requirements, operational constraints, and so on

Constrained by equipment or firmware requirements, only a fraction of the nodes in DSs will be feasible as candidate nodes for MER allocation and connection. Next, we give a more detailed introduction on utilizing MGs in system restoration. Then, a two-stage framework for realizing this operation strategy is introduced.

9.3.1 Microgrid-Based Restoration

9.3.1.1 Features of Microgrid-Based Restoration

Essentially, the MG-based restoration problem is to dispatch MGs in the power grid (normally, the DS) to restore critical loads under emergency situations. MERs can be utilized in this process. Generally, a MER may serve a single location such as a hospital or a government building. However, after a natural disaster, MERs will have to play a more important role than that. First, most critical and large loads, which are small in number, have backup power access to multiple feeders or self-installed emergency generators. And note that many MERs will be in a state of readiness before a natural disaster. Thus, a considerable number of MERs can be spared to serve small yet critical or less-critical loads, which are numerous. Second, as natural disasters often cause prolonged outages for many customers, MERs can be vital power sources for them for days or even for over a week after a natural disaster strikes. Moreover, MER dispatch and MG-based restoration in response to natural disasters have the following characteristics.

First, natural disasters often cause a complete or partial loss of power supply to DSs from the main grid. Major reasons include transmission system outages, substation faults, broken feeders or laterals of DSs, and so on. Timely electric service recovery of isolated outage regions by conventional restoration can be hindered. These are the major areas of interest for MER dispatch. Nevertheless, areas that still have power access to the main grid should also be considered, as operational constraints may prohibit them from being fully restored.

Second, sustained DS element damage, the resulting prolonged power outages to customers, and RN damage/congestion are quite uncertain prior to a natural disaster. Assessments of these may be completed in minutes or hours, or even days, after the natural disaster strikes. Then, matching between MERs and critical loads is conducted to guide MER dispatch. Matching can be a quite difficult and complicated task, considering grid damage, DS operational constraints, requirements for resource utilization efficiency of MERs, desired timely restoration, and so on.

Third, MERs have to travel on RNs to allocated locations. The outage duration of critical loads to be picked up by MERs is greatly influenced by MERs' travel time. Some critical loads may also require a MER to arrive before their backup power runs out. Note that RNs can be vulnerable to natural disasters, too. Thus, the traffic issue is a critical factor in MER dispatch. To reduce the outage duration of critical loads, proactive predispatch measures (prepositioning, in this chapter) are necessary, and vehicle routing of MERs should be considered.

9.3.1.2 Coordination with Conventional Restoration

The conceptual resilience curve in Panteli and Mancarella [31] is used here for more clear statements. In Figure 9.22, R is an index of system resilience level, t_{MG} the timing of MGs participating into DS restoration.

Associated with an event, a DS has these states: event progress $t_e \sim t_{pe}$, postevent degraded state $t_{pe} \sim t_r$, restorative state $t_r \sim t_{pr}$, postrestoration state $t_{pr} \sim t_{ir}$, and infrastructure recovery $t_{ir} \sim t_{pir}$. The time period $t_r \sim t_{ir}$ is the concern of conventional DS restoration strategies, which are generated by expert systems [32], multiagent systems [33], optimization [34], and so on. However, they may be of limited effect for a DS struck by a natural disaster. That is, the

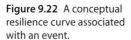

Figure 9.22 A conceptual resilience curve associated with an event.

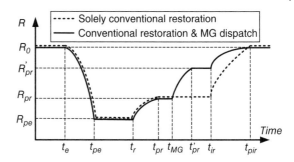

enhancement from R_{pe} to R_{pr} is small. In this case, MERs are desirable resources to enhance the system resilience level from R_{pr} to R'_{pr} in the postrestoration state. The concerned areas of MER dispatch are mainly the isolated outage regions and some areas that cannot be sufficiently restored by the surviving power access. Thus, although the concerned time period of MER dispatch is the same as that of conventional DS restoration, their concerned outage areas are different with limited overlaps. These two kinds of restoration actions are related yet independent to some extent. Generally, we can coordinate them in a straightforward way: For outage areas sufficiently recovered by conventional restoration strategies, apply these strategies; for isolated outage areas without power sources, send MERs and conduct MG formation after MERs arrive; for outage areas insufficiently recovered by conventional restoration strategies, apply conventional strategies first, and then transfer to coordinated restoration strategies after MERs arrive. Note that the coordinated strategies can be generated by solving our real-time allocation problem.

9.3.1.3 Implementation Issues

Figure 9.23, which is self-explanatory, reveals the relationships among data sets and task modules that are necessary to implement MER dispatch of prepositioning and real-time allocation. On the basis of the framework depicted in Figure 9.23, the proposed MER dispatch methodology can also be reduced or modified to address the utility repair truck scheduling problem. Proactive prepositioning and timely real-time response are also preferred in that problem. The detailed DS operation formulation is necessary to optimize the electric service recovery process, too.

To maintain functionality of a MG with two or more generation units, their coordination has to be resolved [35, 36]. This can be a difficult issue not only for temporarily installed MERs but also for preinstalled generators due to communication obstruction and the postevent degraded state of the system after natural disasters. Thus, although allowing two or more MERs in a MG can be more advantageous in matching power with loads and improving voltage security, in this work we choose to form each MG with only one power

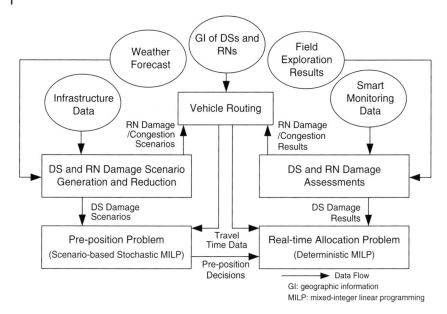

Figure 9.23 Relationships between data sets and task modules.

source following [37]. For some types of MERs equipped with required functional modules, parallel operation of two or three MERs can be conducted by well-trained staff [38]. The premise is that they are connected to the same DS node. Our formulations can accommodate this special case by minor modifications.

Feeder root nodes representing power from substations can also be modeled besides MERs. Other power or flexibility resources, including distributed generations (DGs) and the measures of partial load restoration/curtailment, can also be considered. Specifically, DGs can be treated in the same way as feeder root nodes. In this way, DGs restore critical loads by forming MGs exactly as that in [37]. Partial load restoration/curtailment can be incorporated by adding a continuous variable for each load and some easy modifications of the formulations. It is not always practically feasible. The focus of this chapter is designing a method to better utilize MERs in response to natural disasters.

9.3.2 Demonstration and Practice

9.3.2.1 Existing Practices
MGs are perceived as a viable means to improve distribution system restoration. For resilience measurement, a graph theory-based method was used in Chandra and Srivastava [39] for quantifying and enabling the resilience of

distribution systems. Using this method, the impact of possible control decisions could be analyzed quantitatively to proactively enable resilient operation. For resilience improvement, one promising method is the dynamic formation of MGs after disasters by utilizing available resources for their generation [37, 40]. A MG formation mechanism was proposed in Chen et al. [37] by exploiting DG and automatic remotely controlled switches. In Lei et al. [40], flexible resources, such as trunk-mounted mobile emergency generators, were utilized to form multiple MGs, and a two-stage dispatch framework was established with consideration of the prepositioning and the real-time allocation of these mobile generators. Network configuration for resilience enhancement has also been studied [41, 42]. In Ma et al. [41], an optimal network hardening strategy was proposed by compromising grid hardening investment and load shedding costs during extreme weather events. In Ren et al. [42], a programmable network was integrated into the MGs to provide flexible and an easy-to-manage communication solution. Compared with the conventional system restoration issues, practical considerations of the constraints for MG-driven restoration were incorporated in Xu et al. and Gao et al. [43, 44], and associated resilience-oriented service restoration methods were put forward. The stability of MGs and dynamic performance of DGs during the restoration process were incorporated in Xu et al. [43]. Power generation resource scarcity in MGs was considered using the proposed concept of continuous operating time [44], which determined the availability of MGs for critical load restoration and assessed the service time. In addition to using single MGs, multi-MG coordination was also proposed to enhance distribution system restoration through hierarchical outage management [45]. The functions of different management entities in a multi-MG system can be distributed after outages, and unused resources can be coordinated with consideration of possible power transfer among MGs for feeding unserved loads. In Chen and Zu [46], the interactions between MGs were captured based on a game-theoretic framework, and a resilient and fully distributed algorithm was proposed for MGs to update their control strategy using renewable energy generation (Figure 9.24).

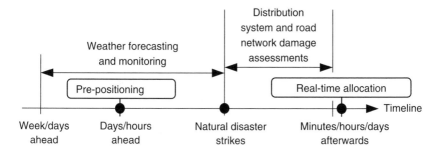

Figure 9.24 Timing of prepositioning and real-time allocation.

9.3.2.2 A Two-Stage Dispatch Framework

The proposed operation strategy for MG formation with MERs is fulfilled by a proposed two-stage dispatch framework comprising prepositioning and real-time allocation.

In the first stage, that is, prior to a natural disaster, prepositioning is conducted. Resource prepositioning is a common proactive measure undertaken by electric utilities. Before a natural disaster, they allot resources (including repair crews and restoration equipment) across their staging locations to ensure the earliest possible response after the natural disaster strikes [47]. We determine that prepositioning of MERs, that is, placing utilities' MERs in staging locations for earliest future response, is also necessary for several interrelated reasons:

- The earliest possible electric service recovery is desired.
- MERs' travel time to allocated places can be saved.
- The influence of RN damage/congestion can be reduced.

In the second stage, that is, after the natural disaster strikes, real-time allocation is optimized. MERs are sent from staging locations to allocated locations. On arrival, they are connected to the grid and form MGs to pick up critical loads.

Currently, we set capacities of available MERs as parameters. In some cases, dispatchers have to or are allowed to select a budgeted number of MERs among different types. Our formulations can accommodate these cases by listing enough numbers of MERs of each type and letting the optimizer choose. A budget constraint limiting the amount of selected MERs should also be added. In this way, the capacities of MERs can be co-optimized with other decisions. Note that this may introduce extra computational burden. Experience-based heuristics may help to relieve it. For example, an initial solution of good quality may be set at first by the dispatchers based on experience.

Generally, the candidate nodes for MER connection are selected based on the following. (i) Site requirements. Their locations should have appropriate space to install and operate a MER, free of potential risks such as flooding [38]. (ii) Access requirements. Their locations should be reachable by the truck-mounted MERs and fuel trucks via the RNs [38]. (iii) Facility requirements. A node with connection panel to interface with plug-terminated cables of MERs is preferred [44]. An underground fuel tank at its location is also a plus [38, 48]. (iv) Some other considerations. One example is the noise. If necessary, dispatchers can further reduce the number of candidate nodes based on their preferences or other factors such as distance.

9.3.2.3 Scenario Generation for Natural Disaster Damages

On one hand, statistical data fitting models use power grid data and environmental data to estimate outages and damages. Measurements of fitting goodness are also studied for evaluation. For example, models such as Bayesian

Additive Regression Trees are assessed in estimating the number of damaged DS poles in Guikema et al. [49]. On the other hand, simulation-based models make predictions based on physical mechanisms of damages. For example, in McClure et al. [50] the mechanism of localized high intensity wind damaging overhead line is studied. Interested readers may refer to others [47, 51, 52] for more detailed reviews. Thus, applying simulation-based or statistical models, scenario generation for natural disaster damages itself is a critical, challenging, and active research topic. A potential topic is to consider graph-theoretic metrics here. In this work we generally assume that extensive expert experience or mature tools for scenario generation are available and do not investigate too much detail on this topic since it is not the focus. Specifically, we generate scenarios following Ouyang and Dueñas-Osorio, and Winkler et al. [53, 54], that is, comparing failure probabilities of vulnerable components with a random variable uniformly distributed in the interval (0, 1). And we reduce scenarios by three rules. First, prioritize scenarios in which MERs can restore more critical loads, since it means little for the MER dispatch problem to consider scenarios in which MERs will not be so helpful. Second, aggregate scenarios with the same real-time allocation solution, as they tend to impact the prepositioning decisions in a similar way. Third, prioritize scenarios with higher probabilities of occurrence. The travel time of edges in RNs is assumed to have a lognormal distribution.

9.3.2.4 Applied Algorithms

Dijkstra's shortest-path algorithm for the vehicle routing (VR) problem: As mentioned before, the traffic issue can influence optimal MER dispatch decisions, as MERs have to spend time traveling on RNs to allocated locations. Thus, a VR module is employed, which finds the shortest or fastest route from an origin to a destination via RNs. VR can be realized by Dijkstra's algorithm [55], the Floyd–Warshall algorithm [56], and so on. We apply Dijkstra's algorithm here, as only routes from staging locations to candidate nodes are of interest. For each scenario, the travel time from each staging location to each candidate node is derived via the VR module. Then the data are used both in the prepositioning and real-time allocation optimization problems. Thus, the traffic issue is considered when optimizing MER dispatch to reduce the outage duration of critical loads to be picked up by MERs. One might also consider MERs' exploration roles in damage assessment of DSs and RNs by assigning must-pass locations.

Scenario decomposition (SD) algorithm for prepositioning: The scenario-based two-stage stochastic optimization problem of prepositioning has a block-diagonal structure. It can be recast in a compact form as follows:

$$\min\left\{\sum_n f_n\left(\mathbf{x},\mathbf{y}_n\right): \mathbf{x} \in \Lambda, \mathbf{y}_n \in \Omega_n\right\}, \tag{9.10}$$

where $\mathbf{x}$, $\mathbf{y}_n$, $f_n(\mathbf{x},\mathbf{y}_n)$, Λ, and Ω_n denote the first-stage prepositioning variables, second-stage real-time allocation variables under scenario n, and second-stage objective function for scenario n, respectively. Generally, the SD algorithm is to transform Eq. (9.10) into Eq. (9.11):

$$\min\left\{\sum_n f_n\left(\mathbf{x}_n,\mathbf{y}_n\right) : \mathbf{x}_n \in \Lambda, \mathbf{y}_n \in \Omega_n, \sum_n \mathbf{A}_n\mathbf{x}_n = \mathbf{h}\right\}. \tag{9.11}$$

where the last constraint is the non-anticipativity constraints enforcing x1 = x2 = , ..., = xn. Thus the problem's block-diagonal structure can be taken advantage of. Implementation of the SD algorithm is described in Algorithm 3 [57].

Algorithm 3   Scenario decomposition algorithm

1) Set *upper_bound* = $+\infty$, *lower_bound* = $-\infty$, $\mathbf{S} = \varnothing$, $\mathbf{x}^* = \varnothing$
2) Set $\lambda = \mathbf{0}$
3) Solve $\alpha_n = \min\{f_n(\mathbf{x}_n,\mathbf{y}_n) + \lambda^\mathrm{T}\mathbf{A}_n\mathbf{x}_n : \mathbf{x}_n \in \Lambda\backslash\mathbf{S}, \ \mathbf{y}_n \in \Omega_n\}$ and obtain $\mathbf{x}_n^*$ for all n
4) If consensus criteria are not met, update λ and go to Step 3, otherwise Step 5
5) Update *lower_bound* = $\sum_n u_n \alpha_n - \lambda^\mathrm{T}\mathbf{h}$, $\mathbf{S}' = \cup_n\{\mathbf{x}_n^*\}$, $\mathbf{S} = \mathbf{S}\cup\mathbf{S}'$
6) For all $\mathbf{x}^0 \in \mathbf{S}'$, do
7) Solve $\beta_n = \min\{f_n(\mathbf{x}^0,\mathbf{y}_n) : \mathbf{y}_n \in \Omega_n\}$ for all n
8) If *upper_bound* $\geq \sum_n u_n \beta_n$, update *upper_bound* = $\sum_n u_n \beta_n$, $\mathbf{x}^* = \mathbf{x}^0$
9) If (*upper_bound*-*lower_bound*)/ *lower_bound* $\geq$ *convergence_tolerance*, go to Step 2; otherwise, terminate

It is essentially a modified version of the progressive hedging algorithm [58], which can be classified as an augmented Lagrangian relaxation algorithm. Specifically, $f_n(\mathbf{x}_n,\mathbf{y}_n) + \lambda^\mathrm{T}\mathbf{A}_n\mathbf{x}_n$, that is, the objective function in Step 3, can be rewritten in a more detailed form as follows:

$$\min_{\mathbf{x}_n,\mathbf{y}_n}\sum_l \left[\mu(l)+\lambda(l)-\rho\bar{\mathbf{x}}_n(l)+0.5\rho\right]\mathbf{x}_n(l)+\vartheta(l)\mathbf{y}_n(l), \tag{9.12}$$

where $\bar{\mathbf{x}}_n$ denotes the weighted average of $\mathbf{x}_n$ over all scenarios $n \in \mathbf{N}$ in the previous iteration; μ and ϑ are cost coefficient vectors, that is, problem parameters; $(\bullet)(l)$ denotes the lth element of the vector $(\bullet)$; ρ is an algorithm parameter set as 10000 in this work. Note that, other than fixed ρ strategies, there exist variable ρ strategies. Interested readers can refer to references [59–61], which discuss both of them. And λ is updated by

$$\lambda(l) \leftarrow \lambda(l)+\rho\left[\mathbf{x}_n(l)-\bar{\mathbf{x}}_n(l)\right], \tag{9.13}$$

which is a generally applied updating rule. The consensus criterion in Step 4 is set as follows:

$$td = \sum_{l,n} \frac{\left| \mathbf{x}_n(l) - \overline{\mathbf{x}}_n(l) \right|}{|\mathbf{N}|} \le \underline{td}, \tag{9.14}$$

where td is defined as the average per-scenario deviation from the average; $\underline{td}$ denotes a threshold set as 0.5; and $|\mathbf{N}|$ is the number of scenarios. The consensus criteria can also be set based on other metrics such as normalized average per-scenario deviation from the average [59] and overall cost discrepancy [60]. As for the actual real-time allocation optimization problem to be solved after the natural disaster strikes, in our case it can be directly solved by a solver such as Gurobi with acceptable efficiency.

References

1 Xiang, D., Ran, L., Bumby, J. et al. (2006). Coordinated control of an HVDC link and doubly fed induction generators in a large offshore wind farm. *IEEE Trans. Power Delivery* 21 (1): 463–471.

2 Foster, S., Xu, L., and Fox, B. (2008). Control of an LCC HVDC system for connecting large offshore wind farms with special consideration of grid fault, in *Power and Energy Society General Meeting, 2008*.

3 Jia, X., Liu, B., Olguin, R.E.T., and Undeland, T. (2010). Grid integration of large offshore wind energy and oil & gas installations using LCC HVDC transmission system, in *SPEEDAM, 2010 International Symposium*. pp. 784–791.

4 Sun, X., Liu, Z., Gao, L., and Ding, Y. (2009). Practice and innovation in the ±800 kV UHVDC demonstration project. *Proc. CSEE.* 29 (22): 35–45.

5 Fan, L., Miao, Z., and Osborn, D. (2009). Wind farms with HVDC delivery in load frequency control. *IEEE Trans. Power Syst.* 24 (4): 1894–1895.

6 Huang, H. (2012). *HVDC and Power Electronics*. CIGRE.

7 Zhou, H., Yang, G., and Geng, H. (2008). Grid integration of DFIG-based offshore wind farms with Hybrid HVDC connection, Electrical Machines and Systems, in *ICEMS October 2008*

8 Andersen, B.R. and Xu, L. (2004). Hybrid HVDC system for power transmission to island networks. *IEEE Trans. Power Delivery.* 19 (4): 1884–1890.

9 Flourentzou, N., Agelidis, V.G., and Demetriades, G.D. (2009). VSC-based HVDC power transmission systems: an overview. *IEEE Trans. Power Electron.* 24 (3): 25–30.

10 Du, C. (2007). *VSC-HVDC for industrial power systems*, Ph.D. dissertation, Department of Energy and Environment, Chalmers University of Technology, Göteborg, Sweden, 2007.

11 Kim, C.K., Sood, V.K., Jang, G.S. et al. (2009). *HVDC Transmission: Power Conversion Applications in Power Systems*. Wiley.

12 Bahrman, M. and Bjorklund, P.E. (2014). The new black start: system restoration with help from voltage-sourced converters. *IEEE Power Energ. Mag* 12 (1): 44–53.

13 Dong, D., Li, J., Boroyevich, D., Mattavelli, P., Cvetkovic, I., and Xue, Y. (2012). Frequency behavior and its stability of grid-interface converter in distributed generation systems, in *Proceedings 2012 IEEE Applied Power Electronics Conference Expo.*, pp. 1887–1893.

14 Guide for the Development of Models for HVDC Converters in a HVDC Grid, *CIGRÉ WG* B4-57, pp. 53–83, 2014.

15 Chaudhuri, N.R., Majumder, R., and Chaudhuri, B. (Feb. 2013). System frequency support through multi-terminal DC (MTDC) grids. *IEEE Trans. Power Syst.* 28 (1): 347–356.

16 Kundur, P.C. (1994). *Power System Stability and Control*, Vol. 7, pp. 598–599. New York; Toronto: McGraw-Hill.

17 Adibi, M.M. and Fink, L.I I. (1994). Power system restoration planning. *IEEE Trans. Power Syst.* 9 (1): 22–28.

18 Li, S., Zhou, M., and Liu. Z. (2009). A study on VSC-HVDC based black start compared with traditional black start, in *2009 International Conference on Sustainable Power Generation and Supply*.

19 Adibi, M.M., Martins, N., and Watanabe, E.H. (2010). The impacts of FACTS and other new technologies on power system restoration dynamics, in *Power and Energy Society General Meeting, 2010 IEEE*, pp. 1–6.

20 NERC. Real-time application of synchrophasors for improving reliability [Online]. Available: http://www.nerc.com/docs/oc/rapirtf/RAPIR%20final%20101710.pdf.

21 NASPI. Actual and potential phasor data applications [Online]. Available: http://www.naspi.org/resources/2009_march/phasorappstable_20091201.pdf.

22 IEEE, IEEE Standard for Synchrophasors for Power System Std C37.118-2005.

23 Nuqui, R.F. and Phadke, A.G. (2005). Phasor measurement unit placement techniques for complete and incomplete observability. *IEEE Trans. Power Delivery* 20 (4): 2381–2388.

24 Emami, R. and Abur, A. (2010). Robust measurement design by placing synchronized Phasor measurements on network branches. *IEEE Trans. Power Syst.* 25 (1): 38–43.

25 Aminifar, F., Lucas, C., Khodaei, A., and Fotuhi-Firuzabad, M. (2009). Optimal placement of phasor measurement units using immunity genetic algorithm. *IEEE Trans. Power Delivery* 24 (3): 1014–1020.

26 Hou, Y., Liu, C.-C., Sun, K. et al. (Jun. 2011). Computation of milestones for decision support during system restoration. *IEEE Trans. Power Syst.* 26 (3): 1399–1409.

27 Hou, Y. and Liu, C.-C. (2001). Reducing duration of system restoration based on GRM algorithms. Annual Report for EPRI (Contract EP-P35424/C16059), November 2010.

28 NERC, Hurricane Sandy event analysis report (2014). [Online]. Available at: http://www.nerc.com/pa/rrm/ea/Oct2012HurricanSandyEvntAnlyssRprtDL/ Hurricane_Sandy_EAR_20140312_Final.pdf.

29 Campbell, R.J. (2012). Weather-related power outages and electric system resiliency, Congress Res. Service, Washington, D.c., Tech. Rep. R42696 [Online]. Available at: www.fas.org/sgp/crs/misc/R42696.pdf.

30 EPRI (2013). Enhancing distribution resiliency: opportunities for applying innovative technologies, [Online]. Available at: www.epri.com/abstracts/ Pages/ProductAbstract.aspx?ProductId=000000000001026889.

31 Panteli, M. and Mancarella, P. (May 2015). The grid: Stronger, bigger, smarter? *IEEE Power Energ. Mag.* 13 (3): 58–66.

32 Liu, C.C., Lee, S.J., and Venkata, S.S. (May 1988). An expert system operational aid for restoration and loss reduction of distribution systems. *IEEE Trans. Power Syst.* 3 (2): 619–626.

33 Nguyen, C.P. and Flueck, A.J. (Jun. 2012). Agent based restoration with distributed energy storage support in smart grids. *IEEE Trans. Smart Grid* 3 (2): 1029–1038.

34 Khushalani, S., Solanki, J.M., and Schulz, N.N. (May 2007). Optimized restoration of unbalanced distribution systems. *IEEE Trans. Power Delivery* 22 (2): 624–630.

35 Qi, H., Wang, X., Tolbert, L. et al. (Dec. 2011). A resilient real-time system design for a secure and reconfigurable power grid. *IEEE Trans. Smart Grid* 2 (4): 770–781.

36 Katiraei, F. and Iravani, M.R. (2006). Power management strategies for a microgrid with multiple distributed generation units. *IEEE Trans. Power Syst.* 21 (4): 1821–1831.

37 Chen, C., Wang, J., Qiu, F., and Zhao, D. (Mar. 2016). Resilient distribution system by microgrids formation after natural disasters. *IEEE Trans. Smart Grid* 7 (2): 958–966.

38 Iwai, S., Kono, T., Hashiwaki, M., and Kawagoe, Y. (2009). Use of mobile engine generators as source of back-up power, in *Proceedings of the IEEE 31st International Telecommunications Energy Conference, October 2009*, pp. 1–6.

39 Chanda, S. and Srivastava, A.K. Defining and enabling resiliency of electric distribution systems with multiple microgrids. *Smart Grid, IEEE Trans. on* (accepted and early access).

40 Lei, S., Wang, J., Chen, C., and Hou, Y. (May 2018). Mobile emergency generator pre-positioning and real-time allocation for resilient response to natural disasters. *Smart Grid, IEEE Trans.* 9 (3): 2030–2041.

41 Ma, S., Chen, B., and Wang, Z. (March 2018). Resilience enhancement strategy for distribution systems under extreme weather events. *Smart Grid, IEEE Trans.* 9 (2): 1442–1451.

42 Ren, L., Qin, Y., Wang, B. et al. (Nov. 2017). Enabling resilient microgrid through programmable network. *Smart Grid, IEEE Trans.* 8 (6): 2826–2836.

43 Xu, Y., Liu, C.C., Schneider, K.P. et al. (Jan. 2018). Microgrids for service restoration to critical load in a resilient distribution system. *Smart Grid, IEEE Trans.* 9 (1): 426–437.

44 Gao, H., Chen, Y., Xu, Y., and Liu, C.C. (Nov. 2016). Resilience-oriented critical load restoration using microgrids in distribution systems. *Smart Grid, IEEE Trans.* 7 (6): 2837–2848.

45 Farzin, H., Firuzabad, M.F., and Aghtaie, M.M. (Nov. 2016). Enhancing power system resilience through hierarchical outage management in multi-microgrids. *Smart Grid, IEEE Trans.* 7 (6): 2869–2879.

46 Chen, J. and Zhu, Q. (Jan. 2017). A game-theoretic framework for resilient and distributed generation control of renewable energies in microgrids. *Smart Grid, IEEE Trans.* 8 (1): 285–295.

47 Whipple, S.D. (2014). Predictive storm modelling and optimizing crew response to improve storm response operations, M.B.A. and M.Sc. Dissertation, MIT, Cambridge, MA.

48 Nightingale, C.R. (1983). The design of mobile engine driven generating sets and their role in the British telecommunications network, in *Proceedings of the 5th International Telecommunications Energy Conference, October 1983*, pp. 144–150.

49 Guikema, S.D., Quiring, S.M., and Han, S.-R. (Dec. 2010). Prestorm estimation of hurricane damage to electric power distribution systems. *Risk Anal.* 30 (12): 1744–1752.

50 Mcclure, G., Langlois, S., and Rogier, J. (2008). Understanding how overhead lines respond to localized high intensity wind storms, in *Proceedings of the Structures Congress, Vancouver, BC, Canada, 2008* [Online]. Available at: http://ascelibrary.org /doi/abs/10.1061/41016(314)192.

51 Han, S.-R. (2008). Estimating hurricane outage and damage risk in power distribution systems, Ph.D. dissertation, Texas A&M University, College Station, TX.

52 Wang, Y., Chen, C., Wang, J., and Baldick, R. (May 2015). Research on resilience of power systems under natural disasters – a review. *IEEE Trans. Power Syst.* 31 (2): 1604–1613.

53 Ouyang, M. and Dueñas-Osorio, L. (2012, Art. no.). Time-dependent resilience assessment and improvement of urban infrastructure systems. *Chaos* 22 (3): 033122.

54 Winkler, J., Dueñas-Osorio, L., Stein, R., and Subramanian, D. (2010). Performance assessment of topologically diverse power systems subjected to hurricane events. *Reliab. Eng. Syst. Saf.* 95 (4): 323–336.

55 Dijkstra, E.W. (Dec. 1959). A note on two problems in connexion with graphs. *Numer. Math.* 1 (1): 269–271.

56 Cormen, T.H., Stein, C., Rivest, R.L., and Leiserson, C.E. (2001). *Introduction to Algorithms*. McGraw-Hill Higher Education.

57 Ahmed, S. (2013). A scenario decomposition algorithm for 0-1 stochastic programs. *Oper. Res. Lett.* 41 (6): 565–569.

58 Rockafellar, R.T. and Wets, R.J.-B. (1991). Scenarios and policy aggregation in optimization under uncertainty. *Math. Oper. Res.* 16 (1): 119–147.

59 Watson, J.-P., Woodruff, D.L., and Strip, D.R. (2007). Progressive hedging innovations for a stochastic spare parts support enterprise problem. *Naval Research Logistics.*

60 Watson, J.-P. and Woodruff, D.L. (Nov. 2011). Progressive hedging innovations for a class of stochastic mixed-integer resource allocation problems. *Comput. Mag. Sci.* 8 (4): 355–370.

61 Crainic, T.G., Fu, X., Gendreau, M. et al. (2011). Progressive hedging-based meta-heuristics for stochastic network design. *Networks* 58 (2): 114–124.

10

Black-Start Capability Assessment and Optimization

10.1 Background of Black Start

Black-start (BS) capability is important for system planners to prepare the power system restoration (PSR) plan. To achieve a faster restoration process, installing new BS generators can be beneficial in accelerating system restoration. While additional BS capability does not automatically benefit the restoration process, power systems have to update the PSR plan and quantify the benefit based on appropriate criteria. In this chapter, a decision support tool is utilized to provide a quantitative way for assessing BS capability.

10.1.1 Definition of Black Start

A BS is the process of restoring a power station to operation without relying on external energy sources. Following an outage of the power system, power stations usually rely on the electric power provided from the station's own generators. For example, small diesel generators can provide electric power to start larger generators (of several MW capacity), which in turn can be used to start the main power station generators. However, steam turbine generators require station service power of up to 10% of their capacity for boiler feed water pumps, boiler forced-draft combustion air blowers, and fuel preparation [1]. It is not economical to provide such a large standby capacity at each station, so BS power must be provided over the transmission network from other stations. If part or all of the plant's generators are shut down, station service power is also supplied from the grid.

After a partial or complete system blackout, dispatchers rely on off-line restoration plans and available BS capabilities to restore system back to normal operation conditions. The typical BS scenario includes BS generating units providing power to start large steam turbine units located electrically close to these units; the supply of auxiliary power to nuclear power stations and off-site

Power System Control Under Cascading Failures: Understanding, Mitigation, and System Restoration, First Edition. Kai Sun, Yunhe Hou, Wei Sun and Junjian Qi.
© 2019 John Wiley & Sons Ltd. Published 2019 by John Wiley & Sons Ltd.
Companion website: www.wiley.com/go/sun/cascade

power to critical service load, such as hospitals and other public health facilities; military facilities; transmission lines that transport the cranking power to NBS units or large motor loads; and transformer units, including step-up transformers of the BS units and steam turbine units, as well as auxiliary transformers serving motor control centers at the steam plant [1].

According to the start-up power requirement, generating units can be divided into two groups: BS generators and non-black-start (NBS) generators. A BS generator, for example, a hydro or combustion turbine unit, can be started with its own resources, while NBS generators, such as steam turbine units, require cranking power from outside. The typical BS generators are

Hydroelectric generating units: These generators need very little initial power to open the intake gates, and have fast response characteristics to provide power to start fossil fueled or nuclear stations.

Diesel generating units: Diesel generators usually require only battery power and can be started quickly to supply the power to start up larger generating units. They are small in size and generally cannot be used to pick up any major transmission system elements.

Gas turbine generating units: Aero-derivative gas turbine generators can be started remotely with the help of local battery power. Large gas turbine generators are coupled with on-site diesel generator sets, which are started and used to energize plant auxiliary buses and start either the gas turbine or steam turbine. Gas turbine generators can be started and pick up load within a short time. Time to restart and available ramping capability are functions of the duration when the unit was off-line.

The typical generators that are contracted for BS service are 10–50 MW small hydro or gas turbine units, and in a few cases even 200–400 MW steam units. The bus voltage values can be 6.9 kV for hydro units, 12.8 or 13.8 kV for gas turbine units, and 22 kV for steam turbine units. However, not all generating plants are suitable for BS service. For example, mini-hydro or micro-hydro plants rely on a power network connection for frequency regulation and reactive power supply. Therefore, BS units must be stable when operated with the large reactive load of a long transmission line. Also, traditional high-voltage direct current (HVDC) converter stations cannot operate without the commutation power from the system at the load end.

10.1.2 Constraints During BS

Power system restoration following a blackout begins with the BS units and then restores the system outward toward critical system loads. Among various BS restoration steps, the key concern is the control of voltage and frequency, both of which must be kept within a tight band around nominal values to guarantee no equipment failure will severely hinder the restoration process.

Voltage stress is a major concern for a BS during power system restoration. BS units are required to be able to absorb the produced reactive power from charging current by energizing the unloaded generator step-up transformer and transmission lines. When energizing a transmission line, the produced charging currents can be large enough to result in the BS generating unit absorbing reactive power, which may cause self-excitation. The self-excitation will result in an uncontrolled rise in voltage or equipment failure. Thus, it is important to verify the reactive power capacity of the BS unit when operated at a leading power factor. The installation of shunt reactors, synchronization of generating units as a block and minimizing the time between paralleling units online will help to reduce the probability of exposing a generating unit to the condition of self-excitation.

During the BS process, starting up generators and picking up large blocks of load will perturb the system frequency, which can be prevented by picking up loads in increments that can be accommodated by system inertia and response of already started-up generators. However, when reenergizing the load that has been de-energized for several hours or longer, the generated inrush current can be as high as 8–10 times higher than normal, known as the phenomenon of cold load pickup. The large inrush current brought by picking up loads need to be carefully considered in a BS plan.

During the implementation of the BS plan, voltage, rotor angle, and frequency stability have to be maintained as important components in the stability assessment of the plan. Therefore, BS plans must be validated by tests or simulation in terms of both steady state and transient operating conditions. A step-by-step simulation is required to verify the BS plan's feasibility and compliance with required operational limits on voltage and power flows. Also, the robustness of the BS plan needs to be verified to ensure its ability to compensate for some equipment unavailability.

10.1.3 BS Service Procurement

BS service is an ancillary service that is procured for power system restoration after a complete or partial outage. These BS resources must be able to energize buses and have on-site diesel or gas turbine generators to provide power for the auxiliary systems of the generating unit, which then can be used to start the unit. In North America, independent system operators (ISOs) identify and contract resources with BS capability and form financial agreements with them to provide this obligatory service. Traditionally, BS service costs were rolled into a broad tariff for cost recovery from ratepayers. In the deregulated environment, some ISOs, for example, the Electric Reliability Council of Texas (ERCOT), have shifted this cost-of-service provision to a competitive procurement. There are three methods of procuring BS service.

- The most common one is cost of service, in which generating units are identified for BS resources and the costs are rolled into a tariff for cost recovery. This simple and traditional method is currently used by the California Independent System Operator, the PJM Interconnection, and the New York Independent System Operator.
- The second method is a new methodology that uses a flat rate in $/kWyr to increase BS remuneration to encourage provision. Then the monthly compensation paid to a generator is determined by multiplying this flat rate by the unit's monthly claimed capability for that month. The new method is aimed at simplifying the procurement process and incentivizing the provision of BS, which is currently used by the Independent System Operator of New England.
- The last method is a competitive procurement as used in ERCOT, which runs a market for BS services. In the market, interested participants submit an hourly standby cost in $/h, which is named an availability bid that is unrelated to the capacity of the unit. Then, based on various criteria, ERCOT evaluates these bids and the selected units are paid as bid. Each BS unit must be able to demonstrate its ability to start up another unit.

There are other procurement methods. The New Zealand System Operator procures the BS service through a competitive tender. Other jurisdictions also have some sort of competitive procurement, although not as structured as ERCOT. Alberta Electric System Operator and Independent Electric System Operator of Ontario use a long-term "Request for Proposals" approach, which is similar to but not as structured as in ERCOT.

In ERCOT, BS service is awarded through a competitive annual bidding process where market participants submit bids for hourly standby price for their generators to provide the service. ERCOT then selects the capable resources of the providing BS service that meets the BS selection reliability criteria at a minimum cost. The criteria include the required amount of load to be recovered and the minimum time to recover that load. The general process in each year is as follows [2]:

1) On April 1st, ERCOT posts the BS Request for Proposal on the website.
2) In the following two months, market participants formulate and submit their bids for entering BS service market.
3) Based on the criteria listed as before, ERCOT evaluates and analyzes the bids to develop a list of preliminary BS units.
4) The preliminary units must undergo physical tests in a "real" BS scenario, including Basic Start test, Line Energizing test, Load Carrying test, and Next Start Resource test, to prove their ability to provide BS service.

5) On successful completion of the BS tests, BS resources are awarded the BS service contracts for the next calendar year.

6) At the same time, the complete BS plan is formulated with the regional transmission operators and is made available to the system operator for training and use.

After BS service resources are selected, ERCOT pays an hourly standby fee at their bid price, with an adjustment for reliability based on a six-month rolling availability equal to 85% in accordance with the BS Agreement.

In the deregulated environment, Independent Power Producers (IPPs) are able to provide BS service, which is not possible in the previous regulated power system. Generators owned by IPPs are usually located near or within industrial areas. These resources can quickly supply power to adjacent users [3]. While more and more BS resources are available, it is important for system dispatchers to provide adequate but not redundant BS capabilities considering BS costs. The BS capability assessment is required to assist PSR plans.

10.1.4 Power System Restoration Procedure

Power system restoration is a complex problem involving a large number of generation, transmission and distribution, and load constraints [4]. A common approach to simplify this task is to divide the restoration process into stages (e.g., preparation, system restoration, and load restoration stages) [5]. According to these restoration stages, PSR strategies can be categorized into six types [5], that is, build-upward, build-downward, build-inward, build-outward, build-together, and serve-critical. Nevertheless, one common thread linking each of these stages is the generation availability at each restorative stage for stabilizing the system, establishing the transmission path, and restoring load [6], as shown in Figure 10.1.

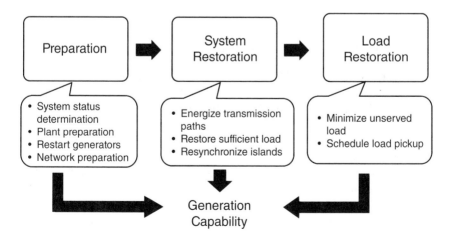

Figure 10.1 Illustration of generation availability connecting three restoration stages.

Following a system blackout, some fossil units may require cranking power from outside in order to start the unit. Some units may have time constraints within which the unit can be started successfully or else they have to be off-line for an extended period of time before they can be restarted and resynchronized to the grid. As a result, it is important that, during system restoration, available BS generating units provide cranking power to NBS generating units in such a way that the overall available generation capability is maximized [7]. Given limited BS resources and different system constraints on different generating units, the maximum available generation can be determined by finding the optimal start-up sequence of all generating units in the system.

10.2 BS Capability Assessment

10.2.1 Installation Criteria of New BS Generators

To achieve a faster restoration, additional BS generators might be useful to accelerate the restoration process. After new BS generating units are installed, system restoration steps, such as generator start-up sequence, transmission path, and load pick-up sequence, will change. Then each system restoration stage will adjust the restoration steps to accommodate the additional BS capability.

1) In the *Preparation* stage, there are three tasks for updating restoration plans:
   - **Task 1**: Update generator start-up sequence. With the help of additional BS generating units, more cranking power can be provided to start up NBS generators earlier to increase system generation capability.
   - **Task 2**: Update the transmission path to deliver the cranking power. With the updated generator start-up sequence, a new transmission path search is required to implement this updated sequence.
   - **Task 3**: Update critical load pickup sequence. Since the critical load needs to be picked up to maintain system stability, the critical load pickup sequence will be updated according to the new generator start-up sequence.

   By completing these three tasks, the updated restoration plan will proceed to the next stage.
2) In the *System Restoration* stage, three tasks needed to be performed to update restoration plans:
   - **Task 4**: Update transmission paths to energize and build the skeleton of the transmission system. The critical restoration actions, such as

energization of high voltage lines and switching actions, need to follow the updated transmission path search.

- **Task 5**: Update the dispatchable load pickup sequence. Sufficient loads need to be restored to stabilize generation and voltage. Larger or base-load units are prepared for load restoration in the next stage.
- **Task 6**: Update the resynchronization of electrical islands. Many system parameters, such as voltage stability, VAR (voltage ampere reactive) balance, and voltage/frequency response, need to be checked and monitored to synchronize islands in a reliable way.

3) In *Load Restoration* stage, there is one task required for restoration plans:

- **Task 7**: Updated load pickup sequence. This is different from load pickup in the previous two stages, which are aimed at stabilizing the power system. The objective in this stage is to minimize the unserved load according to the total system generation capability.

The seven tasks in three restoration stages help to update the restoration plan with installation of additional BS capabilities. This is illustrated in Figure 10.2.

While restoration plans are updated by installing new BS generators, the benefits of additional BS capabilities need to be evaluated. There are multiple options for installation sizes and locations. Each option will lead to a different restoration time. Therefore, the restoration time can be used as a criterion to

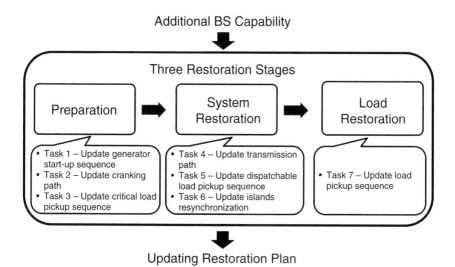

Figure 10.2 Updated restoration plan with additional BS capability.

quantify the benefit. However, there is a point where the benefits will not increase further. Therefore, power systems need to evaluate the strategy of both placement and size of new BS generators and quantify the benefits with the appropriate criteria.

In the literature, there are various objectives or criteria to develop PSR strategies. For example, to maximize the total system generation capability [7], to minimize the unserved energy [8], to maximize the total (or certain percentage of) restored load within the given restoration time [9], to minimize the total restoration time [10], and so on. However, there are obstacles for the direct applications of previous work on the BS capability assessment. The knowledge-based system (KBS) restoration tool [11] has been developed to integrate both dispatchers' knowledge and computational algorithms for system analysis. However, KBSs require special software tools and, furthermore, the maintenance of large-scale knowledge bases is a difficult task. The Critical Path Method [12] is able to estimate system restoration time, which requires the preselected restoration strategies. However, for different installation strategies of new BS generators, it is difficult to provide and compare the updated PSR strategies. The mixed-integer linear programming (MILP)-based optimal generator start-up strategy [7] is able to provide the overall system generation capability and update the solution throughout the BS process. Without considering system topology and power flow constraints, the solution will deviate from the actual system generation capability. Therefore, the appropriate criteria and solution methodology are required to assess BS capability to provide the optimal installation strategy of new BS generators.

One of the objectives of system restoration is to restart as much load as possible within the shortest time. After installing additional BS generating units, the reduced restoration time can be obtained from the updated restoration plan. Each installation strategy, including different placement and size of new BS generators, will bring different restoration time. The value of additional BS capability will be evaluated in terms of the system restoration time. However, from the cost–benefit point of view, the cost of installing additional BS capability is another criterion to evaluate the strategy. These two criteria of reduced restoration time and cost of installing BS capability will decide the installation plan, as shown in Figure 10.3. These two criteria together provide information on the benefit based on the cost of installation; for example, installing a certain amount of BS capability will reduce the restoration time. Then power system operators can develop their best installation strategies according to their own perspectives. Therefore, based on the criteria of the total restoration time and installed additional BS capability, power systems can make decisions on the optimal installation strategy of both location and amount of BS capability.

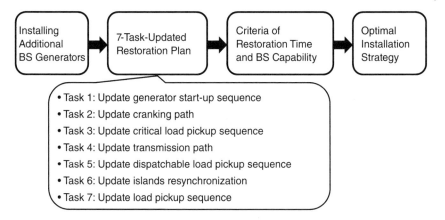

Installing Additional BS Generators ➡ 7-Task-Updated Restoration Plan ➡ Criteria of Restoration Time and BS Capability ➡ Optimal Installation Strategy

- Task 1: Update generator start-up sequence
- Task 2: Update cranking path
- Task 3: Update critical load pickup sequence
- Task 4: Update transmission path
- Task 5: Update dispatchable load pickup sequence
- Task 6: Update islands resynchronization
- Task 7: Update load pickup sequence

Figure 10.3 Criteria of restoration time and BS capability.

10.2.2 Optimal Installation Strategy of BS Capability

The seven tasks will provide the updated restoration plan. However, these tasks will change with different systems or different installation strategies. Among different PSR strategies, there are several general actions to perform these seven restoration tasks, such as

- Generator start-up sequencing
- Transmission path search
- Load pickup sequencing
- Optimal power flow check

Then restoration tools – for example, a generic restoration milestone (GRM)-based restoration tool – can be utilized to provide the algorithms of generic restoration actions to perform the seven tasks, as shown in Figure 10.4.

In this way, a method will be developed to determine the optimal sizes and locations of new BS units:

- **Step 1**: Select the installation location and amount for the additional BS capability.
- **Step 2**: Based on restoration tools, the seven tasks are performed to obtain the updated restoration plan.
- **Step 3**: Based on the criteria of restoration time and BS capability, the installation choice is evaluated.
- **Step 4**: Update the installation strategy, continue the previous steps to obtain the optimal installation strategy.

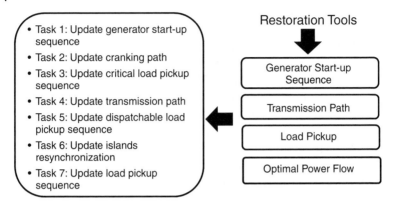

Figure 10.4 Restoration tool for updating restoration plans.

10.2.3 Examples

After installing new BS generators, the restoration process needs to be analyzed using the restoration tool. The System Restoration Navigator (SRN), developed by EPRI (Electric Power Research Institute), is able to compute the total restoration time and provide detailed restoration steps, which facilitate the investment strategy of additional BS capability. In the following example, this GRMs-based restoration tool is utilized to compute the restoration time and update the restoration plan.

Example 10.1 In this example, the PJM 5-bus system [13] is used for illustration of the proposed model and solution methodology, as shown in Figure 10.5. There are four generators, five buses, and six lines. The generator and transmission system information are given in Tables 10.1 and 10.2. The scenario of a complete shutdown is assumed. Unit G4 is a BS unit (BSU), while G1–G3 are

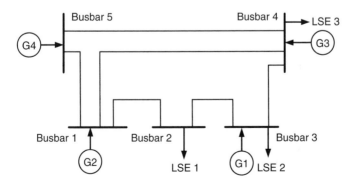

Figure 10.5 PJM five-bus system.

Table 10.1 Data of generator characteristics.

Gen.	T_{ctp} (h)	T_{cmin} (h)	T_{cmax} (h)	Rr (MW/h)	P_{start} (MW)	P_{max} (MW)	Connected Bus
G1	2	N/A	5	2	1	8	3
G2	1	5	N/A	4	1	12	1
G3	2	N/A	4	4	2	20	3
G4	1	N/A	N/A	1	N/A	3	5

Table 10.2 Data of transmission system.

Branch	From Bus	To Bus	Reactance X	Limit
1	1	2	0.0281	2.50
2	2	3	0.0108	3.5
3	3	4	0.0297	2.4
4	4	5	0.0297	2.4
5	5	1	0.0064	4
6	1	4	0.0304	1.5

Table 10.3 Restoration actions.

Time	Action
$t = 0$	Start G4
$t = 1$	N/A
$t = 2$	Energize Bus 5
$t = 3$	Energize Bus 4, Line 4
$t = 4$	Energize Bus 1, Line 5, Line 6; start G3
$t = 5$	Energize Bus 2, Line 1; start G2
$t = 6$	Energize Bus 3, Line 2, Line 3
$t = 7$	Start G1

NBS units (NBSUs). The restoration actions are checked and updated every 10 minutes.

The restoration actions are shown in Table 10.3. In the base case, there is only 1 BSU, G4, on Bus 5, with ramping rate of 1 MW per hour and maximum generation output of 3 MW. When increasing the BS capability by twice the

Table 10.4 Comparison of restoration times and system generation capabilities under different installation strategies.

	$t_{start}1$ (h)	$t_{start}2$ (h)	$t_{start}3$ (h)	$t_{start}4$ (h)	Total Restoration Time (h)	System Generation Capability (MWh)
Base Case	7	5	4	0	13	210.5
Install at Bus 5	5	5	2	0	11	259
Install at Bus 4	4	5	3	0	10	284
Install at Bus 3	3	5	4	0	11	273
Install at Bus 2	4	5	4	0	11	266
Install at Bus 1	5	5	4	0	11	259

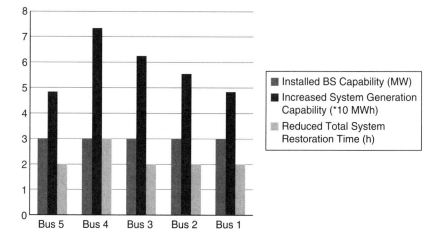

Figure 10.6 Comparison of different installation strategies.

base case, the generator starting sequence and time will change. However, if one continues increasing the BS capability, there will be no further improvements. Therefore, the optimal amount of new BS capability will be 3 MW. The comparisons of generator starting time after installing new BS capability at different buses are shown in Table 10.4.

Also, the comparison of installed BS capabilities, increased system generation capability, and reduced total restoration time with three different installation strategies of BS capabilities are shown in Figure 10.6. The comparison of system generation capability curves under different installation strategies of BS capability is shown in Figure 10.7. The optimal installation strategy will be to install 3 MW BS capabilities at Bus 4 to reduce restoration time by 2 h and increase system generation capability by 63.5 MWh.

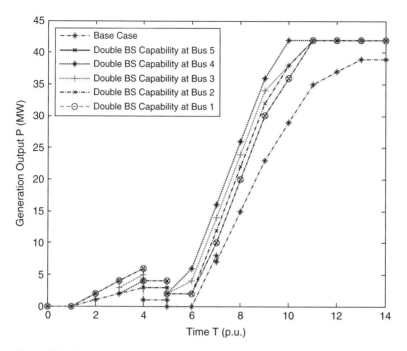

Figure 10.7 Comparison of system generation capability curve under different installation strategies.

10.3 Optimal BS Capability

10.3.1 Problem Formulation

Based on the developed MILP modules of generation and transmission system restoration, the optimal installation of BS capabilities is formulated as the mixed-integer bilevel programming (MIBLP) problem.

10.3.1.1 Optimal Generator Start-Up Sequence Module

During system restoration, it is critical to utilize the available BSUs for providing cranking power to NBSUs in such a way that the overall system generation capability will be maximized. Given limited BS resources and various system constraints on different generating units, the maximum available generation capability can be determined by finding the optimal start-up sequence of all generating units in the system. Therefore, the optimal generator start-up sequencing problem can be formulated as the following MILP problem [7].

First, define the following parameters, sets, and decision variables:

1) Parameters

T represents specified restoration time, N_{bus} is total bus number, N_{line} is total line number, N_{BSU} is total BSU number, N_{NBSU} is total NBSU number, N_{ALLU} is total generator number, $T_{ic\,max}$ is maximum intervals constraint of NBSU i, $T_{ic\,min}$ is critical minimum intervals constraint of NBSU i, Rr_i is ramping rate of generator i, $P_{i\,max}$ is maximum generator output of generator i, P_{istart} is start-up power requirement of NBSU i, T_{ictp} is cranking time for generator i to begin to ramp up and parallel with system, B_{mn} is susceptance of line mn, F_{mn} is thermal limit of line mn, M represents arbitrarily large number.

2) Sets

$\Omega_t = 1, \ldots, T$, $\Omega_{t-1} = 1, \ldots, T-1$, and $\Omega_{t-2} = 2, \ldots, T$ are sets of time, $\Omega_{bus} = 1$, $\ldots, N_{bus}$ is the set of bus numbers, $\Omega_{line} = 1, \ldots, N_{line}$ is the set of line numbers, $\Omega_{BSU} = 1, \ldots, N_{BSU}$ is the set of BSU numbers, $\Omega_{NBSU} = 1, \ldots, N_{NBSU}$ is the set of NBSU numbers, $\Omega_{ALLU} = 1, \ldots, N_{ALLU}$ is the set of all generator numbers, Ω_{line-m} is the set of lines connected with Bus m, and $\Omega_{bus-BSU}$ is the set of buses connected with BSU.

3) Decision variables

$u_{bus\,n}^t$ is a binary decision variable of bus n at time t, 1 means energized and 0 means de-energized; $u_{line\,mn}^t$ is a binary decision variable of line mn at time t, 1 means energized and 0 means de-energized; P_n^t is a linear decision variable of generation output of generator n at time t, u_j^t is a binary decision variable of status of generator j at time t, $j \in \Omega_{NBSU}$, $t \in \Omega_t$; t_{jstart} is the starting time of NBSU j, $j \in \Omega_{NBSU}$; f_{mn}^t is the power flow on line mn at time t; θ_m^t is the bus m voltage angle; $t_{i1}^{t_{kstart}}, t_{i2}^{t_{kstart}}, t_{i3}^{t_{kstart}}$ are linear decision variables to define generator capability function and start-up power function, $i \in \Omega_{ALLU}$, $k \in \Omega_{NBSU}$; $w_{i1}^{t_{kstart}}, w_{i2}^{t_{kstart}}, w_{i3}^{t_{kstart}}$ are binary decision variables to define generator capability function and start-up power function, $i \in \Omega_{ALLU}$, $k \in \Omega_{NBSU}$; $v_{j1t}^{t_{kstart}}, v_{j2t}^{t_{kstart}}, v_{j3t}^{t_{kstart}}$ are binary decision variables, $j \in \Omega_{NBSU}$, $t = \Omega_t$, $k \in \Omega_{NBSU}$.

10.3.1.1.1 Objective Function

The objective is to maximize the overall system MW generation capability during a specified system restoration period.

$$
\underset{t_{istart}}{Max} \left\{ \sum_{i \in \Omega_{ALLU}} \left[\left(P_{i\,max} \right)^2 / \left(2Rr_i \right) + P_{i\,max} \left(T - T_{ictp} - P_{i\,max}/Rr_i \right) \right] \right.
$$
$$
\left. - \sum_{j \in \Omega_{NBSU}} P_{jstart} T \right\} - \left(\sum_{i \in \Omega_{ALLU}} P_{i\,max} t_{istart} - \sum_{j \in \Omega_{NBSU}} P_{jstart} t_{jstart} \right) \tag{10.1}
$$

The first component (in braces) is constant, then Eq. (10.1) is equivalent to

$$\underset{t_{jstart}}{Min} \sum_{j \in \Omega_{NBSU}} \left(P_{j\max} - P_{jstart} \right) t_{jstart}. \tag{10.2}$$

10.3.1.1.2 Constraints

1) Constraints of critical maximum and minimum time intervals

$$\left.\begin{array}{l} t_{istart} \leq T_{ic\max} \\ t_{istart} \geq T_{ic\min} \end{array}\right\} i \in \Omega_{NBSU} \tag{10.3}$$

2) Constraints of MW start-up requirement

$$\sum_{i \in \Omega_{ALLU}} Rr_i \left(t_{kstart} - t_{i1}^{t_{kstart}} - t_{i2}^{t_{kstart}} \right) - \sum_{j \in \Omega_{NBSU}} w_{j3}^{t_{kstart}} P_{jstart} \geq 0$$

$$k \in \Omega_{NBSU} \tag{10.4}$$

3) Constraints of generator capability function

$$w_{i1}^{t_{kstart}} T_{ictp} \leq t_{i1}^{t_{kstart}} \leq T_{ictp}, \quad i \in \Omega_{BSU}, k \in \Omega_{NBSU} \tag{10.5}$$

$$\left(T + 1 + T_{jctp} \right) w_{j1}^{t_{kstart}} - \sum_{t \in \Omega_t} v_{j1t}^{t_{kstart}} \leq t_{j1}^{t_{kstart}} \leq t_{jstart} + T_{jctp}$$

$$j \in \Omega_{NBSU}, k \in \Omega_{NBSU} \tag{10.6}$$

$$w_{i2}^{t_{kstart}} P_{i\max}/Rr_i \leq t_{kstart} - t_{i1}^{t_{kstart}} - t_{i2}^{t_{kstart}} \leq w_{i1}^{t_{kstart}} P_{i\max}/Rr_i$$

$$i \in \Omega_{ALLU}, k \in \Omega_{NBSU} \tag{10.7}$$

$$t_{i2}^{t_{kstart}} \leq w_{i2}^{t_{kstart}} \left(T - T_{ictp} - P_{i\max}/Rr_i \right)$$

$$i \in \Omega_{BSU}, k \in \Omega_{NBSU} \tag{10.8}$$

$$t_{j2}^{t_{kstart}} \leq \sum_{t \in \Omega_t} v_{j2t}^{t_{kstart}} - w_{j2}^{t_{kstart}} \left(T_{jctp} + P_{i\max}/Rr_i + 1 \right)$$

$$j \in \Omega_{NBSU}, k \in \Omega_{NBSU} \tag{10.9}$$

$$w_{i2}^{t_{kstart}} \leq w_{i1}^{t_{kstart}}, \quad i \in \Omega_{ALLU}, k \in \Omega_{NBSU} \tag{10.10}$$

4) Constraints of generator start-up power function

$$w_{j3}^{t_{kstart}} T - \sum_{t \in \Omega_t} v_{j3t}^{t_{kstart}} \leq t_{j3}^{t_{kstart}} \leq t_{jstart} - 1$$

$$j \in \Omega_{NBSU}, k \in \Omega_{NBSU}$$

(10.11)

$$w_{j3}^{t_{kstart}} \leq t_{kstart} - t_{j3}^{t_{kstart}} \leq \sum_{t \in \Omega_t} v_{j3t}^{t_{kstart}}$$

$$j \in \Omega_{NBSU}, k \in \Omega_{NBSU}$$

(10.12)

5) Constraints of decision variables

$$t_{jstart} = \sum_{t \in \Omega_t} \left(1 - u_j^t\right) + 1, \quad j \in \Omega_{NBSU}$$

(10.13)

$$u_j^t \leq u_j^{t+1}, \quad t \in \Omega_{t-1}, \quad j \in \Omega_{NBSU}$$

(10.14)

$$\left.\begin{aligned} v_{jht}^{t_{kstart}} &\geq w_{jh}^{t_{kstart}} + u_j^t - 1 \\ v_{jht}^{t_{kstart}} &\leq w_{jh}^{t_{kstart}} \\ v_{jht}^{t_{kstart}} &\leq u_j^t \end{aligned}\right\} \begin{aligned} &h = 1,2,3, t \in \Omega_{t-1}, \\ &k \in \Omega_{NBSU}, j \in \Omega_{NBSU} \end{aligned}$$

(10.15)

10.3.1.2 Optimal Transmission Path Search Module

By assigning each bus or line one integer decision variable, the optimal transmission path search problem can be formulated as a MILP problem to find the status of each bus or line at each time. The detailed formulations are shown ahead.

10.3.1.2.1 Objective Function

The objective is to maximize the total generation output.

$$\underset{u_{busn}^t, u_{linemn}^t}{Max} \sum_{t \in \Omega_t} \sum_{n \in \Omega_{NBSU}} u_{busn}^t P_n^t$$

(10.16)

10.3.1.2.2 Constraints

1) If both connected buses are de-energized/energized, then the line is de-energized/energized.

$$-M u_{busm}^t \leq u_{linemn}^t \leq M u_{busm}^t$$

$$-M u_{busn}^t \leq u_{linemn}^t \leq M u_{busn}^t$$

$$t \in \Omega_t, mn \in \Omega_{line}, m \& n \in \Omega_{bus}$$

(10.17)

2) If both connected buses are de-energized at t, then the line is de-energized at $t + 1$; if the line is energized at $t + 1$, then at least one of the connected buses is energized at t.

$$-M\left(u_{busm}^{t} + u_{busn}^{t}\right) \leq u_{linemn}^{t+1} \leq M\left(u_{busm}^{t} + u_{busn}^{t}\right)$$
$$t \in \Omega_{t-1}, mn \in \Omega_{line}, m \,\&\, n \in \Omega_{bus} \tag{10.18}$$

3) For any bus not connected with BSU, if all lines connected with this bus are de-energized, then this bus is de-energized.

$$-M \sum_{mn \in \Omega_{line-m}} u_{linemn}^{t} \leq u_{busm}^{t} \leq M \sum_{mn \in \Omega_{line-m}} u_{linemn}^{t}$$
$$t \in \Omega_{t-2}, m \in \Omega_{bus}/\Omega_{bus-BSU} \tag{10.19}$$

4) Once a bus or line is energized, it will not be de-energized again.

$$u_{linemn}^{t} \leq u_{linemn}^{t+1}, \quad t \in \Omega_{t-1}, mn \in \Omega_{line}$$
$$u_{busm}^{t} \leq u_{busm}^{t+1}, \quad t \in \Omega_{t-1}, m \in \Omega_{bus} \tag{10.20}$$

5) All the lines' initial states are de-energized; all the buses' initial states are de-energized, except the ones connected with BSU are energized.

$$u_{linemn}^{t} = 0, \quad t = 1, mn \in \Omega_{line}$$
$$u_{busm}^{t} = 1, \quad t = 1, m \in \Omega_{bus-BSU}$$
$$u_{busm}^{t} = 0, \quad t = 1, m \in \Omega_{bus}/\Omega_{bus-BSU} \tag{10.21}$$

6) Transmission line thermal limit constraint

$$-\left(1 - u_{linemn}^{t}\right)M \leq f_{mn}^{t} - B_{nm}\left(\theta_{m}^{t} - \theta_{n}^{t}\right) \leq \left(1 - u_{linemn}^{t}\right)M$$
$$-u_{linemn}^{t}\overline{F}_{mn} \leq f_{mn}^{t} \leq u_{linemn}^{t}\underline{F}_{mn}$$
$$t \in \Omega_{t}, m \in \Omega_{bus}, mn \in \Omega_{line} \tag{10.22}$$

10.3.1.3 Adding the Time of GRAs into the Optimization Modules

The concept of generic restoration actions (GRAs) is proposed in Fink et al. [5] to generalize various restoration steps in different system restoration strategies. The time to take restoration actions should be considered to achieve a more accurate estimation of total restoration time. The (fictitious) time to complete each GRA is given in Table 10.5.

Table 10.5 Time to complete generic restoration actions (GRAs).

GRA	Time (min)
GRA1: start_black_start_unit	$T_{GRA1} = 15$
GRA2: find_path	$T_{GRA2} = $ N/A
GRA3: energize_line	$T_{GRA3} = 5$
GRA4: pick_up_load	$T_{GRA4} = 10$
GRA5: synchronize	$T_{GRA5} = 20$
GRA6: connect_tie_line	$T_{GRA6} = 25$
GRA7: crank_unit	$T_{GRA7} = 15$
GRA8: energize_bus	$T_{GRA8} = 5$

Then each GRA time can be integrated into the optimization modules in the following way:

1) BSU Module
   - $t = 0$, start BSU
   - $t = \max\{T_{ctp}, T_{GRA1} = 15\}$, BSU is ready
   - $t = \max\{T_{ctp}, T_{GRA1} = 15\} + 5$, bus connected with BSU is energized
2) NBSU Module
   - $t = 0$, crank NBSU from bus
   - $t = 15$, synchronize NBSU with bus
   - $t = 15 + \max\{T_{ctp}, T_{GRA5} = 20\}$, NBSU is ready
3) Bus/Line Module
   Each line needs the following time to energize
   $(T_{GRA3} = 5 \text{ minutes}) + (T_{GRA8} = 5 \text{ minutes}) = 10 \text{ minutes}$
4) Load Module
   - t = 0, pick up load from bus
   - t = 10, load is ready
5) If necessary, add $T_{GRA6} = 25$ minutes to connect tie line.

10.3.1.4 Connections of Optimization Modules

The optimization modules of generator start-up sequencing and transmission path search, considering the time to take restoration actions, can be integrated by adding the following constraints:

1) NBSU will be started after the connected bus is energized.

$$u_{busm}{}^t \geq u_j^t, \quad m \in \Omega_{bus-NBSUj}, j \in \Omega_{NBSU}, t \in \Omega_t \qquad (10.23)$$

2) The starting time of NBSU equals the time when the bus connected with NBSU being energized plus the time of GRA7.

$$\sum_{t\in\Omega_t}\left(1-u_j^t\right)+1=\sum_{t\in\Omega_t}\left(1-u_{busm}{}^t\right)+1+T_{GRA7}$$

$$m\in\Omega_{bus-NBSUj}, j\in\Omega_{NBSU}$$

(10.24)

3) The time when the bus connected with BSU being energized equals the sum of the starting time of BSU (zero), the time of GRA8, and the larger value between the time of GRA1 and T_{ctp} of BSU.

$$\sum_{t\in\Omega_t}\left(1-u_{busm}{}^t\right)+1=\max\left\{T_{GRA1},T_{ctp_i}\right\}+T_{GRA8}+0$$

$$m\in\Omega_{bus-BSUi}, i\in\Omega_{BSU}$$

(10.25)

Then change the constraint (Eq. (10.21)) so that all buses are de-energized at $t = 1$.

10.3.1.5 Optimal Installation Strategy of BS Capability

Based on the developed MILP modules of generation and transmission system restoration, the optimal installation of BS capabilities is formulated as the following MIBLP problem.

$$\underset{P_{i\max},Rr_i,t_{istart}}{\text{Max}}\quad\left\{\sum_{i\in\Omega_{ALLU}}\left[\left(P_{i\max}\right)^2/\left(2Rr_i\right)+P_{i\max}\left(T-T_{ictp}-P_{i\max}/Rr_i\right)\right]\right.$$

$$\left.-\sum_{j\in\Omega_{NBSU}}P_{jstart}T\right\}-\sum_{j\in\Omega_{NBSU}}\left(P_{j\max}-P_{jstart}\right)t_{jstart}$$

$$s.t.\quad\underset{t_{istart}}{\text{Min}}\quad\sum_{j\in\Omega_{NBSU}}\left(P_{j\max}-P_{jstart}\right)t_{jstart}$$

s.t.   Critical Time Intervals Constraint $-$ Eq. (10.3)

MW Start-up Requirement Constraint $-$ Eq. (10.4)

Generator Capability Function Constraints $-$ Eqs. $(10.5-10.10)$

Generator Start-up Power Function Constraints $-$ Eqs. $(10.11-10.12)$

Decision Variables Constraints $-$ Eqs. $(10.13-10.15)$

Transmission Path Search Logic Constraints $-$ Eqs. $(10.17-10.21)$

Transmission Line Thermal Limit Constraint $-$ Eq. (10.22)

GRAs Time and Modules Connection Constraints $-$ Eqs. $(10.23-10.25)$

(10.26)

By solving this problem, system generation capability equals the optimal objective value, and the estimated total restoration time can be obtained as follows:

$$t_{sys} = \max_{i \in \Omega_{ALLU}} \left(t_{istart} + T_{ictp} + P_{i\max} / Rr_i \right) \tag{10.27}$$

10.3.2 Solution Algorithm

In the literature, there are a few methods for a restricted class of bilevel linear programming (BLP) problems. For example, no integer decision variable is involved in the lower level problem. There are no direct applications to MIBLP problems from the previous work. The solution algorithm based on Benders decomposition and transformation procedure [14] is described as follows:

The MIBLP problem can be written in a compact format, that is

$$
\begin{aligned}
\max_{x,y} \quad & C1^* x + d1^* y \\
s.t. \quad & A1^* x + B1^* y \le b1 \\
& y \in \arg\max_{w} \quad C2^* x + d2^* w \\
& \qquad s.t. \qquad A2^* x + B2^* w \le b2 \\
& \qquad\qquad\qquad w \ge 0 \quad w_I \text{ integer} \\
& x \ge 0 \quad x_J \text{ integer}
\end{aligned} \tag{10.28}
$$

Step 1: Divide the MIBLP problem into one restricted master problem (RMP) (Eq. (10.37)) and several slave problems (SPs) (Eq. (10.29)) by fixing binary variables.

1) Fix the values of the binary variables $z = \bar{z}$, and solve the BLP problem:

$$
\begin{aligned}
\max_{x,y,z} \quad & D1^* x + E1^* y + F1^* \bar{z} \\
s.t. \quad & A1^* x + B1^* y \le b1 - C1^* \bar{z} \\
& \max_{y,z} \quad D2^* x + E2^* y + F2^* \bar{z} \\
& \quad s.t. \quad A2^* x + B2^* y \le b2 - C2^* \bar{z} \qquad (w1) \\
& \qquad\qquad -x \le 0 \qquad\qquad\qquad\qquad (w2) \\
& \qquad\qquad -y \le 0 \qquad\qquad\qquad\qquad (w3)
\end{aligned} \tag{10.29}
$$

where $w1, w2, w3$ are the dual variables of the constraint.

2) In the initial step, RMP will only have the objective, and constraints are added in future iterations from the cut of solving SP. Then in each step, the solution of RMP will provide an upper bound of the original problem, and the solution of SP will provide a lower bound. It is because RMP is a relaxation of the original problem whereas the SP represents a restriction.

Step 2: Transform the SP problem Eq. (10.29) to a linear problem with complementarity constraints (LPCC) Eq. (10.30), and solve it by "θ-free algorithm" [15].

1) Based on Karush–Kuhn–Tucker (KKT) conditions, replace the lower level problem of Eq. (10.29) by complementarity constraints and add them to the upper level problem. Then get the following LPCC:

$$
\begin{aligned}
\max_{x,y} \quad & D1^* x + E1^* y + F1^* \overline{z} \\
s.t. \quad & A1^* x + B1^* y \le b1 - C1^* \overline{z} \\
& 0 \le b2 - C2^* \overline{z} - A2^* x - B2^* y \quad \perp \quad w1 \ge 0 \\
& 0 \le A2^* w2 - D2 \quad\quad\quad\quad\quad\;\; \perp \quad x \ge 0 \\
& 0 \le B2^* w3 - E2 \quad\quad\quad\quad\quad\;\; \perp \quad y \ge 0
\end{aligned}
\tag{10.30}
$$

2) Solve Eq. (10.30) using the "θ-free algorithm."
SP is the restricted MIBLP, and it provides a lower bound. The decomposition technique allows parallel computation of solving multiple SPs.

Step 3: From the solution of the LPCC problem, construct the LP problem Eqs. (10.31 and 10.32).

1) From the solution of Eq. (10.30), it is known which constraint in Eq. (10.30) is active. Then formulate the following linear programming problem by removing the optimality constraint of the lower level problem and set the active constraint (here randomly assume one for illustration purpose) to be the equality:

$$
\begin{aligned}
\max_{x,y} \quad & D1^* x + E1^* y + F1^* \overline{z} \\
s.t. \quad & A1^* x + B1^* y = b1 - C1^* \overline{z} \\
& A2^* x + B2^* y \le b2 - C2^* \overline{z} \\
& x, y \ge 0
\end{aligned}
\tag{10.31}
$$

2) Introduce slack variables to transform Eq. (10.31) into the following compact form:

$$
\begin{aligned}
\max_{s} \quad & H^* s \\
s.t. \quad & M^* s = b - C^* \overline{z} \quad (u) \\
& s \ge 0
\end{aligned}
\tag{10.32}
$$

The dual of Eq. (10.32) is

$$
\begin{aligned}
\min_{u} \quad & u^* \left(b - C^* \overline{z} \right) \\
s.t. \quad & u^* M \ge H \\
& u \quad \text{free,}
\end{aligned}
\tag{10.33}
$$

where u is the dual variable of constraint in Eq. (10.31).

3) Based on Farkas' Lemma, the necessary and sufficient condition for Eq. (10.32) to have at least one nonempty solution for z is Eq. (10.32) has a solution s if and only if $u^*(b - C^*\overline{z}) \geq 0$ and $u^*M \geq 0$ for all u.

When choosing z arbitrarily, there is a finite number of possibilities: z_1, z_2, ..., z_n. For each z_i, there is a corresponding inequality constraint. Then to make sure Eq. (10.32) has at least one nonempty solution, solve the master problem Eq. (10.34). Since the values of all z_i are obtained during the iteration process, only some constraints in Eq. (10.34) are known explicitly.

$$\max \quad \xi$$

$$s.t. \quad \begin{cases} u_1^{z_1}*(b-C^*z_1) \geq \xi \\ \cdots \\ u_P^{z_1}*(b-C^*z_1) \geq \xi \end{cases} \begin{cases} u_1^{z_2}*(b-C^*z_2) \geq \xi \\ \cdots \\ u_K^{z_2}*(b-C^*z_2) \geq \xi \end{cases} \cdots \begin{cases} u_1^{z_n}*(b-C^*z_n) \geq \xi \\ \cdots \\ u_R^{z_n}*(b-C^*z_n) \geq \xi \end{cases}$$

$$\begin{cases} u_1^{z_1}*(b-C^*z_1) \geq 0 \\ \cdots \\ u_P^{z_1}*(b-C^*z_1) \geq 0 \end{cases} \begin{cases} u_1^{z_2}*(b-C^*z_2) \geq 0 \\ \cdots \\ u_K^{z_2}*(b-C^*z_2) \geq 0 \end{cases} \cdots \begin{cases} u_1^{z_n}*(b-C^*z_n) \geq 0 \\ \cdots \\ u_R^{z_n}*(b-C^*z_n) \geq 0 \end{cases} \quad (10.34)$$

$$\xi \text{ free}, \quad z \text{ binary}$$

4) Solve Eq. (10.32), and based on the solution add the following cut:
   i) If the dual problem Eq. (10.33) is unbounded, add the following *Feasibility Cut* to RMP and go to Step 4:

$$\overline{u}^*(b-C^*\overline{z}) \geq 0, \quad (10.35)$$

   where $\overline{u}$ is the extreme ray of the dual problem (Eq. (10.33)).

   ii) If the optimal value of the dual problem (Eq. (10.33)) is bounded, which provides a lower bound, and restrict RMP ($\overline{u}^*(b-C^*\overline{z}) < \xi$), add the following *Optimality Cut* to RMP, and go to Step 4:

$$\overline{u}^*(b-C^*\overline{z}) \geq \xi. \quad (10.36)$$

   iii) If the optimal value of dual problem Eq. (10.33) is bounded, which provides a lower bound, and does not restrict RMP ($\overline{u}^*(b-C^*\overline{z}) \geq \xi$), but the difference between the upper and lower bound exceeds the threshold, add the following *Integer Exclusion Cut* to RMP and go to Step 4:

$$\sum_{i \in P} z_i - \sum_{j \in Q} z_j \leq |P| - 1, \quad (10.37)$$

   where P is the set of z's value is 1, and Q is the set of z's value is 0. $|P|$ is the number of variables z that has value of 1.

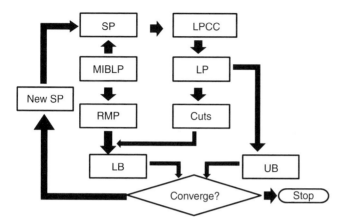

Figure 10.8 Flow chart of algorithm.

Step 4: Solve RMP with an added cut, and obtain an updated upper bound. Find the difference between upper bound and lower bound. If it is within the tolerance, stop; otherwise, update the SP by setting the constraint of the current binary variable z and go back to Step 2.

The flow chart of the proposed algorithm is shown in Figure 10.8. The proposed solution methodology leads to an optimal solution in an efficient way. The decomposition technique is essential for large-scale problems, and the transformation procedure validates the use of KKT conditions and transforms the MIBLP into two single-level problems. Therefore, the proposed algorithm outperforms traditional enumeration or reformulation techniques in both quality and computational efficiency.

10.3.3 Examples

Example 10.2 In this example, the IEEE RTS 24-bus test system [16], as shown in Figure 10.9, is used to illustrate the proposed installation strategy of new BS generators. In this system, there are 1 BSU and 9 NBSUs. Based on the generator data in PECO system [6], the characteristics of generators are shown in Table 10.6.

In the base case, the time to energize branch or transformer (GRA3) is set as two minutes. By utilizing the SRN tool, the time to restore all the generators is $T_{restore} = 36$ (minutes).

The steps to restore all the generators are shown in Table 10.7.

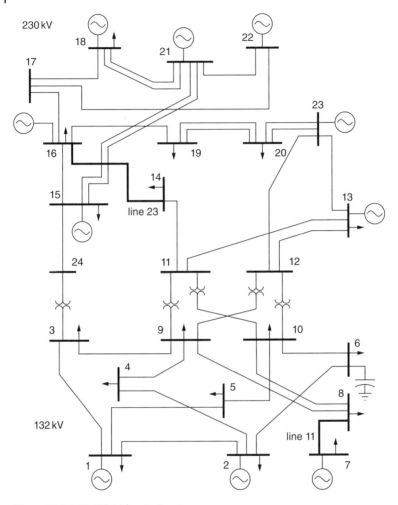

Figure 10.9 IEEE RTS 24-bus test system.

Study 1: Install New BS Generating Units at Bus 22

If a new BS generator is to be installed, which is the same as the current BS unit at Bus 22, the benefit of installing this new BS unit will be analyzed using the SRN tool. The system now has 2 BS units and 9 NBS units. The time to restore all generators is

$$T_{restore} = 28 \, (\text{minutes}).$$

The steps to restore all the generators are shown in Table 10.8:

After increasing BS capability at Bus 22, it provides more cranking power to first start NBSU on Bus 15 and then to crank NBSU on Bus 18, both at earlier

Table 10.6 Data of generator characteristics.

Gen	Bus	BSU/NBSU	T_{ctp} (min)	T_{cmin} (h)	T_{cmax} (h)	R_r (MW/h)	P_{start} (MW)	P_{max} (MW)
1	22	BSU	0	N/A	N/A	138	0	138
2	21	NBSU	30	N/A	N/A	120	6.6	300
3	18	NBSU	30	N/A	N/A	346	13.2	660
4	16	NBSU	100	N/A	N/A	157	12	600
5	15	NBSU	120	N/A	N/A	150	30	252
6	13	NBSU	160	N/A	N/A	30	2.7	135
7	23	NBSU	100	N/A	3.3	120	6	300
8	7	NBSU	120	N/A	3.5	100	9	300
9	2	NBSU	30	N/A	4	148	12	345
10	1	NBSU	0	N/A	N/A	120	0	302

Table 10.7 Sequence of restoration actions.

Restoration Action	Time (min.)	Path	Dispatchable Loads
Restart BSU on Bus 22	0	N/A	N/A
Crank NBSU on Bus 18	12	22-17-18	N/A
Pick up critical loads on 19	12	17-16-19	N/A
Pick up critical loads on Bus 14	12	16-14	N/A
Crank NBSU on Bus 15	20	22-21-15	N/A
Pick up critical loads on Bus 9	20	14-11-9	N/A
Crank NBSU on Bus 16	20	N/A	N/A
Crank NBSU on Bus 2	22	9-4-2	N/A
Crank NBSU on Bus 7	26	9-8-7	N/A
Crank NBSU on Bus 21	26	N/A	N/A
Crank NBSU on Bus 23	30	19-20-23	N/A
Crank NBSU on Bus 13	32	11-13	N/A
Crank NBSU on Bus 1	36	11-10-5-1	4,10,11,17

times than in the base case. Compared with NBSU on Bus 18, NBSU on Bus 15 requires more cranking power but has a higher ramping rate and generation capacity. With the help of additional BS capability, it is started at an earlier time and helps to start other NBS generating units. It is shown that the installation of additional BS generators can benefit system restoration by shortening the total restoration time.

Table 10.8 Different sequence of restoration actions compared with base case.

Restoration Action	Time (min)	Path	Dispatchable Loads
Restart BSU on Bus 22	0	N/A	N/A
Crank NBSU on Bus 15	8	22-21-15	N/A
Pick up critical loads on 19	8	22-17-16-19	N/A
Pick up critical loads on Bus 14	8	16-14	N/A
Crank NBSU on Bus 18	12	17-18	N/A

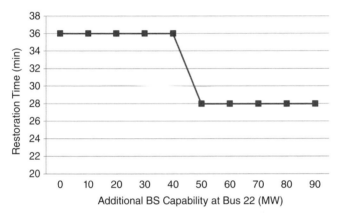

Figure 10.10 Comparison of restoration times under different BS capabilities on Bus 22.

By gradually increasing the total BS capability on Bus 22, the SRN tool is used to calculate the restoration time. The comparison of restoration time under different BS capability is shown in Figure 10.10. It can be seen that when increasing somewhat the BS capability, the system restoration will not benefit. There is a maximum amount under which restoration time is reduced due to the additional BS capability. However, beyond the maximum amount, the extra BS capability will not help to reduce the restoration time further. In this case study, the threshold at Bus 22 is to install additional 50 MW BS capabilities to reduce restoration time by 8 minutes, about 20% of the restoration time in the base case. Therefore, when making decisions on installing additional BS units, the benefit analysis needs to be conducted to find the optimal capability of installed BS units.

Study 2: Install One New BS Unit at Different Buses

When installing new additional BS units, system restoration steps will change. At different installation locations, the new restoration strategy can be made using a GRMs-based restoration tool. In this study, one new BS unit, same

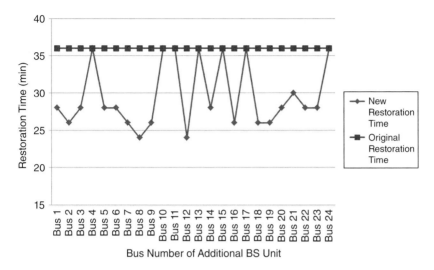

Figure 10.11 Comparison of restoration times with installation of one new BS unit at different buses.

as the BS unit on Bus 22, is installed at different buses. The times to restore all the generators are shown in Figure 10.11. From the comparison, it is shown that after installing one new BS generating unit at different buses, some benefit from the additional BS capability to reduce the restoration time, while others have the same restoration time. In this case study, the optimal locations for installing a new BS generating unit are Bus 8 and Bus 12. The restoration steps of installing an additional BS generating unit at Bus 8 are shown in Table 10.9. Compared with the restoration steps by increasing BS capability at Bus 22, NBSU on Bus 7 and Bus 21 are started earlier, which brings a shorter restoration time.

Study 3: Optimal Installation Strategy of Additional BS Capability

The installation of additional BS generating unit does not automatically benefit system restoration. In Studies 1 and 2, it is shown that when installing new BS generators, both the location and amount of BS capability need to be decided to obtain the optimal installation strategy. In this study, the developed GRMs-based algorithm is utilized to calculate the optimal installation location and amount. Increasing the amount of additional BS capability at each bus, the restoration times are obtained. The comparisons of restoration times are shown in Figure 10.12.

It can be seen that, after installing new BS generating units, the restoration time decreases in the most cases, while it stays the same at some buses. For each situation with reduced restoration time, there is a threshold at which restoration time reaches the minimum value. However, for some buses, such as Bus 2 and Bus 16, there are multiple break points. They provide multiple choices for increasing BS capability to reduce the restoration time.

Table 10.9 Sequence of restoration actions.

Restoration Action	Time (min)	Path	Dispatchable Loads
Restart BSU on Bus 22	0	N/A	N/A
Restart BSU on Bus 8	0	N/A	N/A
Crank NBSU on Bus 15	6	22-21-15	10
Pick up critical loads on 19	6	22-17-16-19	4,10
Pick up critical loads on Bus 14	6	16-14	4,10
Pick up critical loads on Bus 9	6	14-11-9	4,10
Crank NBSU on Bus 18	12	17-18	4,10
Crank NBSU on Bus 16	12	17-16	10,11,17
Crank NBSU on Bus 2	14	9-4-2	10,11,17
Crank NBSU on Bus 7	14	9-8-7	4,10
Crank NBSU on Bus 21	14	N/A	4,10
Crank NBSU on Bus 23	18	19-20-23	4,10
Crank NBSU on Bus 13	20	11-13	4,10
Crank NBSU on Bus 1	24	11-10-5-1	4,10,11,17

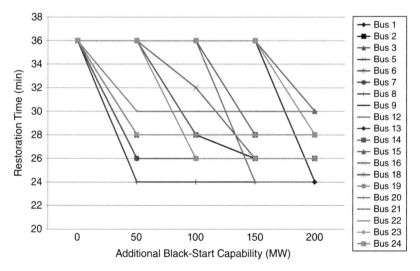

Figure 10.12 Comparison of restoration times at different buses with additional BS capability.

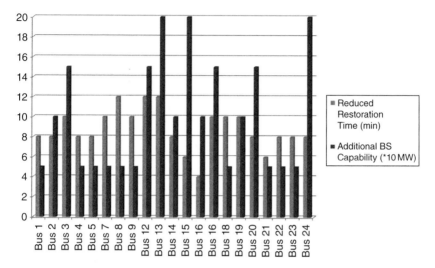

Figure 10.13 Comparison of reduced restoration times and installed BS capability at each bus.

However, the optimal installation strategy of additional BS capability needs to consider both the reduced restoration time and installed BS capability. The comparisons are shown in Figure 10.13. When installing 50 MW additional BS capabilities at Bus 8, the restoration time is reduced by 12 minutes. When installing 100 MW additional BS capabilities at Bus 18, the restoration time is reduced by 10 minutes. When installing 150 MW additional BS capabilities at Bus 12, the restoration time is down by 12 minutes. When installing 200 MW additional BS capabilities at Bus 13, the restoration time is down by 12 minutes. Therefore, in this system, the optimal installation strategy is to install an additional 50 MW BS generating unit at Bus 8 to reduce restoration time by 12 minutes, or 33% of the restoration time in the base case.

Example 10.3 In this example, the IEEE 39-bus system is used for illustration of the proposed MIBLP model and its solution methodology.

1) *Only considering GSS*

Increase system BS capability, and get the generator start-up time, total restoration time, and system generation capability, as shown in Table 10.10. The comparison of increased system generation capability with different BS capabilities is shown in Figure 10.14.

It is shown that when the increased amount is small (10 MW) for the BS capability, the system will benefit from the extra BS capability significantly.

Table 10.10 Comparison of restoration time and system generation capability with different BS capabilities.

BS Capability	G1	G2	G3	G4	G5	G6	G7	G8	G9	Total Restoration Time	System Generation Capability
10	5	4	2	7	3	2	3	3	4	26	94 002
20	4	2	2	7	4	3	3	3	3	26	97 995
30	4	2	2	7	4	3	3	3	3	26	100 346
40	4	2	2	7	3	2	2	3	3	26	103 785
50	4	3	3	7	2	2	2	2	3	26	106 969
60	4	2	2	7	2	2	2	3	3	26	109 769
70	4	2	3	7	2	2	2	3	2	26	112 480
80	4	2	2	7	2	2	2	2	3	26	115 287
90	4	2	2	7	2	2	2	3	2	26	117 805
100	4	2	2	7	2	2	3	2	2	26	120 438
110	4	2	2	7	2	2	2	2	2	26	123 322
120	4	2	2	7	2	2	2	2	2	26	125 672

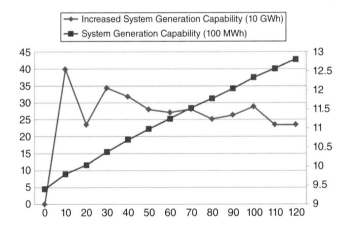

Figure 10.14 Comparison of increased system generation capability with different BS capabilities.

Although the total restoration time does not change, due to the time constraint of G4, system generation capability increases until the amount of installed BS capability exceeds the threshold, which is 110 MW in this case. And the more installed BS capability, the smaller increasing rate of system generation capability.

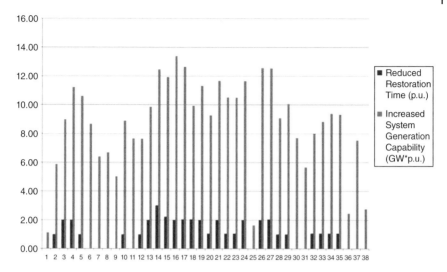

Figure 10.15 Comparison of reduced restoration time and increased system generation capability with installed BS capability at different buses.

2) *Considering GSS, TPS, and GRAs*

Install 10 MW BS capabilities at different buses, and calculate the total restoration time and system generation capability. Continue increasing the BS capability, and it is observed that there is no more improvement. The comparisons are shown in Figure 10.15.

It is shown that several optimal installation strategies are available. Installing at Bus 14 will bring the maximally reduced restoration time of 3 p.u. time and second largest increased system generation capability. Installing at Bus 15 will bring the third largest reduced restoration time and second largest increased system generation capability. Installing at Bus 17 will bring the third largest reduced restoration time and the largest increase in system generation capability. System planners can utilize the information provided by the proposed method to make optimal installation decisions.

If installing at Bus 14 to reduce restoration time by 3 p.u. time, which saves 30 minutes of power system outage, it contributes substantially to the enhancement of power system reliability. Reliability is a high priority in power system operation. Following an outage, it is critical to restore the system back to a normal operating condition as efficiently as possible. According to the Mid-Atlantic Area Council (MAAC) Reliability Principles and Standards, "sufficient megawatt generating capacity shall be installed to ensure that in each year for the MAAC system the probability of occurrence of load exceeding the available generating capacity shall not be greater, on the average, than one day in ten years" [17]. Based on this "one day in ten years" loss-of-load expectation principle, the North American Electric Reliability Corporation (NERC) standard of "Planned Resource Adequacy Assessment" is established to analyze and assess the resource adequacy for load in the Reliability First Corporation (RFC) region [18]. Therefore, it is important to reduce the duration of outages in order to meet this NERC reliability standard. The reduced outage time is an essential performance index of a power system's resilience.

Moreover, from the economic point of view, saving 30 minutes could save millions of dollars. The cost increases exponentially during the scale and duration of an outage. A major power outage makes a significant impact on people's daily lives and the economy. The blackout on August 14, 2003 in the US and Canada affected an area with an estimated 50 million people and 61 800 MW of load. The duration of the outage was about two days, and the estimated total cost ranged between $4 billion and $10 billion [19]. After the blackout, industrial sectors, public transportation, and financial and other physical systems were severely affected.

The NERC reliability standard requires a system to establish a Black Start Capability Plan to ensure that the quantity and location of system BS generators are sufficient to provide a BS service [20]. Therefore, it is mandatory for power systems to have sufficient BS capability to increase its resilience against disturbances or outages. Installing new BS capability can reduce the total restoration time and achieve an efficient restoration process.

Typical BS generating units are diesel, hydro, or combustion turbine units, which are expensive and are often used to serve the peak load. A rolling blackout often happens during the periods of peak energy demand. By installing more BS generators based on the system reliability requirement, these peaking units can serve the peak load and enhance power system security.

Table 10.11 Performance analysis.

	PJM 5-Bus Case	IEEE 39-Bus Case
Number of NBS generators	3	9
Number of all generators	4	10
Number of buses	5	39
Number of lines	6	46
Total restoration time (h)	2	6
Number of decision variables	480	10 642
Number of constraints	1521	36 253
Computational time (s)	0.27	3.20

The performance of the proposed optimization modules is shown in Table 10.11. From the simulation results, it is seen that the computational time is within the practical range for both system restoration planning and online decision support environments.

References

1 Feltes, J.J.W. and Grande-Moran, C. (2008). Black start studies for system restoration, *Proceedings of the IEEE Power and Energy Society General Meeting – Conversion and Delivery of Electrical Energy in the 21st Century, IEEE, 2008*, pp. 1–8.

2 Saraf, N., McIntyre, K., Dumas, J., and Santoso, S. (Nov. 2009). The annual black start service selection analysis of ERCOT grid. *IEEE Trans. Power Syst.* 24 (4): 1867–1874.

3 Lin, Z. and Wen, F. (2007). Power system restoration in restructured power industry, in *Power Engineering Society General Meeting, IEEE*, pp. 1–6.

4 Adibi, M.M., Borkoski, J.N., and Kafaka, R.J. (Nov. 1987). Power system restoration – the second task force report. *IEEE Trans. Power Syst.* 2 (4): 927–932.

5 Fink, L.H., Liou, K.L., and Liu, C.C. (May 1995). From generic restoration actions to specific restoration strategies. *IEEE Trans. Power Syst.* 10 (2): 745–751.

6 Sun, W., Liu, C.C., and Chu, R.F. (2009). Optimal generator start-up strategy for power system restoration, *Proceedings of the 15th International Conference on Intelligent System Applications to Power Systems, Brazil, November 2009*.

7 Sun, W., Liu, C.C., and Zhang, L. (August 2011). Optimal generator start-up strategy for bulk power system restoration. *IEEE Trans. Power Syst.* 26 (3): 1357–1366.

8 Perez-Guerrero, R.E., Heydt, G.T., Jack, N.J. et al. (Jul. 2008). Optimal restoration of distribution systems using dynamic programming. *IEEE Trans. Power Syst.* 23 (3): 1589–1596.

9 Liu, C.C., Vittal, V., Heydt, G.T., Tomsovic, K., and Sun, W. (2009). Development and evaluation of system restoration strategies from a blackout, PSERC Publication 09–08.

10 Feltes, J.W., Grande-Moran, C., Duggan, P. et al. (2006). Some considerations in the development of restoration plans for electric utilities serving large metropolitan areas. *IEEE Trans. Power Syst.* 21 (2): 909–915.

11 Liu, C.C., Liou, K.L., Chu, R.F., and Holen, A.T. (Feb. 1993). Generation capability dispatch for bulk power system restoration: a knowledge-based approach. *IEEE Trans. Power Syst.* 8 (1): 316–325.

12 Adibi, M.M. and Milanicz, D.P. (1999). Estimating restoration duration. *IEEE Trans. Power Syst.* 14 (4): 1493–1498.

13 PJM Training Materials-LMP 101, PJM.

14 Saharidis, G.K. and Ierapetritou, M.G. (2009). Resolution method for mixed integer bi-level linear problems based on decomposition technique. *J. Glob. Optim.* 44 (1): 29–51.

15 Wang, L. (2010). θ-free algorithm for LPCC, IE 634X Computational Optimization, Iowa State University, Spring 2010.

16 Power Systems Test Case Archive [Online]. Available: https://www2. ee.washington.edu/research/pstca/rts/pg_tcarts.htm.

17 MAAC Reliability Principles and Standards. https://www.rfirst.org/standards/ Documents/MAAC%20Documents%20Retired%20After%20082709.pdf.

18 NERC Standard BAL-502-RFC-02, Planned Resource Adequacy Assessment (RFC). http://www.nerc.com/files/BAL-502-RFC-02.pdf.

19 U.S.-Canada Power System Outage Task Force, Final Report on the August 14th Blackout in the United States and Canada: Causes and Recommendations. https://reports.energy.gov/BlackoutFinal-Web.pdf.

20 NERC Standard EOP-007-0, Establish, Maintain, and Document a Regional Black Start Capability Plan. http://www.ncrc.com/files/EOP-007-0.pdf.

Index

a

AC OPA model   58
  fast dynamics of cascading events
    59–60
  slow dynamics of system evolution
    CCDs   61
  critical lines   62
  RTS-96 three-area system   60
adjacency matrix   34

b

backtracking search method   229
Bak–Tang–Wiesenfeld (BTW)
    sandpile model   23–24
Barabási–Albert preferential
    attachment model   29
biggest–load-increment branching
    (BLB)   278
bilevel linear programming (BLP)
    problems   417
binary-decision model   33–34
black start (BS)
  auxiliary load recovery in
    247–248
  capability

generator characteristics,
    data of   409
  increased system generation
    capability   429
  installation criteria, of new black
    start generators   404–407
  optimal black start capability (see
    optimal black start capability)
  PJM 5-bus system   408
  reduced restoration time   427–429
  restoration actions   409
  restoration time and system
    generation capabilities   410,
    427, 428
  transmission system, data of   409
  updating restoration plans,
    restoration tool for   408
  constraints
    self-excitation   401
    voltage stress   401
  definition of   399
  diesel generating units   400
  dynamic analysis of   225
  gas turbine generating units   400
  hydroelectric generating units   400

Power System Control Under Cascading Failures: Understanding, Mitigation, and System Restoration, First Edition. Kai Sun, Yunhe Hou, Wei Sun and Junjian Qi.
© 2019 John Wiley & Sons Ltd. Published 2019 by John Wiley & Sons Ltd.
Companion website: www.wiley.com/go/sun/cascade

black start (BS) (*cont'd*)
  motor start-up, influence
    of   249–250
  performance analysis   431
  power system restoration
    procedure   403–404
  self-excitation criteria   237
  service procurement
    401–403
  VSC-HVDC technology
    367–369
black start unit (BSU)
  IEEE RTS 24-bus test system
    423, 424
  module   415
  restoration actions, sequence of
    423, 425
  restoration times   424–426
Boolean
  matrix theory   177
  N-node graph model   175
  six-bus power system   178
Borel–Tanner distributions   81
branching processes
  continuous data   78–81
  cross-validation   85
  descriptive parameters   73
  Galton–Watson branching
    process   73, 74
  line outage data   82–83
  load shed   74, 84
  multi-type branching processes
    cross-validation for   100–102
    estimated joint distribution
      98–100
    estimated parameters   96–98
    failure propagation
      104–105
    joint probability distribution
      90–91
    number of cascades needed
      94–96

predicting joint distribution
  102–104
two-type branching process
  91–92
offspring distribution   74–75
simulation data   72
statistical insight 81–82

c
California–Oregon Intertie   217
capability assessment, black start
  generator characteristics, data
    of   409
  installation criteria, of new black
    start generators
    critical path method   406
    installation strategies
      410, 411
    KBS restoration tool   406
    load restoration stage   405
    preparation stage   404
    restoration time, criteria of
      406, 407
    system restoration
      stage   404–405
    updated restoration plan   405
  optimal installation strategy
    407–408
  PJM 5-bus system   408
  restoration actions   409
  restoration times and
    system generation
    capabilities, installation
    strategies   410
  transmission system, data
    of   409
CASCADE model   35–37
cascading failures
  blackouts   69
  branching processes
    continuous data   78–81
    cross-validation   85

descriptive parameters 73
efficiency improvement
85–87
Galton–Watson branching
process 73–74
line outage data 82–83
load shed 74, 84
models 69
multi-type branching processes
(*see* multi-type branching
processes)
offspring distribution
74–75
simulation data 72
statistical insight 81–82
electric power grids 1, 69
failure interaction analysis
(*see* failure interaction
analysis)
2012 Indian blackout 2
interaction model 69
2003 Italy blackout 2
large-scale cascading blackouts 1
low-probability high-impact
events 1
mitigation 5–6
modeling and understanding
benchmarking and validation 5
complexities of system 4
computational power 4
emerging problems 5
evolving system 4
external factors 4
mechanisms 4
size of system 3
1965 Northeast blackout 1
self-organized criticality (SOC)
theory 69
blackout data 71–72
finite-dimensional attractor 70
2017 South California
disturbance 3

system restoration
challenges and research
opportunities 12
characteristics 9
control actions 9–10
generation and consumption 9
generation side 9
interconnected components 9
market environment 12–13
optimization and computation
technologies 14–15
reliability guidelines and
standards 11–12
resilience requirements
13–14
restoration phases
10–11
temporal dependency 9
U.S.-Canadian blackout 9
2015 Ukraine blackout 2–3
uncontrolled and controlled system
separations
electromechanical oscillation
modes, 7
high generation-load
imbalance, 7
out-of-step protection
schemes 8
power blackout, 7
WAMS-based CSS scheme 8
2003 U.S.-Canadian blackout 2
1996 Western North America
blackouts 1–2
Chapman-Richards model 54
cold load pickup 250–251
complementary cumulative
distribution (CCD)
127–128
AC OPA simulation 61
basic and improved OPA
models 49
slow process 57, 58

controlled system separation (CSS)
generator grouping
Boolean Algebraic model
175–178
computational complexity
164–167
elementary coherent
groups 158–160
generation coherency and power
balance constraints
178–182
minimum-cut strategies 184
network decomposition, for
parallel processing 173–175
network reduction 167–172
OBDD 182–184
slow coherency analysis
154–158
small disruption
constraint 185–190
heuristic searching algorithms 143
implementation issues 142
locations of separation 142
online decision, separation
strategy
frequency-amplitude
characteristics, of
electromechanical
oscillation 199–204
generator grouping 197
phase-locked loop-based method
204–210
separation timing 197
spectral analysis-based method
198–199
timing of controlled
separation 210–212
ordered binary decision diagram
(OBDD)-based methods 143
power blackout 141
power network, graph models
directed edge-weighted
graph 152–153

undirected node-weighted
graph 149–151
real-time synchrophasor
measurements 143
separation points
actual generation and spinning
reserve 147
adaptive separation points 145
balanced K-way graph partition
161–162
disturbances triggering cascading
failures 145
electrical speed, angular
frequency 144
fixed separation points 145
generation coherency constraint
(GCC) 146
K-cut of edge-weighted graph
model 161, 163
K-cut of node-weighted graph
model 160–161
K-cut with acceptable total
weight 162
K-cut with minimum total
absolution weight 163
minimally imbalanced K-way
graph partition 162
minimum cut constraint 147
power balance constraint
(PBC) 146
real-time algorithm 146
small disruption constraint 147
steady-state constraints 148
transmission capacity constraint
(TCC) 146
transmission capacity
limits 144
slow-coherency-based CSS
techniques 143
system resynchronization 141
timing of separation 142
WAMS-based unified
framework for

offline analysis stage
212–213
online monitoring stage   212,
213, 216–221
real-time control stage   212, 213,
221–222
controlling unstable equilibrium point
(CUEP)   204
coupled map lattice model   34–35
critical path method   406
cross-spectral density (CSD)
198, 199

d
Decision Trees   143
distance (DIST) relays   63
dynamic cascading failure
model   62

e
Electric Reliability Council of Texas
(ERCOT)   402, 403
electromagnetic torque   249, 250
electromagnetic transients, system
restoration   235, 236
generator self-excitation
black start   237
suppression of   237
magnetizing inrush current on
transformer, impact of
245–246
characteristics and identification
methods of   246
suppression   246–247
resonant overvoltage, no-load
transformer   242–243
ferroresonance
overvoltage   243–245
harmonic resonance
overvoltage   243
switching overvoltage   237, 238
controlling
overvoltage   239–240

equivalent circuit   238
evaluating transient voltage,
approach for   240
load-ended line model   241–242
open-ended line model
240–241
overvoltage factor   239
switching operation   238
total closing overvoltage   238
unloaded transformers,
energization of   239
voltage and frequency analysis, load
pickup
auxiliary load recovery, in black
start   247–248
cold load pickup   250–251
induction motor
model   248–249
motor start-up, influence
of   249–250
electromechanical oscillation,
frequency-amplitude
characteristics
frequency-amplitude curve
203–204
damped rotor angle oscillation,
large disturbance   202
FFT algorithm   199
oscillation, nonlinearities
of   202
SEP   200–201
SMIB   200–201
elementary coherent groups (ECGs)
197, 206
angle differences   218–220
angle of CSD   218, 2201
FFT results   198
operating condition   158
phase angle of CSD   199
power system   160
real-time operations   160
real-time oscillations   159
slow coherency analysis   159

elementary coherent groups (ECGs)
(*cont'd*)
squared-coherency index
218, 220
WECC 179-bus system, graph
model of   215
energy management system
(EMS)   46, 212
energy storage, in system restoration
batteries for
generator frequency response
rate   333
one BS unit (CT)   337, 338
one NBS unit (ST)   338
simulation results   334–336
solution algorithm   334
state of charge of batteries   334
two BS Units   338–340
plug-in hybrid electric vehicle
(PHEV)
capacity of aggregation
345–346
charging/discharging sequence
344, 350
definition   341
flexible generation and
distributed energy
storage   342
grid stabilization   342
load pickup process   342
operation and maintenance
costs   341
problem formulation
325–326, 344
pumped-storage hydro units
conventional generators   323
economic dispatch and unit
commitment problems   323
optimal coordination   323
wind uncertainty and
variability   324
simulation results
PSH unit location   329

restoration with/without PSH
Contribution   328–329
wind power
fluctuation   331–332
wind power uncertainty   330
solution algorithm   325–326

f

failure interaction analysis
estimation of interactions
106–108
interaction-based mitigation   105
interaction matrix and interaction
network   119–120
interaction model   105,
134–137
average propagation for original
and simulated cascades
126–127
complementary cumulative
distribution (CCD)
127–128
flow chart   111–112
implementation   112–113
online decision-making
support   113
probability distribution
125–126
similarity indices   129
validation   113–115
key links
and component identification
108–111
mitigation strategies
AC OPA model   130, 134
average propagation   132
implementation   133
link weights   132
power law distribution   131
probability distributions of total
line outages   131, 133
relay blocking strategy   130
risks   129

number of cascades needed
  AC OPA simulation   117
  cause-indistinguishable
    components   119
  lower bound for M, 115
  lower bound for M_u, 115–116
  propagation capacity for M_u, 118
  unnecessary original cascade
    simulation M_{un}, 117
fast Fourier transform (FFT)   198,
  199, 207
feedback path (FP)   201
ferroresonance overvoltage   243–245
flexible AC transmission systems
  (FACTS)
  benefits of   370
  potential applications
    oscillation stability   372–373
    reactive power control   374
    smooth load pickup   375
    standing phase angle control
      373–374
    stored energy for circuit
      breaker   375
    sustained overvoltage   374
    transient overvoltage   374–375
    transient stability   373
    voltage stability   373
  series capacitor (SC)   370
  static synchronous compensator
    (STATCOM)   371
  static var compensator (SVC)   370
  thyristor-controlled series capacitor
    (TCSC)   370
  unified power flow controller
    (UPFC)   371–372

g
Galton–Watson branching
  process   74–75
gas turbines   228
Gaussian filter   207
general cascading failure models

binary-decision model
  average vulnerable cluster size   34
  normalized generating
    function   34
  vulnerable node degree   33
BTW sandpile model   23–24
CASCADE model   35–37
coupled map lattice model
  adjacency matrix   34
  nodes   35
failure-tolerance sandpile model
  cascade sizes   26
  economic loss   27
  mechanisms   25
  optimal control   27–29
  random network   29
  range width   28, 29
  scale-free network   29
influence model
  network influence
    matrix   31–32
  probability vector   31
  realization   32
  stochastic matrix   31
interdependent failure model
  37–38
Motter–Lai model   30
generating units
  acceptable operating point of
    system   267–268
  case examples   278–283
  and critical loads   257
  decision trees   261
  diesel generating   400
  dispatchable loads   260, 268–269
  dynamic programming   259
  energized block   261–263
  gas turbine   400
  generic model   256–257
  hydroelectric   400
  non-black-start unit   260
  operating constraints   260
  picking up critical loads   259

generating units (*cont'd*)
  practical constraints 259
  restoration actions 256
  technical constraints 258
  time interval 260
  transmission paths 263–265
generating unit start-up
  active power balance and frequency
    control 226–227
  backtracking search method 229
  bilevel optimization algorithm 230
  fuzzy analytic hierarchy process 229
  knapsack subproblems 229
  knowledge-based system 229
  reactive power balance and
    overvoltage control 227
  switching transient voltage
    227–228
generation coherency and power
    balance constraints
  Boolean expressions 178–182
  OBDDs 182–184
generator reactive capability
    (GRC) 227
generic restoration actions (GRAs),
    optimization modules
  BSU module 415
  bus/line module 416
  load module 416
  NBSU module 416

h
harmonic resonance overvoltage
    236, 243
hidden failure model 39–40
  blackout size distribution 40
  simulation procedure 39
high-voltage direct current
    (HVDC) 400

i
improved OPA model 46–47
  fast dynamics of cascading events

accidental faults of lines 48
  control center 48
  DC power flow 48
  tripping overloaded/normal
    lines 48
  flow chart 46
  slow dynamics of system evolution
    CCDs 45
    NPGC 49
  transmission line upgrade 48
independent system operators
    (ISOs) 401, 402
induction motor model 248–249
influence model 30–33
interdependent failure model 37–38

k
KarushKuhn–Tucker (KKT)
    conditions 418, 420
knowledge-based system (KBS) 406

l
least mean square error (LMSE) 208
linear problem with complementarity
    constraints (LPCC) 418
line commutated converter-high-
    voltage direct current (LCC-
    HVDC) technology
  black start capability
    AC grid voltage 361
    auxiliary device support 362–363
    control modification 362
  characteristics of 357–359
  inverter controller
    constant extinction angle (γ)
      control 361
    DC current control 361
  rectifier controller
    DC current control 359
    minimum firing angle α_{min}
      control 359–360
    voltage-dependent current-order
      limit (VDCOL) 360

load-ended line model 241–242
load restoration 233–235
  active power balance and frequency
    control 234
  branch-and-cut (B&C) solver
    algorithm 276
    applicability of cutting
      planes 276–277
    assumption 275
  branching methods
    biggest-load-increment
      branching (BLB) 278
    maximum fractional
      branching 278
    pseudo-cost branching 278
  coordinated control 269
  design and implementation 269
  in distribution systems 235
  features 236
  frequency response capacity 270
  industry practices 270
  load characteristics modeling
    234–235
  mixed integer nonlinear load
    restoration (MINLR) model
    cold load effect 274
    cutting planes 277–278
    dispatch actions and
      processes 275
    frequency security
      constraints 274
    objective function 272
    operational constraints 273
    power balance constraints 273
    responsive (dynamic) reserve
      constraints 274
    spinning (synchronous) reserve
      constraints 273–274
    operational region bound 271–272
  optimal strategies
    IEEE 118-bus System 287–291
    RTS 24-Bus System 283–287
  security constraints 269

  subtasks 270
  in transmission systems 235
loop filter (LF) 201

m
magnetizing inrush current,
  transformer 245–246
  characteristics and identification
    methods of 246
  suppression
    grounding resistance 247
    three-phase closing angle,
      controlling 246
    windings distribution of
      transformer 247
Manchester model
  flow chart 41
  time-dependent phenomena
    40–42
maximum fractional branching 278
microgrid (MG)
  conventional restoration 386–387
  demonstration and practice
    algorithms, for vehicle routing
      (VR) problem 391
    existing practices 388–389
    natural disaster damages
      390–391
    scenario decomposition (SD)
      algorithm 391–393
    two-stage dispatch
      framework 389
  features of 385–386
  implementation issues 387–388
Mid-Atlantic Area Council
  (MAAC) 430
mixed-integer bilevel programming
  (MIBLP) problem 410, 417,
  418, 420, 427
mixed-integer linear programming
  (MILP)-based optimal
  generator start-up
  strategy 406

mixed integer nonlinear load
restoration (MINLR) model
cold load effect   274
dispatch actions and
processes   275
frequency security constraints   274
objective function   272
operational constraints   273
power balance constraints   273
responsive (dynamic) reserve
constraints   274
spinning (synchronous) reserve
constraints   273–274
Motter–Lai model   30
multitype branching processes   69
cross-validation   100–102
estimated joint distribution
98–100
estimated parameters   96–98
estimating failure
propagation   104–105
estimation   88–90
joint probability distribution   90–91
number of cascades needed   94–96
predicting joint distribution
102–104
two-type branching process
91–92
validation of
conditional largest possible total
outage (CLO)   93
joint entropy   93
marginal distribution   93

n
network influence matrix   31–32
network reduction
cluster of loads   168
generation and potential out-of-step
modes   167
geographical area   167
IEEE 30-bus power
system   169–170

IEEE 118-bus power system
and graph model
171–172
one-degree node   168–169
simplification   168
zero/small-weight node   168
non-black-start (NBS)   400
non-black-start unit (NBSUs)   409,
410, 416
nonlinear programming (NLP)
model   232
North American Electric Reliability
Corporation (NERC) reliability
standard   430
Northeastern Power Grid of China
(NPGC)   49, 56, 57
number-controlled oscillator
(NCO)   201

o
online decision support, CSS. *See*
controlled system separation
(CSS)
OPA model   42–46
AC OPA model   58
fast dynamics of cascading
events   59–60
slow dynamics of system
evolution   60–62
improved OPA model   46–47
fast dynamics of cascading
events   43–45, 48
flow chart   46
slow dynamics of system
evolution   48–49
sandpile model   46
fast dynamics of cascading
events   48, 52–54
flow chart   50
slow dynamics of system
evolution   54–58
open-ended line model   240–241
optimal black start capability

generator characteristics   422
GRA time, optimization modules
  BSU module   415
  bus/line module   416
  load module   416
  NBSU module   416
  optimal generator start-up sequence
    module   410, 411
    critical maximum and minimum
      time intervals, constraints
      of   412
    decision variables   411–412
    decision variables, constraints
      of   413
    generator capability function,
      constraints of   412–413
    generator start-up power
      function, constraints of   413
    MW start-up requirement,
      constraints of   412
    objective function   412
    parameters   411
    sets   411
  optimal installation strategy   417
  optimal transmission path search
    module   413–414
    constraints   414–415
    objective function   414
  optimization modules, connections
    of   416
  restoration actions, sequence
    of   422
  solution algorithm
    flow chart of algorithm
      420, 421
    LPCC   418
    MIBLP problem   417, 418
    RMP   418, 420
    slack variables   419
    SPs   418
optimal load restoration strategies
  IEEE 118-bus System   287–291
  RTS 24-bus system

computational settings   284, 285
loss of energy contingency   286
parameters   284
restoration process   284
optimal power flow (OPF)   43, 46, 59
ordered binary decision diagram
  (OBDD)   143, 214, 231
oscillation amplitude (OA)   203, 204
oscillation frequency (OF)   203, 204
overcurrent (OC) relays   63

p
phase detector (PD)   201
phase-locked loop (PLL)-based
  method   204
  flowchart of   206
  implementation of   205
  mode detection   206–207
  mode shape analysis   208–210
  signal tracking   207–208
  simple PLL algorithm, in
    Z-domain   206
  structure of   205
phasor measurement units (PMUs)
  contributions of   381–382
  establish restoration
    strategy   383–384
  global positioning system (GPS)
    radio clock   376
  IEEE standard C38.118, 376
  observability of power systems
    378–381
  restoration-oriented PMU
    placement   382–383
  standards of   377–378
  time-stamped measurements   376
power spectral densities (PSD)
  198, 199
power system cascading failure
  models
  dynamic cascading failure
    model   62
  hidden failure model   39–40

power system cascading failure
models (*cont'd*)
Manchester model 40–42
OPA model 42–46
AC OPA model 58–62
improved OPA model 46–47
slow process 52–54
power system restoration (PSR)
procedure 403–404
pseudo-cost branching 278

r
Reliability First Corporation
(RFC) 430
renewable generators, in system
restoration
doubly fed induction generators
(DFIGs) 295–296
independent system operators
(ISOs) 295
large-scale renewable energy
sources 295
normal operating conditions 295
off-line restoration tool
distribution system operators
(DSOs) 297, 298
mixed-integer linear
programming (MILP) 298
TSOs control and
coordinate 296–297
wind energy resources 296
operation and control
build-down strategy 305
build-up strategy 305
synchronous generators
(SGs) 306
Type 3 WT 306–323
voltage-source converter (VSC)
technology 305–306
power grid resilience 295
problem formulation
generation and transmission
restoration 299

generator output limit 299
generator ramping limit 299
line flow limit 299
load pickup limit 299
load shedding limit 299
power balance 298
solution algorithm 300–301
transmission system operators
(TOS) 296
wind fluctuation 304–305
wind location 302
wind penetration 302–304
restoration
electromagnetic transients
235, 236
generator self-excitation 237
magnetizing inrush current on
transformer, impact
of 245–247
resonant overvoltage, no-load
transformer 242–245
switching overvoltage 237–242
voltage and frequency analysis,
load pickup 247–251
physical constraints
generating unit
start-up 225–230
load restoration 233–235
system sectionalizing and
reconfiguration 230–233
steady-state analysis 225
restricted master problem
(RMP) 418
feasibility cut 420
integer exclusion cut 420
optimality cut 420

s
scale-free network 29
self-excitation of generator
black start 237
strategies 228
suppression of 237

self-organized criticality (SOC)
theory   69
blackout data   71–72
finite-dimensional attractor   70
single-machine-infinite-bus (SMIB)
system   200–202
slave problems (SPs)   418
slow coherency analysis
algorithm   156
assumptions   154–155
N-bus power system   155
nonoscillatory mode   156
six-bus power system   158
torque coefficients   155
slow process, OPA model
fast dynamics of cascading
events   59–60
accidental faults of lines   52
control center   52
DC power flow   49, 52
overcurrent relays, operation and
maloperation   52
tree contact/failure of lines   52
flow chart   50
slow dynamics of system evolution
cause of blackout   56–57
CCDs   58, 61
gradual development   57
NPGC   56
operation of relays   57
process of blackout   57
UVM   54–56
spectral analysis–based method
inter-area modes, identification
of   198–199
limitations of   199–203
separation boundary, prediction
of   198–199
squared linear coherency
index   209
stable equilibrium point (SEP)
200–204
static Var compensator   243

stochastic matrix   31
switching overvoltage   237, 238
controlling overvoltage   239–240
equivalent circuit   238
evaluating transient voltage,
approach for   240
load-ended line model   241–242
open-ended line model   240–241
overvoltage factor   239
switching operation   238
total closing overvoltage   238
unloaded transformers,
energization of   239
switching surges   375
system restoration
electromagnetic transients
235, 236
generator self-excitation   237
magnetizing inrush current on
transformer, impact
of   245–247
resonant overvoltage, no-load
transformer   242–245
switching overvoltage   237–242
voltage and frequency analysis,
load pickup   247–251
generating unit start-up
active power balance and
frequency control   226–227
features   229
reactive power balance and
overvoltage control   227
self-excitation   228–230
switching transient
voltage   227–228
LCC-HVDC technology (*see* line
commutated converter-high-
voltage direct current (LCC-
HVDC) technology)
load restoration   233–235
active power balance and
frequency control   234
in distribution systems   235

system restoration (*cont'd*)
  load characteristics
    modeling   234–235
    in transmission systems   235
  microgrid in
    conventional restoration   386–387
    demonstration and
      practice   388–393
    features of   385–386
    implementation issues   387–388
  network reconfiguration and path
    selection   231–232
  phasor measurement units (PMUs)
    contributions of   381–382
    establish restoration strategy
      383–384
    observability of power systems
      378–381
    restoration-oriented PMU
      placement   382–383
  switching impact   232
  system sectionalizing   230–233

t
temperature (TEMP) relays   63
thyristor-controlled braking
  resistors   375
thyristor-controlled circuit
  breakers   375
thyristor-controlled phase
  regulators   375
thyristor-controlled voltage
  regulators   375
Type 3 wind turbines (WTs)
  autonomous frequency mechanism
    frequency adjustment
      mechanism   317–318
    mathematical performance
      analysis   318–320
    self-maintained frequency
      service   318
  black starting control and
    sequence

proposed control strategy   315
  sequence of   315–317
black start stage   309
control configuration   309
control scheme for   310
grid side converter (GSC)
  control   309
inherent properties and
  defects   309
mathematical model   311
multitimescale controllers   309
one-line circuit model   313
P-f relationship   307–308, 313
prerequisites   308
and Q-V relationship   306, 308,
  313–314
rotor side converter (RSC) current
  control   309
simulation study
  black-starting validation   321
  stand-alone loaded validation
    321–323
stator voltage orientation and
  dual-loop control   309
system frequency performance   306

u
underfrequency load shedding (UFLS)
  relays   63
undervoltage load shedding (UVLS)
  relays   63
utility vegetation management
  (UVM)   54–56

v
voltage source converter-high-voltage
  direct current (VSC-HVDC)
  technology
  auxiliary control of
    dead grid   367
    P-V_{dc} droop control   367
    Q-V_{ac} droop control   367
    V_{dc}-P and f-P droop control   366

black start of 367–369
characteristics of 363
regular control of
  active power control 364
  constant active power and
    reactive power control
    mode 365
  control structures 363, 364
  DC-voltage deviation
    control 365
  dual-loop structure 364
  phase-locked loop (PLL)-based
    vector control 364
  power passive systems 365
  reactive power control 364
  receiving-end station 365
  sending-end station 365

w

wide area measurement system
  (WAMS) 197, 231
measurement, communication and
  actuation systems,
  infrastructures for 212, 213
offline analysis stage 212, 213
  ECGs, determination of
    214, 215
  postseparation control 216, 217
  separation points, determination
    of 214–216
online monitoring stage 212, 213,
  216–221
real-time control stage 212, 213,
  221–222